the
technique
of
**television
production**

*This book is sold subject to the Standard Conditions of Sale of
Net Books and may not be re-sold in the UK below the net price.*

THE LIBRARY
OF COMMUNICATION TECHNIQUES

FILM
THE TECHNIQUE OF DOCUMENTARY FILM PRODUCTION
W. Hugh Baddeley

THE TECHNIQUE OF EDITING 16 *mm* FILMS
John Burder

THE TECHNIQUE OF FILM ANIMATION
John Halas and Roger Manvell

THE TECHNIQUE OF THE FILM CUTTING ROOM
Ernest Walter

THE TECHNIQUE OF FILM EDITING
Compiled by Karel Reisz and Gavin Millar

THE TECHNIQUE OF FILM MUSIC
John Huntley and Roger Manvell

THE TECHNIQUE OF FILM AND TELEVISION MAKE-UP
Vincent J-R Kehoe

THE TECHNIQUE OF THE MOTION PICTURE CAMERA
H. Mario Raimondo Souto

THE TECHNIQUE OS SPECIAL EFFECTS CINEMATOGRAPHY
Raymond Fielding

TELEVISION
THE TECHNIQUE OF LIGHTING FOR TELEVISION AND MOTION PICTURES
Gerald Millerson

THE TECHNIQUE OF SPECIAL EFFECTS IN TELEVISION
Bernard Wilkie

THE TECHNIQUE OF THE TELEVISION CAMERAMAN
Peter Jones

THE TECHNIQUE OF TELEVISION NEWS
Ivor Yorke

THE TECHNIQUE OF TELEVISION PRODUCTION
Gerald Millerson

SOUND
THE TECHNIQUE OF RADIO PRODUCTION
Robert McLeish

THE TECHNIQUE OF THE SOUND STUDIO
Alec Nisbett

the technique of television production

Gerald Millerson

Focal Press · London

Focal/Hastings House · New York

British Library Cataloguing in Publication Data
Millerson, Gerald
   The technique of television production.—10th ed.
   —(The library of communication techniques).
   1. Television—Production and direction—
   Handbooks, manuals, etc.
   I. Title   II. Series
   791.45'0232   PN1992.75

ISBN (excl. USA) 0 240 51002 X (casebound)
ISBN (excl. USA) 0 240 51010 0 (paperback)
ISBN (USA only) 0 8038 7123 6 (casebound)
ISBN (USA only) 0 8038 7124 4 (paperback)

First edition 1961
Second edition 1963
Third Edition 1964
Fourth Edition 1966
Fifth Edition 1967
Sixth Revised and Enlarged Edition 1968 (USA)
Seventh Edition 1969
Eighth Revised Edition 1970
Ninth Revised Edition 1972
Tenth Impression 1973
Eleventh Impression 1974
Twelfth Impression 1975
Thirteenth Impression 1977
Tenth Edition 1979

Printed by M. & A. Thomson Litho Ltd., East Kilbride.
Bound by Hunter & Foulis Ltd., Edinburgh.

Contents

Preface   13

1 Studio tour   15

The television studio   15
The television studio in action   16
The production control room   19
Master control   23

2 How television works   24

The video signal   24
Color television   24
Picture detail   29
Tonal contrast   29
Tonal gradation   30

3 What your camera can do   31

The lens   31
The lens system   31
Focal length and lens angle   32
Perspective distortion   36
Long-focus lens (narrow-angle lens)   37
Short-focus lens (wide-angle lens)   39
Supplementary lenses   41
The zoom lens   41
Focus controls   43
Lens aperture—f-stop   46
Neutral density filters   48
Camera-tube sensitivity   48
Mountings for lightweight cameras   49
Static mountings   50
The rolling tripod   50
The studio pedestal   50
The small crane   52
Larger cranes   53
Motorized dolly   54
Low dollies (creepers)   55
Hydraulic platforms (cherry-pickers; Simon crane)   55
Remotely controlled cameras   55
Basic camera operation   56

4 The persuasive camera   59

Defining the shot   59
Selecting the length of shot   60
The very long shot (vista shot)   61
The long shot   61
Medium shots   62
The close-up   62
Deep-focus techniques   63
Shallow-focus techniques   64
Moving the camera head   66
Panning the camera   66
Tilting the camera head   69
Camera height   70
Moving the camera   72
Using the zoom lens   78

5 Composing the picture   81

Composition by design   81
Composition by arrangement   81
Composition by selection   81
The director and composition   82
Composition principles   82
The effect of the picture frame   82
Proportions   83
Framing   85
Pictorial balance   87
Emotional influence of tone   90
Scale   90
Subject prominence   91
Subject attitude   92
Picture shape   93
Unifying interest   93
Speed of composition lines   94
Continuity of centres of interest   94
Composition continuity in multicamera production   95
Color and the picture   98
Dynamic composition   103
Potentials of the still picture   103
Potentials of the moving picture   103
A theory of dynamic composition   104
Accepted maxims   107
Mirror images   109

6 Editing   110

The nature of editing   110
Editing and the TV director   110
Television editing techniques   111
Continuity cutting   112
Relation cutting   112
Types of editing   113
Dynamic cutting   113
The cut   113
The fade   116
The mix (dissolve, lap dissolve)   118
The wipe   120
Superimpositions (half-lap dissolves, supers)   121
Order of shots   124
Duration of shots   126
Cutting rate   127
Cutting rhythm   128
The production switcher   128
Basic production switcher functions   131

7 Lighting   133

Influence of lighting   133
Why is lighting necessary?   133
Light quality—dispersion   134
Direction of the light   136
Light intensity   136
Pictorial quality   136
Color quality of light   137
Lighting flat surfaces   137
Lighting an object   138

Lighting people 140
Lighting areas 145
Pictorial treatment 147
Light sources (luminants) 149
Types of lamp housing—luminaires 151
Lamp supports 153
Light control 156
Lighting control equipment 157
TV lighting problems 157
Electronic problems 157
Sound boom shadows 158
How settings affect lighting treatment 159
Other factors affecting lighting 160
Lighting for color 161
Atmospheric lighting 162
Lighting effects 166

8 Scenery 167

Basic organization 167
Basic scenic forms 170
The flat 170
Set pieces (built pieces/solid pieces/rigid units) 172
Profile pieces (cut-outs) 173
Cyclorama 174
Backgrounds 175
Surface detail and contouring 176
Floor treatment 178
Ceilings 179
Size and shape of sets 180
Shooting off (overshoot) 180
Height and depth in floors 180
Space economies 181
Multiplane techniques 182
The illusion of four walls 182
Partial settings 185
Realism of settings 186
Scrim (scenic gauze) 187
Mobile scenic units 187
Set dressing 188
Electronic and lighting considerations 189
Staging for color 192
Costume (wardrobe) 193

9 Make-up 194

Types of make-up 194
Straight make-up 194
Corrective make-up 194
Character make-up 195
Conditions of television make-up 195
Principles of make-up 196
Make-up materials 197
Surface modeling (prosthetics) 199
Hair 200

10 Audio 202

Sound quality 202
Reproduced sound 204

Acoustics   204
Microphone characteristics   205
Personal microphones   207
Hand microphone (baton mike, stick mike)   208
Stand microphone   209
Desk microphone   209
Slung microphone (hung, hanging mike)   209
Fishpole (fishing rod)   210
Sound booms   211
Omnidirectional characteristics   212
Unidirectional characteristics   213
Dynamic (moving coil) microphone   217
Condenser (electrostatic) microphone   218
Ribbon (velocity, pressure-gradient) microphone   218
Crystal microphone   219
Audio control   219
Dynamic range control   220
Sound balance   221
Sound perspective   222
Sound quality   223
Audio tape—recording and replay   224
Playback (foldback)   226
Music and effects (M & E track)   226
Audio effects   226
Audio effects using tape   227
Synthetic sound   230
Dolby noise reduction   231

11 **Film**   232

Basic filming process   232
Scripting   234
Single-Camera shooting   234
Production decisions   235
Uses of film in TV   236
Sound on film   239
Transmitting negative film   240
Reversal film   240
The mechanics of film reproduction   241
Telecine   242
Video control for telecine   242
Non-standard projection   243
Cuing film   244
Recording color TV on film   244

12 **Videotape**   247

Production flexibility   247
Problems of standards   247
The major advantages of VTR   247
Quadruplex (quad, transverse) scanning   248
Helical (slant track) scanning   249
Magnetic video disc   252
Tape formats   253
The reasons for videotape editing   254
Physical editing   254
Electronic editing   255

Manual edits using electronic splicing 256
Editing controllers 257
Editing using a time code 257
Off-line editing techniques 258
Videotape processors 259
Audio dub 260

13 Graphics 261

The purpose of titles 261
Readability 261
Backgrounds to titles 262
Forms of lettering 263
Character generators (alpha-numeric generators) 264
Forms of graphics 265
Animated graphics 267
Graphics with movement 268
Unseen drawing 271

14 The background of production 275

The director's role 276
Production emphasis 276
Production approaches 276
Routines 278
Ambience 279
Selective tools 279
Selective techniques 279
The screen transforms reality 281
Interpretative techniques 281
Making the contrived arise 'naturally' 282
Gratuitous techniques 282
Production pressures 283

15 Production practices 284

Single-camera treatment 284
Multi-camera treatment 284
Visual variety 286
Shot organization 289
Program opening 295
Subjective and objective approaches 296
Focusing audience attention 297
Creating tension 300
Pace 300
Timing 301
Visual clarity 302
Confusing and frustrating techniques 302
Interest or concentration patterns 303
Visual padding 306
Illusion of time 308
Interscene devices 310
Deliberate disruption 310
Illusion of spectacle 312

16 Production organization 314

Unplanned production 314
Unrehearsed formats 314
Regular formats 315

Complex productions   316
The script   320
Auxiliary information   322
Calculating shots   328

17 In production   332

Pre-studio rehearsal   332
Studio rehearsal   332
Rehearsal procedures   334
Rehearsal problems   336
Floor manager   338
Guiding performers   341
Cuing   342
Prompting   344
Production timing   345

18 Imaginative production   352

Pictorial function   352
Production rhetoric   352
Summary of devices used   353
Examples   355
Imaginative sound   358
Sound elements   358
Sound emphasis   359
Sound usage   360
Off-screen sound   360
Substituted sound   361
Controlling sound treatment   362
Audio analysis   362
Focusing attention   365
Selective sound   366
Audio-visual relationships   367

19 Visual effects   368

Mirror effects   368
Rear (back) projection   370
Front projection   370
Reflex projection   371
Camera filters   371
Lens attachments   375
Camera mattes (gobos)   376
Electronic effects   377
Keyed insertion electronic matting)   380
External key   382
Self-keying (self-matting, internal key)   384
Using keyed insertion   385
Chroma-key insertion techniques   390
Complex insertion   397
Color synthesis   398
Digital systems   399
Temporal effects   400

20 On location

Large mobile control room (location production unit)   403
Remotes van   403
Lightweight truck small van units   403
    404

Station wagon small van units   404
Portable units   404
Program handling   404
ENG Editing   406
Types of field camera   407
Field camera equipment   408

21 Video engineering   411

Video control (shading)   411
Preset control   411
Automatic control   412
Shader   412
Video control operator (video control)   412
Exposure control   413
Black level   413
Further controls   414
Artistic aspects of video control   414
The television camera-tube   416
Synchronization   418

Glossary   423

Bibliography   440

Index   443

To
my wife, Pamela

Preface

You cannot learn TV production direct from any book! It is essentially a very practical process, and there is no substitute for personal experience. But knowledge through experience alone can take a long time to acquire. In any craft you find a blend of proven principles and practices that support the free expression of the artist. Without a systematic study, it is all too easy to work hopefully but ineffectually. The more quickly you master these background facts, the sooner you will build up confidence and expertise.

This is essentially a *practical* book, discussing from first principles the processes behind effective production techniques. It has been designed to be adapted to introductory, intermediate and advanced studies.

Do not be daunted by technicalities! As you will see, the basics are simple enough. Included here are easy reference digests, compact diagrams, and quick-guide tables. You will find guidance rather than 'rules'; advice, rather than instructions. Reminder lists summarize operational 'know-how.'

This book will not only provide you with 'how-to,' but explain 'why.' It discusses not only production *mechanics* and *techniques*, but the *aesthetics* of the medium; the power to persuade and the nature of its audience appeal. This book aims to encourage experiment and individuality, as well as showing how the various facets of production come together.

Over the past decade television has undergone many changes in technology, production techniques, program formats and teaching requirements. Consequently, this book has been totally revised and reset to meet your current and future training needs. Combined with practical experience, it can help you develop a firm understanding of the most important contributory features of TV production.

Throughout the book, the masculine gender rather than the more laborious 'persons' has been used for convenience and consistency, although the various jobs are carried out by women and men.

The author would like to thank the Director of Engineering of the British Broadcasting Corporation for permission to publish this book.

Typical development for a complex production

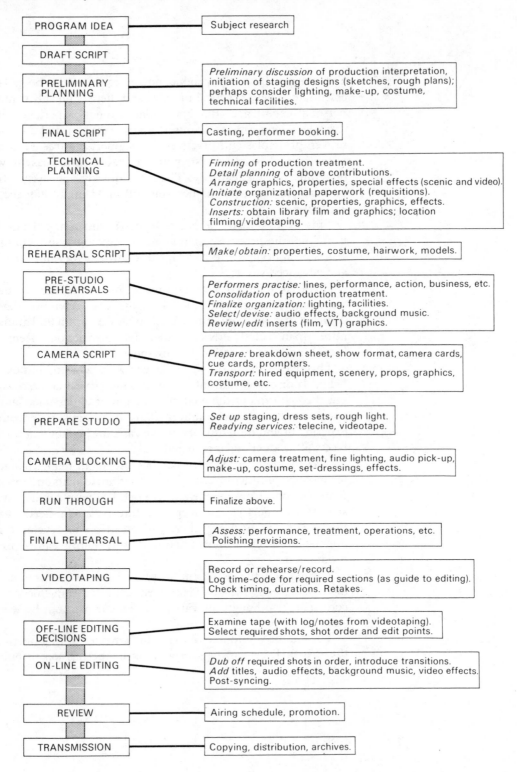

PROGRAM IDEA — Subject research

DRAFT SCRIPT

PRELIMINARY PLANNING — *Preliminary discussion* of production interpretation, initiation of staging designs (sketches, rough plans); perhaps consider lighting, make-up, costume, technical facilities.

FINAL SCRIPT — Casting, performer booking.

TECHNICAL PLANNING — *Firming* of production treatment.
Detail planning of above contributions.
Arrange graphics, properties, special effects (scenic and video).
Initiate organizational paperwork (requisitions).
Construction: scenic, properties, graphics, effects.
Inserts: obtain library film and graphics; location filming/videotaping.

REHEARSAL SCRIPT — *Make/obtain:* properties, costume, hairwork, models.

PRE-STUDIO REHEARSALS — *Performers practise:* lines, performance, action, business, etc.
Consolidation of production treatment.
Finalize organization: lighting, facilities.
Select/devise: audio effects, background music.
Review/edit inserts (film, VT) graphics.

CAMERA SCRIPT — *Prepare:* breakdown sheet, show format, camera cards, cue cards, prompters.
Transport: hired equipment, scenery, props, graphics, costume, etc.

PREPARE STUDIO — *Set up* staging, dress sets, rough light.
Readying services: telecine, videotape.

CAMERA BLOCKING — *Adjust:* camera treatment, fine lighting, audio pick-up, make-up, costume, set-dressings, effects.

RUN THROUGH — Finalize above.

FINAL REHEARSAL — *Assess:* performance, treatment, operations, etc.
Polishing revisions.

VIDEOTAPING — Record or rehearse/record.
Log time-code for required sections (as guide to editing).
Check timing, durations. Retakes.

OFF-LINE EDITING DECISIONS — Examine tape (with log/notes from videotaping).
Select required shots, shot order and edit points.

ON-LINE EDITING — *Dub off* required shots in order, introduce transitions.
Add titles, audio effects, background music, video effects.
Post-syncing.

REVIEW — Airing schedule, promotion.

TRANSMISSION — Copying, distribution, archives.

1 Studio tour

Let us do the obvious and begin our exploration by visiting the TV studio itself. Although its size and facilities will vary considerably, a certain spirit seems to pervade them all; whether on campus, local station or mammoth network centers.

The television studio

As you push through the soundproof entry doors, the empty TV studio has a strangely inhospitable complexity. The stark expanse of smooth floor with lifeless equipment huddled in one corner. The acoustically padded walls bespattered with technical paraphernalia. A ceiling of dormant lamps clustering above. It has the impersonal silence of a deserted warehouse.

But introduce the scenery, switch on those lamps to bathe the settings in light and shade . . . and the atmosphere becomes expectantly transformed. And with this magic comes a feeling of urgency; a purposeful tension that radiates to performers and crew alike.

A hundred and one jobs converge. The furniture, drapes and bric-a-brac begin to clothe the scene as the sets are dressed. Cameras and sound booms are pushed into their opening positions. Lamps are selectively adjusted. Last-minute rearrangements of sets and props . . . By a daily organization miracle, the various pieces of the production combine to provide the complete TV show we have been planning and preparing. The cameras set up

Fig. 1.1 The television studio
A typical view of the empty studio: with uncluttered staging area, surround safety lane, overhead lighting battens and control-room observation window.

their shots. Now, on the picture monitor screens, the illusion begins to live!

Yet, within a time span of hours, the production will have completed rehearsals; the crew will have learned and practiced the director's treatment; performances hopefully reached their peak . . . and the program been recorded or transmitted. And then the whole 'magic cave' will be dismantled. Unbelievably quickly, the settings will be 'struck' and trucked away for stock or junking; equipment stored; and floors cleaned off. The original emptiness will return.

These are the conditions in which our theories and hopes become reality. And it is only when we relate our aspirations to the relentless pressure of practical conditions that we see TV production in its true perspective.

The television studio in action

Turn back the clock now, to earlier in the day when camera rehearsal was in progress for a drama production. The illuminated wall signs show that the studio is in closed-circuit rehearsal. So, although all the equipment is *hot* (operating), its output is only being monitored locally. At scheduled time, the studio video (pictures) and audio (sound) signals will be fed via the cooordinating *master control room* (the central switching/routing control point) to the videotape recorders or immediately to the transmitters for *live* telecasting.

Around the studio you can see examples of the *set designer's* art: the three-walled rooms, a section of a street scene and a summer garden. After surprise at the colorful realism of these sets, the first impression is one of endless lights! Suspended, clamped to

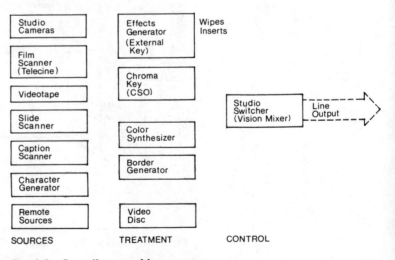

Fig. 1.2   Contributory video sources

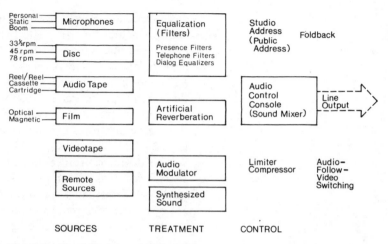

SOURCES   TREATMENT   CONTROL

Fig. 1.3   Contributory audio sources

sets, on floor stands . . . it is hard to appreciate that each has been placed and angled with precision, for a particular purpose.

At the moment, action is in a 'living room' (an *interior*). Despite the number of people working around, everywhere is surprisingly quiet. There is little reverberation, thanks to the dampening effect of the acoustically treated walls. A camera moves silently over the specially leveled floor—where the slightest bumps could judder the picture. Its long cable snakes away to a wall outlet. Through headsets, the studio crew hear the coordinating instructions from the production director. The technical and operation staff overlook the studio from the control room(s).

Outside the 'living room' you see the 'sunny garden' (an *exterior*). Close scrutiny reveals that the convincingly flowering

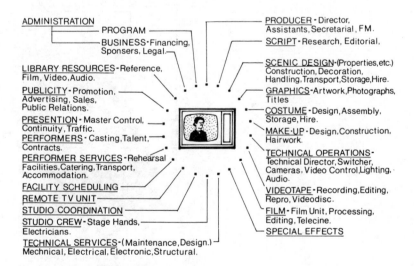

Fig. 1.4   Contributory services

trees are really plastic blossoms wired to lopped-off branches and the turf is grass matting laid over sandbags. But on camera, the effect is idyllic. Quiet birdsong fills the air (from a nearby *playback* or *foldback* loudspeaker).

The *floor manager* (FM), the director's link with the studio floor, hand cues an actress to enter the garden. The sound boom's telescopic arm swings, its microphone following her as she speaks. The cameras within the room pull away quickly and move to the next set.

The director, from behind the tinted glass of the *production control room*, is watching a series of picture monitors (high-quality TV displays) continuously showing each camera's shot. He hears the *program audio* (all sound sources blended together) on an adjacent hi-fi loudspeaker system. His continuous *intercom* (*talkback*) instructions to the crew's headsets are unheard by the actors or the studio microphones. The FM cues performers, guiding and advising them where necessary; diplomatically relaying the director's instructions and observations.

The director has just seen a bad shot, as a foreground actor *masks* (obscures) another. He asks the FM to stop action. 'Hold it please,' calls the FM. Action pauses while he explains the

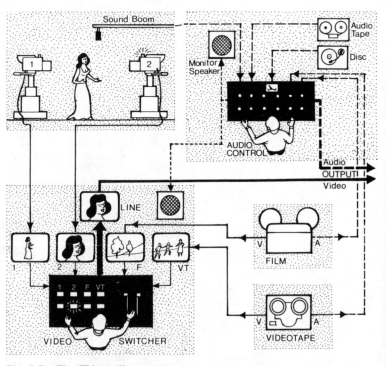

Fig. 1.5 The TV studio system
This simplified diagram shows how video and audio sources are routed to their respective selection points.

problem. A slight position change is crayon-marked on the floor (*toe/location marks*). Actors go through the revised moves, and all is well. But the director feels that the shot now looks too close. 'Can you pull back on Two?' Camera Two's cameraman shakes the camera's head to signal 'No.' 'Can you zoom out a little?' And with a wider lens angle, picture composition improves. On the director's go-ahead, the FM calls for silence. 'Right, back to Shot 54, as Joe enters.' His arm moves cuing action, and the game of make-believe springs back into life . . . as if no mere technicalities had interrupted. The rehearsal continues.

The production control room

Here in the production nerve center, the director and his staff of specialists control the production. There are two basic approaches to control room design: communal and sectionalized layouts. Each has its merits and shortcomings.

■ *The communal layout* This uses open-plan arrangements in either a single-level or split-level forms. Here everyone concerned with production control, sits at a long central desk: the *production director*, his *assistant* (who checks program timing, calls shot numbers and standby cues), the *switcher* (*vision mixer*), the *technical director* and *audio engineer*; and perhaps the *video control engineer* (*shader*). (As you will see the jobs, titles and names vary considerably!)

The desk faces a row of picture monitor screens. The *preview* (*channel*) *monitors* show the outputs of the various video sources and continuously display each camera's output as well as film, videotape, slide scanners and remote sources. Any source may be selected and switched (or faded) by the *production switcher* to the *studio output* (line output to master control.) The *master monitor* (*line, main channel, transmission*) displays the result.

The director's attention is divided between the various preview monitors, the selected output on the master monitor, and the *program audio* from a nearby loudspeaker. Via his headset, the director instructs his team—cuing action, assessing, guiding, explaining, coordinating their various contributions. All members of the team are joined into the communal intercom circuit; this can involve a fair amount of extraneous talk when participants hear others' instructions and exchanges.

For straightforward interswitching, the production director may operate the production switcher (video switching panel, vision mixer)—particularly on remotes (OBs). But for more complex operations, a specialist switcher (vision mixer) or the technical director usually does this job, enabling the production director to concentrate on controlling the many other aspects of the show.

The technical director's responsibilities can vary between one organization and another. He may direct cameras, actively assisting in staging and treatment, and he may carry out video switching. As a supervisory engineer, he may oversee the engineering aspects of the production such as aligning effects, checking shots, ensuring source availability and monitoring quality.

At the far end of the desk the *audio specialist* (*audio engineer, audio control*) sits at his audio console (board, panel) adjusting and blending the various sound sources.

Depending on the show and its organization, others directly involved in its presentation may include:

1. The *set designer* (*scenic designer*) responsible for the *scenic treatment* (*staging*).

2. The *lighting director* who designs and arranges the studio lighting.

3. The *producer, editors* (script, film, videotape), *sponsor*, special assistants, researchers . . . and so on ad infinitum.

The technical quality of pictures is controlled by a *video engineer* (*shader, video operator*) who continually monitors and adjusts video equipment, either within the production control room or in a separate technical area (*apparatus room* or in *master control*). There may be a small adjacent *announce booth* from which announcements and commentaries can be read *out of vision* (*off camera*).

■ *The sectionalized layout* Here the control area is subdivided into separate rooms:

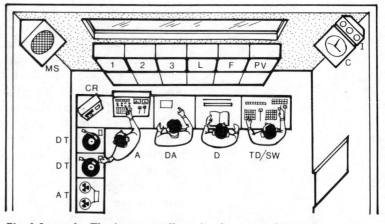

Fig. 1.6 part 1   The 'communal' production control room
All production operations are concentrated at a central desk. Video and lighting may be controlled locally or elsewhere.

1. Production control—where the director, assistant, switcher and technical director are located.

2. Sound control—audio coordination and control.

3. Vision control—video quality and lighting coordination.

Large glass panels and intercom circuits link these separate rooms. This layout has the advantage that their various activities and discussions do not impinge. No one in these control areas wears headsets and the intercom is of the local mike/loudspeaker type. The director's desk mike relays the general (communal) intercom (talkback) which is heard over other control rooms' loudspeakers. Individual specialists have their switched *private wire* (*party line*) circuits, for example from audio console to

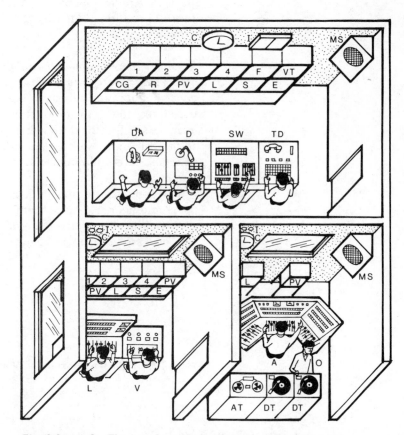

Fig. 1.6 part 2   The 'sectionalized' production control room
Production control, audio control and video/lighting control functions are located in separate rooms to prevent interaction.
Preview picture monitors show: Cameras 1, 2, 3, 4; film F; videotape VT; video effects E; slides S; character generator/titles CG; remote source R; switchable preview monitor PV.
Personnel: Director D; director's assistant DA; switcher (vision mixer) SW; technical director/technical manager TD; lighting controller L; Video controller V; audio controller A; audio operator (disc/tape) O.
Equipment: Audio tape AT; disc turntable DT; monitor speaker MS; clock C; on-air indicator lights I.

sound-boom operator, as well as being able to break into the general intercom system.

■ *Video control—vision control room*   In this area, the picture quality of all the local video sources is controlled and matched (cameras, slide/film scanner). Each studio camera's cable is plugged via a studio wall outlet to its main video apparatus—the *camera control unit* (CCU). These units, one per camera, are usually housed in an engineering area (master control, apparatus room) where they may be controlled directly by a video engineer.

In a more sophisticated arrangement, operation controls are extended to a central *video control console*, where an operator continually adjusts picture quality. Lighting and pictorial effect are closely interrelated, so the video operator usually sits next to the *lighting console* operator and shares the same picture monitors. These operators, guided by the lighting director, can therefore work in harmony.

■ *Audio control—sound control room*   At his *audio console* (*sound desk, mixer, board*) the *audio engineer* (*sound mixer, sound supervisor*) monitors and adjusts sound dynamics and quality, blending the sources to suit the production's artistic and technical requirements. As well as the studio microphones, he is controlling various additional audio sources including discs, audio tapes (reels, cassettes, carts), film and videotape tracks, announce booth and remotes.

The audio control man guides by intercom his sound crew on the studio floor. He helps them to avoid mikes or their shadows appearing in shot, warns them of subject moves or shot changes, etc., and assists them in achieving sound perspective to match

Table 1.1   Typical approximate TV studio sizes

			Typical	Ceiling
Local TV/campus studios	Small	150 m² (1600 ft²)	15×10 m (50×30 ft)	3·5 m (11 ft)
	Medium	216 m² (2400 ft²)	18×12 m (60×40 ft)	7 m (23 ft)
Network studio center	Small general purpose	330 m² (3500 ft²)	22×15 m (70×50 ft)	9 m (30 ft)
	Medium	672 m² (7200 ft²)	28×24 m (90×80 ft)	10 m (33 ft)
	Large	1024 m² (10 000 ft²)	32×32 m (100×100 ft)	13 m (43 ft)

the pictures. Whether he operates disc and tape recorders himself, or is assisted by an operator depends on production complexity and control room layout.

Master control

The *master control room* is the engineering coordination center for a TV station. It provides a master switching point which routes closed-circuit facilities, such as film chains to studios and studios to VTRs; and also program sources to the *continuity control room* for transmission. Because of the complexities of source routing, switching is becoming increasingly automated.

The master control room area may house film chains, VTRs, slide scanners, camera control equipment (CCUs), synchronizing generators, audio equipment, and so on. The continuity (presentation) control room is concerned with station program output continuity, that is program source switching; liaison with contributors concerning source cuing, timing, breakdowns, etc.; station idents; commercials insertion; promotion (trailers); and a local associated studio which covers such items as station announcements, interviews and weather.

2 How television works

Considering its technical complexity, the basic principles involved in producing a TV picture are surprisingly simple.

The video signal

In a monochrome TV system, an image of the scene is focused onto a *camera tube*. A *target* inside the tube builds up a pattern of electrical charges corresponding to light and shade variations in the image. A continuous *electron beam* generated within the camera tube reads across this charge pattern line by line and at the same time discharges it. The net result is a fluctuating voltage which is the *video* or picture signal.

In the receiver, a *picture tube (kinescope)* has a glass screen coated with a fluorescent powder. This glows according to the strength of an electron beam scanning across the screen and is synchronized with the camera tube's scanning process. As this beam's strength fluctuates, controlled by the video signal, it traces a light and shade pattern corresponding with the image of the original scene.

The brief period during which the scanning process returns (re-traces) from bottom-right to top-left of the field (the vertical interval/field blanking period) contains no picture information. It is used for various identification, test and control signals (perhaps including *teletext* information).

Color television

Any color can be analyzed into various proportions of red, green and blue light. Therefore to televise a scene in color, we analyze it into these three *primaries*. Although color TV cameras may use one, two or three camera pick-up tubes to do this, the principles of operation remain the same. The single lens image is split into three identical light paths which incorporate red, green and blue filters respectively. A normal monochrome camera tube behind each filter responds to the light and shade in its particular filtered

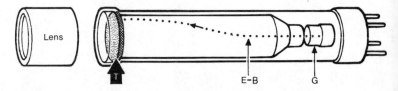

Fig. 2.1   The camera tube (Photoconductive types)
The inside of the camera tube's glass face-plate is covered with a fine metallic coating (signal plate), on which a layer of photoconductive material is deposited. The scene's image is focused onto this target (T). The electrical resistance of the layer varies at each surface point, with the light and shade of the scene, so building up a corresponding charge-pattern. The constant electron-beam (E-B) from the tube's gun (G) scans across the target in a series of close parallel lines, neutralizing each point (element) in turn. The resultant re-charging currents constitute the camera's video signal.

image, producing a corresponding video signal. So the camera provides information about the red, green and blue content of the scene (*chrominance*), and the brightness (*luminance*) at each point.

In the TV receiver (or a picture monitor) the color picture tube screen has a close pattern of different phosphors emitting red, green or blue light when scanned by three electron beams (one for each color). Again, they glow according to their relative video signals, and the eye blends and interprets these colors as a reproduction of the original scene.

Some color receivers include special compensatory circuits that

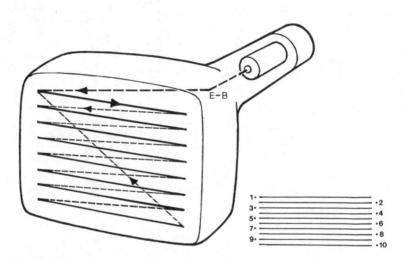

Fig. 2.2   The monochrome picture tube
A gun in the picture tube produces an electron beam (E-B), which fluctuates in intensity with the video signal. As the beam scans across the screen in a series of close parallel lines, it causes the fluorescent coating to glow correspondingly, building up the light and shade pattern of the original image. The picture (frame) is scanned in 525 lines (485 active) 30 times each second (Europe—625 lines, 50 per sec). To conserve bandwidth and decrease flicker effects, the odd lines (*odd field*) are scanned, and then the even lines (*even field*) fitted between them (*interlacing*). In simple TV systems, this interlacing varies (*random interlace*) and definition deteriorates.

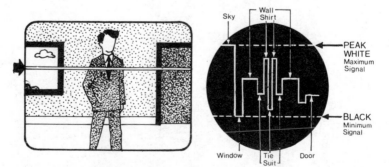

Fig. 2.3   The picture and its video
Showing a typical scanned line and its corresponding video signal.

automatically correct for signal distortions arising between the transmitter and the set. These circuits adjust tint and screen color automatically to the levels specified by the station's VIR (vertical interval reference) signals.

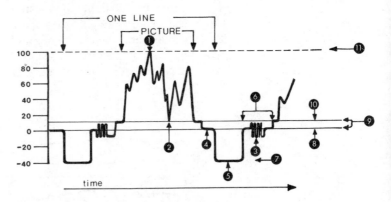

Fig. 2.4   The parts of the video signal
1, Peak white (peak modulation). 2, Peak black. 3, Color burst. 4, Front porch. 5, Horizontal (line) sync pulse. 6, Back porch. 7, Synchronizing level. 8, Blanking level. 9, Set-up (pedestal). 10, Reference black (black level). 11, Reference white (white level).
The picture information (video) is suppressed (blanking) at the end of each line (and each field) and a synchronizing pulse inserted—this ensures that the entire TV system from camera to picture tube scans in step. The brief color burst (of sub-carrier frequency) provides a color-reference signal.

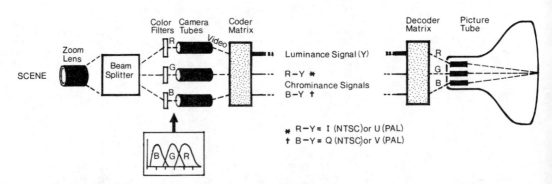

Fig. 2.5   Color TV principles
The focused image is split by a prismatic block (or dichroic filter-mirrors) into three light paths—covering the red, green, and blue regions of the spectrum. Camera tubes (3, 2, or 1) produce corresponding video signals. These signals could be transmitted in this form, and recombined to recreate the original color image (as in color printing processes). However, to provide a *compatible* color TV picture (that will reproduce on monochrome receivers) a complex inter mediate system is necessary—NTSC, PAL or SECAM. This involves *coding* to derive special *luminance* (brightness) and *chrominance* (hue, color) video signals. (Monochrome receivers respond to the luminous component of the signal.)
In the color picture-monitor or receiver, the separate RGB components are recovered by a *de-coding matrix* (decoder), each video signal being used to control their respective electron-beam strengths in the picture tube.

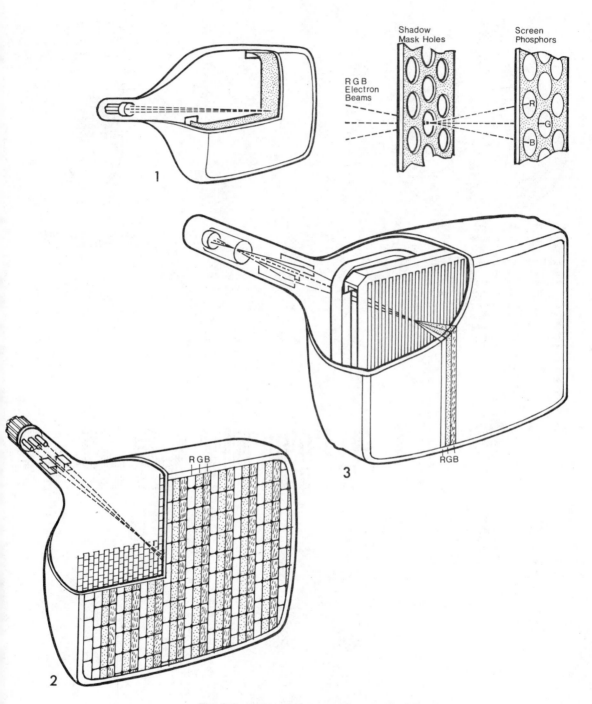

Fig. 2.6 Color picture-monitor or receiver
The tube screen comprises three patterns of phosphors (dots or stripes) that glow red, green, or blue when energized by their associated electron beam. The eye merges these tiny RGB patterns, seeing them as color mixtures; when proportionately activated they appear white to gray.
The color picture tube may take three forms: 1, Delta-gun dot mask (shadow mask); 2, Self-converging precision in-line (PIL) slot-mask design, or 3, Single-gun (three beams) aperture-grille type (Trinitron). The last two avoid the convergence color-fringing problems of 1.

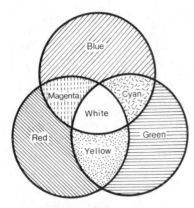

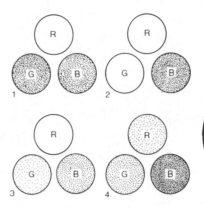

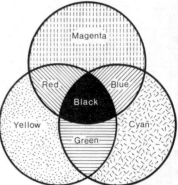

Fig. 2.7 Light mixtures— Additive primaries

Light can be analyzed into red, green, and blue components. Suitable mixtures of these *additive primaries* can reproduce most colors. Equal proportions of two such primaries produces a *secondary* color—magenta (purple), cyan (blue-green), yellow; also known as *subtractive primaries*. This mixture is often termed a 'minus' of the third primary hue: hence yellow is *minus blue*, cyan is *minus red*, and magenta is *minus green*.

Fig. 2.7 part 2

Hue derives from the *proportions* of red, green, and blue present. *Luminance* (brightness) results from their *quantities*. 1, If only one phosphor is activated, the screen appears of primary hue. 2, Two or three phosphors produce additive color mixtures (e.g. R+G=Yellow). 3, 4, If three phosphors are combined, various shades or degrees of color dilution (saturation) result (e.g. Diluted red= pink; Dim orange or yellow=brown).

Fig. 2.7 part 3 Pigment mixtures—Subtractive primaries

When pigments are used (ink, paint, dyes), the apparent surface color is actually the light remaining after surface absorption. So the mixing process becomes a subtractive one.

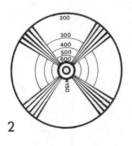

Fig. 2.8 Picture detail

The TV system's ability to resolve detail (its resolution) is checked with standard test charts. These contain *gratings* (1) or *resolution wedges* (2). The finest information that can be discerned, reveals the system's maximum resolution. (The calibrations indicate the 'number of picture lines' that could be resolved in the picture height.) In a 525 line system, typically '320 lines' (4 MHz) represents maximum transmitted detail. Maximum closed-circuit resolution may be '640 lines' (8 MHz).

Picture detail

The finest detail that can be resolved in a TV picture is determined by how rapidly the system can change from one tone to another during the scanning process. This rate of change is measured as frequency (4 MHz for a 525 line system; 5 MHz for 625 lines).

All systems (video and film) have a maximum detail limit (*definition, resolution*). In TV this is initially influenced by the number of transmitted picture lines. But ultimately it depends also on the performance of the whole video chain, from lens to receiver picture tube. Various factors affect picture clarity including lighting contrast, lens flares, dirty lens or prompter, focusing, video equipment design, alignment, and so on. Hence an unsharp picture can have many causes!

Tonal contrast

The tonal range available in the televised picture (*contrast range*) really lies between the picture tube's maximum brightness and its unlit tube face. This is often a range of about 10:1—depending on the design and adjustment of the receiver (or picture monitor), and ambient light spilling onto the screen. Under good conditions adjacent tone contrast of 20:1 (30:1 on widely spaced areas) can be achieved.

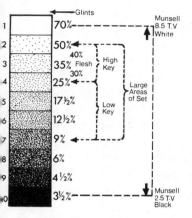

Fig. 2.9 The gray scale
The standard gray scale takes the maximum-to-minimum tonal range readily reproduced by the TV system (overall 20:1 contrast), and divides it into 10 steps. (Relative light reflectance 70–3½%) Each step is √2 times brightness of the next; the logarithmic scale looking linear to the eye. (What appears mid-gray, is only 17½% reflectance; not the 50% you might expect.)

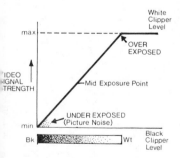

Fig. 2.11 Exposure curve
At very low inputs (dark tones) the video signal's strength is comparable with inherent system noise. At high inputs, the video reaches system limits and any variations block off to white. Only within the straight part of the curve (with unity gamma) are relative tonal values reproduced accurately.

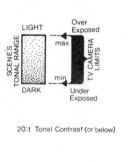
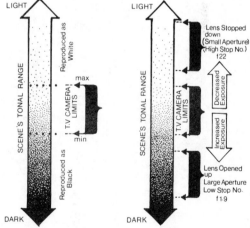

Fig. 2.10 Exposure
The camera's contrast range is limited (e.g. 20:1). Tones within these limits are reproduced proportionally. If scenic tones are relatively restricted (within 20:1 range), the f-stop can be adjusted to accommodate them.
If scenic contrast (subject tones, lighting contrast) exceeds the TV camera's range, detail and gradation beyond these limits are lost. (Above max, reproduces white; below min, reproduces black.) Tones within the system's limits, reproduce well.
Adjusting lens aperture (f-stop) moves camera's exposure bracket up or down the scenic tonal scale, determining which tonal segment is to be reproduced successfully. *Opening up* (e.g. f2) progressively improves shadow detail, but increasingly over-exposes light tones (blocked off to white). *Stopping down* (e.g. f22) progressively improves detail in lightest tones, but increasingly under-exposes dark tones (crushing to black).

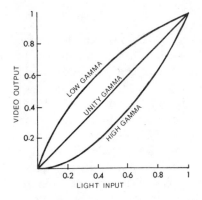

Fig. 2.12 Gamma
The *gamma curve* reveals how a video system reproduces tonal gradation. Where a device is linear (gamma of 1) its output is proportional to the input. A *low gamma* device accepts a wide contrast range, compressing it into the system's limits; resulting in reduced tonal contrast (thin). A high gamma device accepts only limited subject contrast, expanding it to fit the reproduction limits; giving exaggerated coarsened tonal values.

When you shoot a scene involving a considerable contrast range, due to extreme scenic tones and lighting contrast, you must recognize that although the eye can accommodate about 100:1, the camera tube can only handle a range of some 20:1 to 30:1. The reproduced picture therefore has its limitations.

Where you can control contrast of subjects (scenic tones, costume, lighting), it is best to keep these within the camera's tonal limits, unless you are prepared to lose extreme tones (highlights or shadows).

Tonal contrasts between subjects and backgrounds are important, too. If they are of fairly similar tone (their *contrast ratio* is small), they will tend to merge and the picture appear flat and lifeless.

Tonal gradation

For most pictorial subjects, we are concerned not only with a picture's tonal extremes, but how well the system reproduces intermediate tones (gray scale values) between these limits. And this is directly affected by camera channel design and operation, and by the receiver or picture monitor's performance.

Table 2.1   Color terms

Hue	The predominant sensation of color i.e. red, blue, yellow, etc.
Saturation/chroma/purity/ intensity	The color strength; how far it has been 'diluted'. *Desaturation* produces pastel colors, such as pink.
Luminance/brightness/value	A color's apparent brightness (lightness or darkness).
Tint	A hue diluted with white.
Tone	A grayed hue.
Shade	A hue mixed with black.
Achromatic/gray scale	Intermediate tonal steps (*values*) from black to white.
Monochrome	Generally refers to 'black-and-white' (achromatic) reproduction. (Strictly means varying brightnesses of *any* hue.)

The terms Hue, Chroma, and Value are used in the Munsell system of color notation.

3 What your camera can do

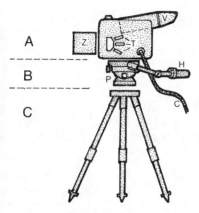

Fig. 3.1   The television camera
The TV camera consists of: A. *The camera head*—Z, Zoom lens. T, Camera tubes. V, Viewfinder. C, Camera cable (taking technical supplies to the camera; and the resultant video to the camera control unit/CCU). B. *The panning head* (pan head)— This mechanism (P) enables the camera head to tilt and turn (*pan*) smoothly. These movements can be restrained by deliberately introduced friction (*drag*), and can be *locked off* to hold the head rigid. *Tilt balance* positions the head horizontally, preventing nose or tail unbalance. An attached *panning* (*pan*) *handle*(*s*) (H) enables the cameraman to direct and control the camera head. C. *The camera mounting*—This can take many forms e.g. camera-clamp, monopod, tripod, pedestal, crane.

Whether your TV camera produces color or monochrome pictures, is hand-held or fitted to a giant crane, the basic operating principles remain the same.

The lens

The camera's lens is actually made up of a series of individual lenses, referred to as *elements*, set within a tubular barrel. These elements have been carefully designed and combined for optimum overall performance, minimizing various inherent optical defects (aberrations). This lens arrangement focuses a small sharp image of the scene on to the camera's pickup tube(s).

There are two basic lens types: those with a fixed focal length and those with an adjustable focal length (*zoom lens*). Fixed focal length lenses are fitted to most photographic cameras, to film and TV cameras of simpler design and to units with rotatable lens turrets. The fixed focal length lens (focus and *f*-stop are adjustable) is designed to cover a certain segment of the scene—known as the *angle of view* or *lens angle*. The zoom lens on the other hand, has variable coverage.

The impressions of distance, space and size that your picture conveys, will be strongly influenced by the angle of view of the lens used. While a 'normal' or 'standard' lens will show the scene in reasonably natural proportions, other lens angles will modify effective perspective. While a *wide-angle* ('*short-focus*') lens seems to exaggerate distance, a *narrow-angle* ('*long-focus*') lens appears to compress it.

The lens system

Your camera's lens system will influence how you use the camera.

■ *A fixed focal length lens*   This provides you with a certain image size at a given distance. If you want a larger image of your subject you must move the camera closer or bring the subject nearer.

■ *A lens turret system*   This provides you with a progressive selection of fixed lenses, each with a different coverage of the scene. You can change the picture (while off-shot) by turning to the nearest suitable lens angle—moving the camera slightly perhaps to trim the framing exactly.

■ *A zoom lens*   This has a variable focal length, and so can be continuously adjusted to provide any shot size you wish within its range, without interrupting the picture or needing to move the camera at all.

So you can select the camera lens angle to suit the size of shot required, instead of having to move the subject or the camera. But as you will see, this technique has its disadvantages because it can introduce perspective distortion and handling problems for the cameraman.

Focal length and lens angle

When we want to identify which particular lens (or zoom setting) is being used, we can quote either its *focal length* (*F*) or its *lens angle* (horizontal angle of view).

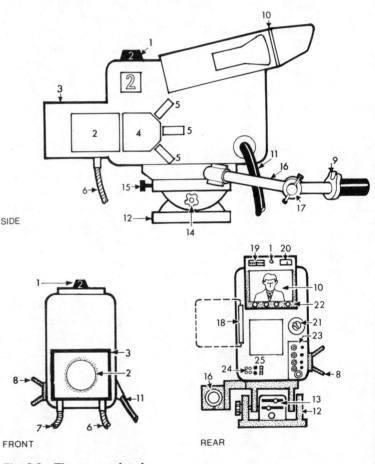

Fig. 3.2   The camera head
1, Tally light (on-air selected) with camera number. 2, Zoom lens. 3, Lens hood (sun shade, ray shade). 4, Beam-splitter prism block. 5, Camera tubes. 6, Zoom-control cable. 7, Zoom focus cable. 8, Zoom focus control (integrated in camera head). 9, Zoom control (thumb lever). 10, Viewfinder (monochrome) with magnifier. 11, Camera cable. 12, Panning head. 13, Pan head controls (Pan and tilt lock. Adjust Pan drag/friction. 14, Adjust tilt drag/friction. 15, Tilt balance. 16, Panning handle (pan bar). 17, Adjust pan-handle angle. 18, Clip for camera card. 19, Zoom lens indicator meter (For lens-angle). 20, Lens aperture (*f*-stop) indicator. 21, Lens range-extender. 22, Viewfinder controls (including hi-peaker crispening detail). 23, Shot box (integrated in camera head or on pan bar). 24, Headset jack-points (intercom and program), with volume controls. 25, Mixed viewfinder feeds switch.

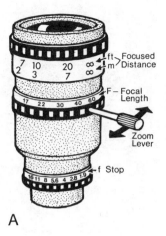

A

B

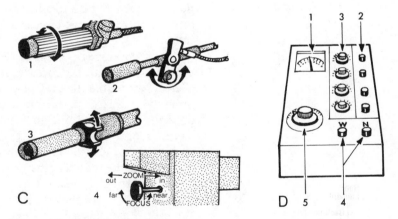

C

D

Fig. 3.3 Camera controls
Camera controls for focus, zoom and f stop (lens-aperture) can take several forms: *Direct lens adjustment*—turning a sleeve ring, lever, or handle. *Pushrod system* (through the camera head)—turn to focus, push-pull to zoom. *Direct cable control*—Flexible cables from lens to pan handle; focus control on left, zoom control on right. (*Iris control*: lens barrel, remotely motorized, or automatic.) *Motor control*—Small motors directly adjust lens (local or remotely controlled). *Servo control*—Adjustment of a control unit, produces corresponding change in slave unit. Turning pan-bar twist-grip or thumb-lever adjusts zoom direction and speed (increases with angle of twist). With direct manual control, adjustment often coarsens at longer focal lengths. Compensated servo systems maintain a similar 'feel' irrespective of focal length. A, *Lightweight zoom lens* showing focus (distance) control, zoom (focal length) lever, f stop (lens aperture). B, *Forms of focus control*—1, Spoked capstan control; 2, Twist-grip; 3, Camera-head control. C, *Forms of zoom control*—1, Twist-grip; 2, Hand crank; 3, Thumb-lever, 4, Push rod (also focuses). D, *Shot box* (provides pre-selected zoom settings, overriden by thumb control)—1, Zoom meter (shows focal length or lens angle); 2, Angle-select buttons; 3, Preset angle (for button selection); 4, Select widest/narrowest lens angle; 5, Zoom-speed adjustment. The shot may be attached to panning handle, or integrated into camera-head.

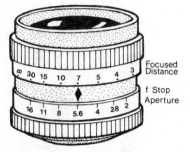

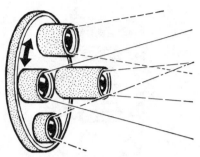

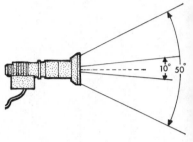

Fig. 3.4 part 1 Fixed optics lens
Adjustable focus (distance) and lens stop. Fixed lens angle (focal length).

Fig. 3.4 part 2 Turret lens
Selection of fixed lenses of different focal lengths fitted to a rotated disc (manual or electrical control). Typically, focal lengths of 12·5 mm (wide angle), 25 mm (normal), 75 mm (narrow), 100 mm ($\frac{1}{2}$, 1, 3, 4 inch).

Fig. 3.4 part 3 Zoom lens
Adjustable focus, stop, and focal length (lens angle). Altering focal length varies the field of view. Usually set at a pre-selected angle. Changing the angle in shot produces 'zooming in' (as angle narrows, F increases), or 'zooming out' (angle widens, F decreases). Typically 12–120 mm focal length adjustments.

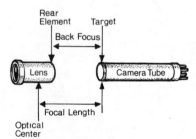

Fig. 3.5 Basic lens characteristics
Focal length—The distance from the lens' optical center (rear nodal point) to the tube target when focused at infinity. *Back focus*—The distance from the back lens surface to the target, when focused at infinity.

■ *Focal length* For many years, film and TV cameramen using fixed lenses have found the focal length (F) inscribed on each lens in millimetres of inches a very convenient method of identification. (Focal length is defined as the distance from the camera tube target to the lens' optical center when the lens is focused on infinity.)

A lens' coverage (its angle of view) and the subject image size it provides vary with the focal length of the lens. If you change to a lens with double the previous focal length, the subject now appears twice as large in the frame, but the shot covers only half the previous scene width/height. A lens of half this original focal length reproduces the subject as half the size with the scene width/height being double the original.

Knowing the focal length of any lens, you can check with tables for your particular *format* (e.g. for 35 mm film or 25 mm [1 inch] camera tube) to see the area it will show at the required distance. Where the lens has a focal length equal to the diagonal length of the format used, it will provide 'natural' perspective and be termed a *normal* lens.

Providing you continually use cameras of the same format, the operational practice of identifying lenses by their focal length is very convenient. You become accustomed, for example, to the subject distance at which a 25 mm (1 inch) lens gives a 'waist shot'. But what happens if you then use a camera with a different size camera tube? Now you may find that a 25 mm (1 inch) lens at this distance provides a 'large-head shot', and your previous associations between shot size and focal length are upset! This is because the lens' coverage or angle of view depends on both the focal length *and* the film or camera tube format. As there are now several formats in use, these shot variations can be confusing, especially as lens designs of identical focal length can have dis-similar coverage anyway!

■ *Lens angle* To overcome such problems, some TV organiza-tion refer to their lenses operationally by the *horizontal angle of view*.

This convention has several practical advantages. The *camera-man* gets used to the shots various lens angles provide, and these will be the same whatever the camera system's format. All '25° lenses' give the same shots; whereas a '25 mm (1 inch) lens' gives different shots with different size camera tubes.

For the *director*, there are many bonuses. You do not need

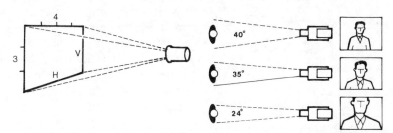

Fig. 3.6 part 1   Angle of view
The TV camera lens 'sees' 4 by 3 proportions. The vertical angle of view is therefore 3/4 of its horizontal angle.
The angle a lens covers, depends on its *focal length* relative to the size of camera-tube used.
For a given camera position, the longer the lens focal length (narrower lens angle) the closer the subject appears; but less of the scene is visible. The shorter the focal length (wider lens angle), the more distant the subject appears.

Fig. 3.6 part 2   Lens angles
Changes are proportional as the lens-angle alters. Using a lens of three times the present angle (i.e. $\frac{1}{3}$F), the subject now appears $\frac{1}{3}$ of former size, and ×3 of scene width is now visible. The effect is that of increasing the camera distance by 3 times. Changing to a narrower angle has the opposite effect.

Fig. 3.7   Viewing distance
Ideally, a photograph should be viewed so that the image subtends a similar angle to that of the original camera lens. The perspective (relative distances, sizes, depths, proportions) will then appear *natural*. Viewing distance, therefore, should ideally depend on the screen size.

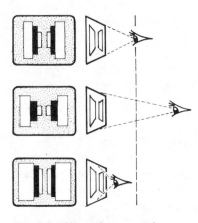

Fig. 3.8 Natural perspective
When the viewing angle equals the camera's lens angle our impression of the picture's perspective will be similar to that of the original scene. Viewing the picture from too far away or using a camera lens of too wide angle for normal viewing conditions gives a false impression of perspective. Depth and distance are exaggerated; distant subjects unduly small. Viewing the picture too closely or using a too narrow angle lens for normal viewing conditions again gives a false impression. Depth and distance are compressed; distant subjects unduly large.

tables to determine shot size. Simply by drawing a 'V' of the chosen angle on a scale plan, or better still using a protractor, you can judge exactly what the camera has in shot. You can determine whether unwanted subjects will be seen, the shot size, relative proportions in the shot, and so on.

Perspective distortion

The strange pictorial effects arising from perspective distortion have become familiar to most TV audiences. Examples of this are grotesquely distorted closeups, squashed-up cardboarded images of distant subjects, and the bent architecture portrayed by the fisheye lens. Used deliberately for a special purpose, they can have dramatic impact. But how do you control this distortion and use it selectively?

Perspective distortion arises whenever your camera's lens angle differs considerably from the audience's *viewing angle*, which is around 24° from the eye to either side-edge of the screen. Their impression of perspective—space, distance, proportion—will be inaccurate.

■ *Normal perspective* In practice, 'normal' perspective is a fairly arbitrary concept. We are least likely to notice inaccuracies in shots of open spaces, such as seascapes or deserts, and are most critical of perspective for familiar architectural surroundings.

Although the ideal distance is where line detail just disappears, people watch TV screens of various sizes at various distances. Even a photograph shot on a 'normal' lens will produce different perspective effects as the print is enlarged.

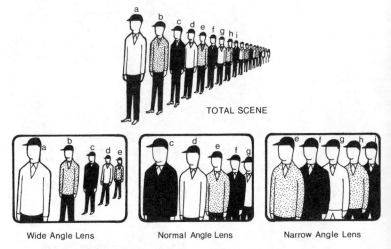

TOTAL SCENE

Wide Angle Lens   Normal Angle Lens   Narrow Angle Lens

Fig. 3.9 Perspective distortion
Reducing the lens angle (increasing F) provides a magnified image of part of the scene. This emphasizes perspective effects normally associated with distant subjects, and produces an illusion of space compression. (See GLOSSARY.)

Table 3.1   Typical lens coverages

Focal length (F) mm (in)	Hori- zontal angle	Vertical angle	Horizontal field of view at distances of		
			1·5 m (5 ft) m (ft)	6 m (20 ft) m (ft)	30 m (100 ft) m (ft)
12·5 (0·5)	54°	40°	1·53 (5·0)	6·1 (20·0)	30·5 (100·0)
25 (1·0)	27	20	0·76 (2·5)	3·1 (10·0)	15·3 (50·0)
50 (2·0)	13·5	10	0·40 (1·3)	1·5 (5·0)	7·6 (25·0)
75 (3·0)	9	7	0·24 (0·8)	1·0 (3·3)	5·2 (17·0)
150 (6·0)	4·5	3·5	0·12 (0·4)	0·5 (1·6)	2·5 (8·3)
17 (0·6)	50°	37·5°	1·46 (4·8)	5·80 (19·0)	29·0 (95)
35 (1·5)	26	19	0·64 (2·1)	2·56 (8·4)	12·8 (42)
52 (2·0)	17	13	0·49 (1·6)	1·95 (6·4)	9·8 (32)
90 (3·5)	9·5	7	0·28 (0·9)	1·10 (3·6)	5·5 (18)
135 (5·3)	6·5	5	0·18 (0·6)	0·73 (2·4)	3·7 (12)
215 (8·5)	4	3	0·12 (0·4)	0·49 (1·6)	2·4 (8)

25 mm (1 in) Plumbicon and Vidicon (first block)
32 mm (1·25 in) standard Plumbicon (second block)

However, for most purposes, if we use a lens around 20–27° the results are natural enough. In practice this means for a:

30 mm (1·25 in) camera tube a lens focal length of about 20–25 mm (0·8–1·0 in) gives a 'normal' lens angle.

25 mm (1·00 in) camera tube a lens focal length of about 16–20 mm (0·6–0·8 in) gives a 'normal' lens angle.

17 mm (0·67 in) camera tube a lens focal length of about 11 mm (0·4 in) gives a 'normal' lens angle.

Long-focus lens (narrow-angle lens)

The long-focus (telephoto) or narrow-angle lens enables you to select small parts of the subject from a distance. This is a particular advantage when you cannot move the camera nearer the subject because of obstructions, terrain, static camera, etc. Outside the studio, at remotes or outside broadcasts, these lenses often prove the only means of getting 'detailed' shots.

■ *Distortion effects*   Purists argue that no lens *distorts* perspective, but this is only true providing the viewer adjusts his distance from the screen to compensate for perspective changes. Under

Table 3.2   Camera tube sizes

	Diameter mm (in)	Image size mm (in)	Diagonal of image mm (in)
Image Orthicon	75 (3·00) or 113 (4·5)	32·5×24·4 (1·28×0·96)	40·6 (1·6)
Vidicons/Plumbicons	30 (1·25)	16·0×12·0 (0·63×0·47)	20·0 (0·84)
	25 (1·00)	12·8× 9·6 (0·50×0·38)	16·0 (0·63)
Plumbicons	18 (0·67)	8·8× 6·6 (0·35×0·26)	11·0 (0·43)
Saticon	Small lightweight formats		

practical conditions we find when using a narrow-angle lens, that depth appears remarkably compressed and foreground to background distance seems much shorter than it really is. Solid objects look 'depth-squashed'. People seem to take an interminable time, even when running, to move toward or from the camera. Distant objects are shown disproportionately large:

Seen end-on, a jumbo jet looks only a few meters long.
At the race track, horses gallop towards us but appear to cover little ground.
Flat cramped figures of a rowing eight, squat in a foreshortened boat.
Shots along piano keyboards appear strangely foreshortened.

Even in the studio, at much shorter distances using lenses around 5–10°, people can appear quite distorted as 'closeups' show unpleasantly flattened features, flat chests and reduced modeling (contouring). This is particularly noticeable in three-quarters face and full-face views.

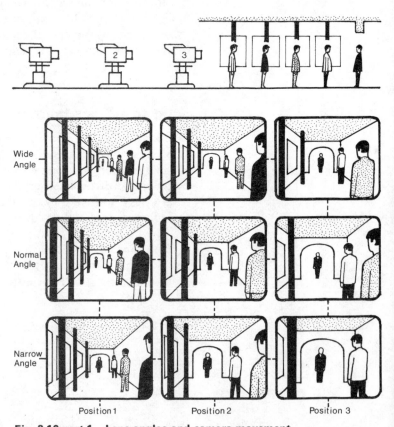

Fig. 3.10 part 1   Lens angles and camera movement
Here we see the effect of moving the camera (positions 1, 2, 3) with wide angle, normal, narrow-angle lenses.

■ *Camera handling*   Even if you chose to disregard such effects productionally, the cameraman cannot overlook that camerawork becomes more difficult as the lens angle narrows (focal length increases). Even slight camera movement, caused by a wavering hand-grip or dollying over uneven floor, produces disturbing picture judder. On remote (OBs) where lenses of $\frac{1}{2}$–$5°$ become essential to see otherwise inaccessible action, such problems become acute. The camera may even have to be rigidly *locked-off* on its panning head to ensure picture steadiness, rather than held ready to tilt and pan.

In hot weather, the diffraction due to heat haze may cause an overall shimmering on 'close' shots of distant subjects. The only remedy is to move closer to the subject.

Focusing too, can become a problem as the available depth of field becomes quite shallow on long-focus (narrow-angle) lenses.

Short-focus lens (wide-angle lens)

This lens can provide you with a wide overall view of a scene, even when the camera is relatively close. A boon when shooting in confined or congested surroundings, where you cannot move away sufficiently to obtain such a shot on a normal lens, or where foreground objects would intrude.

■ *Distortion effects*   However, the wide-angle lens appears to *exaggerate* perspective, making subjects seem further away than they really are. A small room may look quite spacious—even impressively extensive when shot on this type of lens. This is a great facility when we want to emphasize space or make smaller settings appear large. But in naturally small surroundings, such as a cell or a hovel, this can destroy scale and realism.

Shooting people on wide-angle lenses results in bizarre distortions; puffed-up bulbous heads, large protruding noses, stumpy legs, and enlarged chests. Fine for grotesques but totally unacceptable for flattering portraiture.

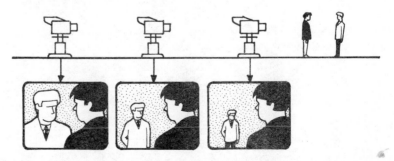

Fig. 3.10 part 2   Maintaining foreground subject size
If lens is changed, and camera distance altered to maintain the same foreground subject size, proportions change too. 1, Narrow; 2, Normal; 3, Wide angle lens.

Very wide-angle lenses can produce geometric distortion that is particularly noticeable on graphics, music and printing. When the camera pans, straight verticals bend as they approach the picture edge.

The wider the lens angle (shorter focal length), the more pronounced will all these effects be. The fisheye lens (between 140° and 360°) can only be used for special applications.

The wide-angle lens exaggerates speed of movement along the lens axis, so that camera dollying looks faster than it actually is and people coming up to camera seem to cover distance remarkably rapidly.

■ *Camera handling* On the plus side, smooth camerawork is much easier when using a wide-angle lens, for any movement or unevenness is less liable to cause picture judder. Moreover, focusing is less critical as the available depth of field is much greater.

Table 3.3   Lens angles

	Advantages	Disadvantages
'Normal' lens *(20°–27°)*	Perspective appears natural.	Inadequate width-coverage in confined spaces.
	Space/distance appears consistent when intercutting 'normal' lenses. Camera handling feels natural and is relatively stable. Focused depth is generally satisfactory, even though fairly shallow for closer shots.	Camera must move close to subject for detail; so may be seen by other cameras taking longer shots of the same subject.
Narrow-angle long-focus lens *(less than 20°)*	Brings distant subjects closer. Permits 'closer' shots of inaccessible subjects. Can flatten or reduce emphasis on surface contours. Enables objects spaced-out away from the camera to be compressed into a more cohesive group (e.g. a string of automobiles). Can suggest aborted effort in subjects moving quickly to/from the camera (e.g. running). Defocuses distracting background.	Depth appears compressed. Heat haze may shimmer the picture. Flattening of subject modeling can appear unnatural. Camera handling is considerably more difficult (pan, tilt, smooth dollying). Restricted depth of field makes focusing critical.
Wide-angle short-focus lens *(greater than 30°)*	Can obtain wide views at closer distances, especially in confined space. · Camera handling causes negligible judder. Because of extended depth of field, focusing is less critical. Enables modeling and space to be exaggerated. Impressive spatial effects possible. Useful for grotesque or dramatic shots. Provides considerable depth of field.	Depth appears exaggerated. (Confined spaces cannot be shown in correct proportions.) Pronounced geometric distortion, at widest angles. Susceptible to lens flares. On close-ups, problems with camera shadows falling on subject.

Supplementary lenses

By attaching a supplementary lens to the front of your lens, you can modify its focal length and hence its angle of view. Although mainly used for fixed lenses, these add-on lenses are employed in flip-in *range extenders* on zoom lenses.

A *positive supplementary lens* reduces the effective focal length and therefore widens the lens angle. It also means that the lens can be focused on closer subjects than normal. Held at arm's length, the positive supplementary lens shows a small inverted image.

A *negative supplementary lens* has the opposite effect, recognizable by its tiny upright image when held at arm's length. When used normally this lens will therefore narrow the lens angle. However, if placed beyond the main lens' focal length, the result is an increase in lens angle. The power of a supplementary lens is quoted in diopters.

The zoom lens

The practical disadvantage of a fixed lens is that it covers only one particular angle of view. If you want to see more or less of the scene you must either alter your distance from the subject or replace the lens with another of more suitable focal length. Replacement involves either unscrewing it, or using a *lens-turret* system. Before the advent of zoom lenses, these were the only possible solutions.

■ *Zoom range* Ingenious optical designs have enabled lenses of *variable focal length* to be constructed—the familiar 'zoom lens'. This can be adjusted to any lens angle within its operating range, either on shot ('zooming') or to a chosen angle. By turning a control we can fill the screen with a person's eye, or broaden the view until he becomes a dot in the landscape. The *zoom range/ zoom ratio*, which is the ratio of the longest to shortest focal length, may vary from some 3:1 to over 40:1. The zoomed-in

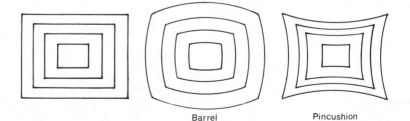

Barrel   Pincushion

Fig. 3.11   Geometrical (curvilinear) distortion
Optical distortion in which lines parallel to the frame edges appear progressively bent outwards (barrel distortion) or inwards (pincushion distortion).

image (narrowest angle) then appear forty times as large as when fully zoomed-out (widest angle). A 10:1 zoom is widely used in studio operations.

■ *Zoom performance* The zoom lens has certain limitations: higher cost, greater complexity, increased light losses (due to absorption and interlens reflections), with greater bulk and weight. Overall performance may well be below that achievable by fixed lenses. But technically and operationally, the zoom lens has many merits.

■ *Zoom advantages* Any chosen lens angle can be selected quickly (manually or by push-button—*shot box*) without losing picture. (Turret selection or lens-replacement methods involve delays.) Lens angle can be adjusted specifically to frame a subject without moving the camera.

Adjusting the lens angle alters the image size, simulating camera movement to/from the subject. For *flat* subjects, the effect is identical to dollying. But for three-dimensional scenes, the effect is artificial owing to the absence of spatial displacement movements between planes (*parallactic movement*). Smooth zooming is easier to achieve than smooth camera moves; particularly over long distances, or for fast shot changes.

■ *Zoom disadvantages* Unfortunately, it is all too easy to select lens angles indiscriminately, so that perspective becomes inappropriately distorted and intercut shots give different spatial

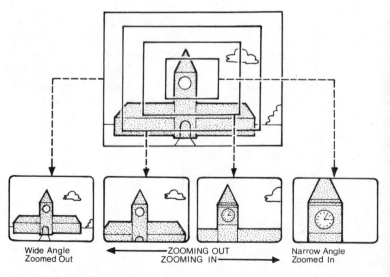

| Wide Angle
Zoomed Out | ←————ZOOMING OUT
ZOOMING IN————→ | Narrow Angle
Zoomed In |

Fig. 3.12 Zooming
Zooming in progressively fills the screen with a smaller section of the scene. (Narrowing lens angle, lengthening F.)

impressions. Shot sizes may not be repeated accurately, unless the cameraman notes his exact position and the angle used.

Remember, a zoom lens handles like a fixed-angle lens of the focal length you have selected. Consequently, at a *narrow-angle* (long focal length) setting, camerawork is more difficult and focusing is more critical. At a *wide-angle* (short focal length) setting, operation is easy and focusing less critical. The 'feel' of a zoom lens continually alters therefore, over its range.

Many zoom lenses do not have a constant performance throughout their range—focus may not remain sharp and the image brightness may vary (aperture losses).

Some zoom lens systems do not have meters or reliable indicators showing the selected angle (or focal length). So the zoom lens has to be used entirely empirically and production standards fall.

■ *Zoom range extender*   Continuous zooming over a wide range poses considerable lens design problems, so some systems introduce a cable-controlled flip-in supplementary *extender lens* to change the normal range by a factor of $1\frac{1}{2}$, 2 or 3 times. To zoom over its entire available range therefore requires resetting the system midway; often with a short overlap in between. Unfortunately, such extenders can cut down light and degrade picture definition.

Close-up attachment lenses are sometimes added similarly, to reduce the minimum focusing distance (closest point at which an object can be focused).

Focus controls

When you focus a camera, you adjust its lens to produce the sharpest possible image of your subject. With a fixed focal length lens this involves altering the distance between the lens and the camera tube; in a zoom lens, focusing is achieved by internal readjustments of lens elements.

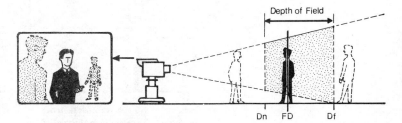

Fig. 3.13   Depth of field
Within the depth of field, subjects appear sharp; although definition is optimum at the *focused distance* (FD). Beyond the furthest acceptable distance (Df) and closer than the nearest limits (Dn), subjects are defocused.

■ *Focusing methods*   Lens focus can be operated by:

1. A *focus ring* on the lens barrel (screwing the lens in/out).
2. A *knob* at the side of the camera head.
3. A *twist grip* on a panning handle.

   Some cameras use a mechanical focusing system, in which the focus control is directly linked to the lens. Although simple, this method has disadvantages. Adjustment feels critical when focusing at narrow lens angles, while at wide angles appreciable control movement is needed before differences in focus are apparent.
   Where electrical *servo systems* are used, the focus control directs

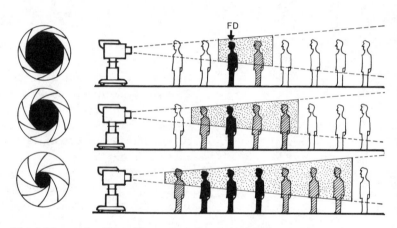

Fig. 3.14 part 1   The effect of aperture on depth of field
Keeping the focal length and focusing distance (FD) constant, the depth of field increases as lens aperture is reduced (stopped-down).

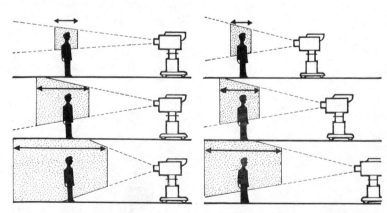

Fig. 3.14 part 2 The effect of subject distance on depth of field
Keeping the focal length and aperture constant, the depth of field increases as the camera gets further from the subject; but image size decreases.

Fig. 3.14 part 3 The effect of focal length on depth of field
Keeping the aperture and subject distance constant, the depth of field increases as focal length decreases (lens-angle is widened), but image size decreases.

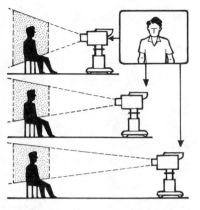

Fig. 3.15 Constant depth of field
If the camera distance is adjusted for each lens angle to keep the same subject size, the depth of field will remain the same (for a given *f* stop).

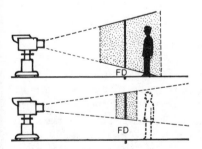

Fig. 3.16 Zoom lens focusing problems
Unless the camera has been pre-focused, the close-up after zooming in may be unsharp. *Zoomed out*—Although the focused distance is incorrect, the subject appears focused due to the considerable depth of field available (wide lens angle). *Zoomed in*—Without re-adjusting focusing, we zoom in. Now the depth of field has become shallower (narrower lens angle) and it no longer includes subject, which appears defocused.

a motor which adjusts the lens focus, and its 'feel' can be maintained irrespective of the lens angle (focal length).

■ *The need to adjust focus* A lens can only provide maximum sharpness at the distance to which it has been focused (focused distance). Sharpness gradually falls off nearer and further than this focused distance, until the image is too blurred for us to accept as sharp or 'in focus'.

The range of distances over which subjects are apparently sharp, is called the *depth of field* (*not depth of focus* which refers to focal plane tolerances). How far sharpness can deteriorate before you notice it, depends on your subject's detail and the system's limits (definition, resolution). Given a vague subject (detailless, plain) or a low-definition system, you have difficulty in seeing what *is* in focus!

Providing a subject remains within this depth of field, you do not need to readjust the focus control. The focused depth may be considerable, especially when the lens is focused at the *hyperfocal distance*; but it can also become embarrassingly shallow, particularly in close-shots.

Depth of field varies with:

1. Focusing distance.
2. Focal length.
3. Lens aperture (*f*-stop).

Changing any of these alters the available focused depth in the scene (Fig. 3.14).

■ *Focusing problems* When using the zoom lens, you may for instance, find focus drifting as you zoom (bad focus tracking) due to design or incorrect setting-up adjustments. But even with a high performance zoom lens there is always a major operational hazard:

The cameraman is zoomed out showing a sharp focused long shot of a crowd (using a wide angle—short focal length). He zooms in to a big closeup of a person (to a narrow angle—long focal length setting) . . . and the shot is now *completely out of focus*. Why?

Focusing with any zoom lens is uncritical when it is zoomed out. This is because the depth of field is considerable so the exact focused distance is not easy to see. Most subjects in the scene appear sharp. But when you zoom *in*, focused depth becomes progressively limited as the lens angle narrows, so you may well find that you have been focused at the wrong distance! Apart from corrective refocusing while on shot (often unavoidable in ENG camerawork), the only solution is anticipatory *prefocusing*—while off-shot, you zoom in, sharp focus, and zoom out to the required wider angle, so that you are now ready for a subsequent zoom in on shot.

45

■ *Minimum focusing distance*   Any lens can be focused over a range from *infinity* (far distance) to a plane a short distance from the camera. Nearer than this *minimum focusing distance* (mfd) the shot remains defocused, the control having reached its limit. This minimum distance varies with the type and design of the lens. While wide-angle (short-focus) lenses can focus down to a few centimeters away, the mfd for a very narrow-angle (long-focus) lens may be several meters from the lens. For a zoom lens therefore, this distance varies with the angle selected.

With *macro-focus zoom lens* design, focusing is possible right up to the front lens surface.

■ *Setting up a zoom lens*   In principle, you can regard the zoom assembly as a fixed-focus lens with a telescope of variable magnification in front of it. While its field of view changes, its back-focus distance (rear-lens to tube target) remains constant and sharply focused.

To ensure that the zoom lens maintains focus as its focal length is changed, the system has to be set up, as follows. First zoom in to distant detail and adjust the normal operational *zoom focus* control (*lens focus*, *rough* or *coarse focus*, *front focus*) for greatest clarity. Then zoom out and adjust the internal preset focus control (*camera focus*, *back* or *fine focus*) for sharpest image. Ideally you can now zoom in/out without focus drift (i.e. correct *focus tracking*). If necessary, repeat this setting-up for optimum results. When the subject's distance changes, you may need to readjust the zoom focus control to maintain sharpness.

Lens aperture—*f*-stop

Looking into a lens, you will see a variable circular diaphragm comprising a number of overlapping metal blades. This device (*diaphragm*, *iris*) alters the effective diameter of the lens opening over a wide range. The *aperture* (*stop*) is adjusted by a marked ring on the lens barrel—either directly or by remote control.

Changing the aperture (stop) does two quite separate things simultaneously:

1. It adjusts the amount of light (image brightness) falling on the camera tube (or film).

2. It alters the depth of field.

■ *Over- and underexposure*   If the camera tube receives excessive light from any part of the scene, those tones become *overexposed* and merge into unmodeled detailless areas (block off, crush out).

Insufficient light results in *underexposure*. Lighter tones are

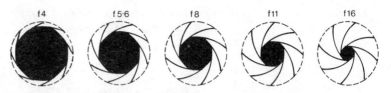

Fig. 3.17 Lens aperture
Lens aperture is adjustable from a maximum (e.g. *f*2 for a fast lens, *f*5·6 for a slow lens) to a minimum (e.g. *f*22). *Large aperture*—Small *f*-number, shallow depth of field, but less light needed. *Small aperture*—Large *f*-number, increased depth of field, but more light needed.

reproduced at too low a level, and darker tones mass as black detailless areas (*clog*, *crush*).

Over- or underexposure can also cause technical defects in the TV picture.

The lens aperture enables you to regulate exposure to suit the lighting intensities and subject tones. *Correct exposure* involves adjusting the aperture for optimum pictorial effect. For bright lighting and light-toned scenes, you reduce the lens aperture (*stop-down*); while for dimly lit surrounding or darker tones, you *open-up* the lens.

■ *f-numbers/f-stops* The lens aperture is graduated in *stops*. These are marked as *f-numbers* (*f-stops*) or *transmission* (*T*) numbers. The amount of light the lens lets through will be doubled for each complete stop you open (halved when stopping down). As a whole stop is an inconveniently large calibration, subdivisional *f*-numbers are normally used.

The standard series of lens markings is:
1·4/2/2·8/3·5/4/5·6/8/11/16/22/32.

The larger the aperture, the smaller will be the *f*-number.

To calculate the difference in light passed at two apertures, you use the formula:

$$\frac{(\text{First } f \text{ no.})^2}{(\text{Second } f \text{ no.})^2} = \text{light change}$$

Thus, from *f*4 to *f*8 the image brightness changes to $\frac{1}{4}$. If, therefore, you stop the lens down from *f*4 to *f*8 to increase the depth of field, you must increase lighting fourfold to maintain the same exposure.

■ *Working apertures* The highest grade lenses will still provide sharp pictures at their maximum aperture (e.g. *f*2) when opened up for low light levels, or to achieve a deliberately shallow depth of field (to isolate a subject). But clarity (*definition*, *resolution*) falls off discernibly towards the frame edges at large apertures.

Many lenses produce optimum picture quality at around *f*5·6 to $f8$. Stopping down to small apertures ($f16$) does not provide maximum definition, although increased focused depth may improve subject clarity.

Neutral density filters

When the scene is too bright for the aperture you want to work at, a *neutral density filter* is used to cut down the overall intensity. These transparent gray-tinted (neutral) discs are usually fitted in a filter wheel behind the camera lens. Typical values range from $1/10$ (10%) to $1/100$ (1%) transmission. A 10% filter will, for example, cut light down by nearly three stops (e.g. $f5\cdot6$ to $f16$).

Neutral density (ND) filters may be used when shooting in very strong sunlight, to prevent overexposure or the need for unduly small lens apertures. Again, should you want to open up the lens of a particular camera to restrict focused depth for artistic reasons, while perhaps other cameras use smaller stops, a ND filter can be used to compensate exposure.

Camera tube sensitivity

All TV cameras require a certain minimum general light level to produce clear, crisp pictures with good tonal gradation. The actual intensities needed depend on the camera tube's inherent sensitivity, scenic tones, the atmospheric effect required, and suitable depth of field (hence the f-stop).

Camera sensitivity varies with tube design and with technical adjustment (line-up). High sensitivity permits shooting under low light conditions and is invaluable for adverse situations when additional lighting is impractical.

In the studio, however, very dim lighting provides depressing and uncomfortable working conditions. Artistic lighting balance becomes difficult, as the eye cannot evaluate light values accurately, or detect extraneous shadows or spill light that are very evident on camera. Instead, when using high sensitivity tubes, we normally find ourselves lighting to comfortable intensities and stopping down or filtering to prevent overexposure.

Table 3.4 Light levels

Monochrome cameras	25 mm (1 in) vidicon	1200–1600 lux $f2\cdot8$
	25 mm (1 in) plumbicon	250 lux $f2\cdot8$
	30 mm ($1\frac{1}{4}$ in) plumbicon	500 lux $f4$
Color cameras	3×25 mm (1 in) plumbicon	750 lux $f2\cdot8$
	3×30 mm ($1\frac{1}{4}$ in) plumbicon*	1000–1600 lux $f4$
	3×17–18 mm ($\frac{2}{3}$ in) plumbicon	2000 lux $f2$

Typical operating levels 1600–2700 lux/150–250 at f4–f8.

Light levels (intensities) are measured in lux (lumens per square meter) or in foot candles (lumens per square foot). 1 fc=10·74 lux. Light levels are only an indication of camera sensitivity when related to the working aperture (f-stop).

Mountings for lightweight cameras

A camera's operational flexibility is considerably influenced by the mounting you use. Over the years a variety of devices have been developed, each providing its particular opportunities and limitations.

Lightweight cameras are of two kinds. Simpler designs are usually employed in closed-circuit TV (CCTV), classroom TV and for demonstration and surveillance purposes. More sophisticated mountings are used for broadcast work—for mobile location shooting and electronic news gathering (ENG).

Lightweight TV cameras have increased production opportunities considerably. They can move within the action, along the touchline, amongst the crowd and generally provide a new immediacy, covering events as they happen.

Lightweight cameras are mainly hand-held, and to avoid undue camera shake, jitter and unsteady handling, a *body-brace (shoulder harness)* is usually worn. Elaborate harnesses (Steadicam, Panaglide) provide outstanding compensatory stabilization, even for violent running and climbing! But underneath it all there is a human cameraman whose arms tire and back aches. Shots waver and standards become difficult to maintain.

Even when capable of the sustained broadcast quality of larger studio cameras, hand-held lightweights do not lend themselves to long days of continuous exacting rehearse/record production techniques. Although ideal for mobility over difficult terrain and confined conditions, the operational problems for high-grade camerawork on sustained static shots or when using narrower lens angles or multicamera treatment, must not be overlooked.

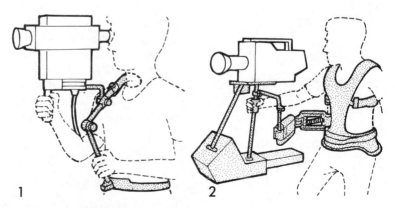

Fig. 3.18   Lightweight camera supports
To increase stability of hand-held cameras, (1) a body brace or (2) an elaborate stabilizing spring harness may be worn (Steadicam, Panaglide).

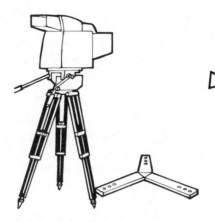

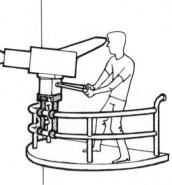

Fig. 3.19 part 1   The tripod
A simple 3-legged stand with in-dependently extensible legs. Height is preset. Unlike other mountings it can be set up on sloping or uneven ground. A *spider* (spreader) may be used to prevent tripod feet slipping.

Fig. 3.19 part 2   Camera clamps
The panning head may be clamped to a firm tubular-rail structure at a vantage point.

Static mountings

For certain applications (e.g. remotes or isolated camera positions) a *totally static* mounting provides the safest and simplest method of securing a camera. Particularly where extreme narrow-angle (long-focus) lenses are used, this ensures complete rigidity for stable shots.

Typically, a fixed tripod is used, alternatively a scaffold clamp fastens the camera panning head to tubular steel scaffolding, or a *camera frame/camera tower* structure.

The rolling tripod

The simplest wheeled camera mounting, known as the *rolling tripod (tripod dolly)*, comprises a collapsible metal or wooden tripod fitted to a three-castered base (*skid*; *skate*). This makes wheeling easy, but not always accurately controlled. The wheels have adjustable guards to prevent overrunning floor cables. Screw-jacks lock-off (immobilize) the mounting when required. The tripod height is preset, and not quickly altered. Suitable for simpler studio operations, the rolling tripod is extremely useful for remotes.

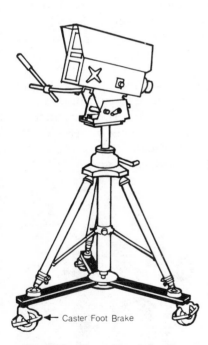

← Caster Foot Brake

Fig. 3.20   The rolling tripod/ tripod dolly
The tripod can be mounted onto a castered base (skid/skate/rolling spider). Height is preset (pneumatic or hand-crank). Easily wheeled on a flat smooth floor (although steering may be imprecise). Caster foot-brakes, and cable guards may be fitted.

The studio pedestal

The *pedestal (ped)* is the most widely used TV camera mounting. Designs range from preset lightweight hydraulic columns on casters to heavyweight monstrosities requiring assisted operation.

The well-designed pedestal is stable, flexible, easy to move and controlled quickly by one person. During complex moves, a cameraman might need to zoom, focus, pan, tilt, change height, push and guide simultaneously; so controlled movement is essential! At his eye level, this can be a difficult feat. At maximum or minimum height, it becomes impractical.

Using the pedestal, the camera height can be adjusted smoothly while on shot and held firmly in position at heights from about 0·9–1·8 m (3–6 ft). A large *ring steering wheel* around the central extendable column, usually serves to raise/lower the camera as

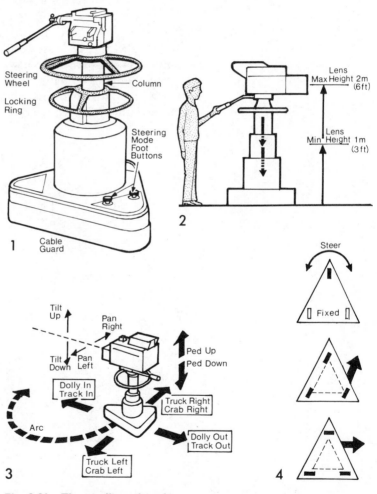

Fig. 3.21 The studio pedestal
1, A surround wheel or ring at the top of the column, is used to adjust column height (hence the lens height) and to steer the pedestal. A second ring may be fitted, to lock-off the column. 2, The lens-height range is typically (3–6 ft). 3, *Camera movements*—The camera can be moved around and repositioned in various ways. 4, The rubber-tired tricycle wheels can be steered in either of two ways (foot-pedal selection). *Dolly* (*tricycle tracking*) *mode*: Single wheel steered, other two fixed. Used for general dolly moves and curved tracks (arcs). *Crab* (*parallel*) *mode*: All three wheels interlinked, steered simultaneously. Used for trucking—crabbing, sideways movements, moving into confined spaces.

well as guide dolly movements. Pedestal designs include pneu
matic, hydraulic, counterbalanced and hand-cranked versions
they can be operated manually or with motor assistance.

The pedestal's three base wheels are guided by the steerin
wheel and/or a handle (*tiller*) in an interlinked or single whee
mode.

The small crane

The small crane (Panoram or Vinten dolly, velocitor, small boom
in various forms is widely used for filming and TV remotes (OBs)
They also have some TV studio use. The crane offers greate
height variations than the pedestal, for example 0·6–2 m (2–7 ft)
and unlike the pedestal, can overreach foreground obstacles
Some versions permit the boom (crane arm, jib) to slew left/righ
for tonguing movements.

However, the small crane requires a crew of two or three and a
fair amount of floor space to move around. The cameraman use

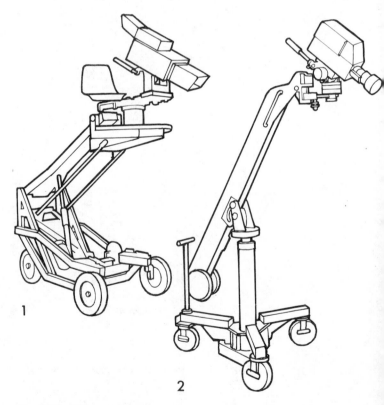

Fig. 3.22   The small crane
Lightweight versions of the small crane are increasingly used. 1, Particularly fo
remotes, the height range (0·6–2 m, 2–7 ft) with 360° seat and camera-mountin
rotation, offer improved shot flexibility (Vinten—'Kestrel'). 2, A counterbalance
crane-arm attached to a pedestal, tripod, or wheeled base again provides
variety of viewpoints for portable and ENG cameras.

finger movements to coordinate assistants (grips, pusher, tracker) in pushing, steering and elevating the mounting.

A level even floor surface is essential to ensure smooth dollying (less critical for pneumatically tired dollies); but special rails or tracks are desirable on rough terrain.

Larger cranes

Large camera cranes range from the Motion Picture Research Council crane (Academy Mole crane) which covers heights from $0.45-2.7$ m ($1\frac{1}{2}-9$ ft), to the truck-mounted giants of the large studio that are used at heights of $1-8.2$ m ($3\frac{1}{2}-27$ ft).

Needing considerable maneuvering space and a three/four man crew, we find cranes come into their own for production numbers, mass spectacles and large-area displays.

In all cranes, the camera and panning-head remain level

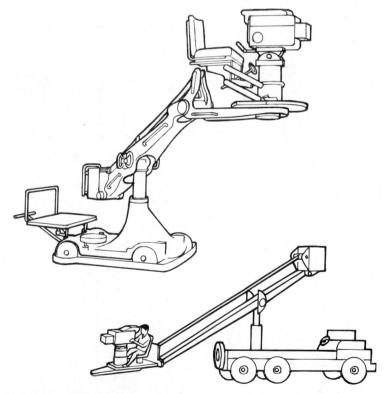

Fig. 3.23   Larger cranes
Heavy-duty cranes are normally used where extensive camera viewpoint changes, high-angle or swooping shots are required for large-scale action. Thanks to precision counter-balancing, the boom can be swung manually by a single operator (guided by a picture monitor and cameraman's finger-signals).

irrespective of the boom (crane-arm) angle. The cameraman can often rotate his seat with the camera through 360°.

Owing to their size, weight and balance problems, cranes require highly skilled operation for smooth safe movement. A heavy moving crane attains speeds up to 40 kph (25 mph) and therefore any misjudgements can be highly dangerous. The cameraman must take all precautions, including use of a helmet and harness, and have total confidence in his crew. He also needs a good head for heights!

Motorized dolly

To relieve the tedium of manual effort and improve coordination, two-man camera cranes with motorized controls were introduced into larger TV studios. Foot pedals for height control and camera-platform direction are operated by the cameraman, while a dolly operator (tracker) on a rear-platform controls speed and direction. During operation, the cameraman must take care not to depress

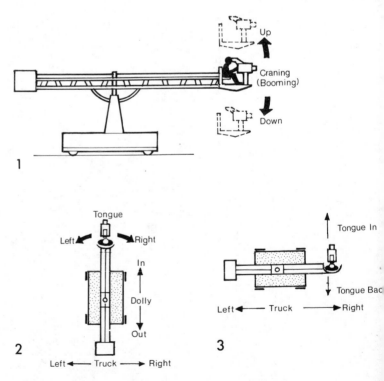

Fig. 3.24 Crane movements
Upward/downward movements of the boom (crane arm, jib) are *craning booming, jibbing*. Swinging the boom sideways is *tonguing left/right*. Swinging the boom to/from action is *tonguing in/back*. The base may be steered, using a tracking line to/from action (*dollying, tracking*), or across the action (*trucking crabbing*).

the jib onto anything beneath its overhang or crush himself against scenery (ceiling, arches).

Low dollies (creepers)

This is basically a steerable wheeled platform. The camera head can be raised/lowered from around 0·45 m to 1·06 m ($1\frac{1}{2}$ ft to $3\frac{1}{2}$ ft) for moving low-angle shots. However, for static set-ups, a mirror periscope attached to a pedestal may suffice instead.

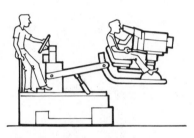 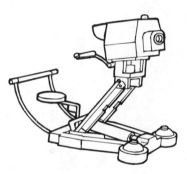

Fig. 3.25 Motorized dolly
A 2-man crane in which all opera-tions are motor-controlled, reducing crewing requirements and opera-tional fatigue.

Fig. 3.26 Low angle dolly
Gives height-mobility (0·1–0·3 m/ 4 in–2 ft) for low angle shots, together with considerable man-euverability.

Hydraulic platforms (cherry-pickers; simon crane)

Truck-mounted hydraulic lifting platforms can raise a camera from 1·8–30 m (6–85 ft) above the ground for high overall shots. It is also useful for positioning microwave dishes (radio links) for remote telecasts or ENG work. If you want to go higher than 30 m (85 ft) then you require the services of a helicopter—with a special camera-stabilizing device to avoid rhythmic vibration.

Remotely controlled cameras

For semi-static presentations, such as newscasts, or from small announce studios or unmanned reporting studios; a remotely controlled pedestal camera can be operated from a master control room several miles away. A keyboard control system can store up to 100 shots with detail variations of pan, tilt, zoom, focus, aperture and pedestal height. Lighting, cuing and equipment switching can be similarly operated.

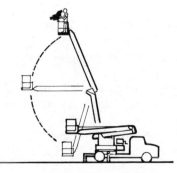

Fig. 3.27 Hydraulic platform
Considerable platform-height varia-tion is possible (even while on shot), giving a high stable viewpoint.

Table 3.5   Camera checkout list

Prerehearsal	Camera switched on (+ 1 hour) and lined up.
1. Camera cable	Camera plug and wall plug tight? Cable secured to mounting? Routing OK? Sufficient cable for moves?
2. Camera-head	Panning handle firmly attached? Unlock tilt. Camera-head tilt balance OK? (Nose or tail heavy?) Adjust vertical drag (tilt-friction). Unlock pan. Action smooth? Adjust horizontal drag (pan-friction).
3. Column	Unlock column. Ped up/down. Vertical movement OK? (Balance is affected by prompter attachments.)
4. Steering	Check steering in dolly and crab modes.
5. Cable guard	Adjusted to prevent cable-overrun or floor scraping.
6. Lens	Uncap lens, both physically and electronically. Check lens is clean.
7. Viewfinder	Shooting lift area, check its focus, brightness, contrast, aspect ratio, cut-off and image-sharpening. Tally and indicator cue-lamps OK?
8. Zoom	Zoom action smooth? Focus at maximum and minimum angles, checking that focus is constant throughout range (tracking). Check zoom meter. Shot box OK? Adjust preset angles.
9. Focus	Check focus control smoothness.
10. f stop	Check lens aperture (adjusted during line-up).
11. Intercom (talkback)	Check general and private wire. Program sound OK?
12. Shot card	Check through list.
Stand down	At conclusion of transmission or recording, inform video engineer, who caps up and switches off. Apply panning-head lock controls (do not tighten drag). Fix any safety chains. Lock pedestal column movement. Affix lens cap. Check general cleanliness, including tires. Report any mechanical, electrical or electronic faults. Remove camera cable and store (in figure-of-eight pile). Push mounting to storage area. Cover with dust sheet.

See Table 3.5

Basic camera operation

Good camerawork is a mixture of know-how, dexterity, observation and anticipation. Here is a digest of the most important points to look out for:

■ *1. Tally light*   Be aware of your tally/cue light. Do not move off shot (*clear*) unless instructed or when light goes out.

■ *2. Posture*   Avoid having a tense stance or gripping controls too tightly.

■ *3. Pan head*   Always be ready to pan or tilt on shot. Only lock off the head during breaks or when respositioning the mounting or when shooting captions and floor titles.

■ *4. Viewfinder*   Ensure it is at a comfortable angle and not diluted with spill light. Make sure you have a sharp and bright image.

■ *5. Lens angle*   Be aware of your lens angle (focal length). This helps you to anticipate handling problems or depth of field limitations.

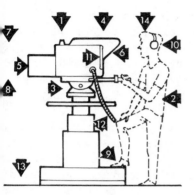

■ 6. *Checking your shot* Make sure you are *focused* on your subject and that you have sufficient depth of field. Be ready to follow focus as subject moves. Always anticipate the need to prefocus lens (for close shot) before zooming in from a wide shot.

Check your *framing* for such things as headroom. Remember to off-center your shot for angled, profile, or moving people ('looking-room').

Recheck your *composition* (for unity, balance, subject importance, clarity, etc.).

It is best to mark the dolly base floor position. This will help to give *consistent* shots.

Always try to match other cameras' shots to maintain similar headroom and height. Where several cameras' shots are combined (for 'supers' or effects) watch floor monitor or use mixed viewfinder feeds.

■ 7. *Spurious items in shot* Check that scenic lines are not 'growing out' of subjects. Take care that the mike (or its shadow), lamps, or other cameras are not visible. Guard against shooting off settings. Avoid strong specular light reflections. If lens flares arise, raise your viewpoint, or stretch tape across top of lens shade. Always be ready to compensate composition, if a performer moves unexpectedly out of position, or others intrude into your shot.

■ 8. *Shot changes* Anticipate subject movements and prevent it passing out of frame or becoming defocused in close shots. Be aware of people 'breaking' (moving apart).

Maintain focus and framing on moving subjects.

Also reframe (recompose) shot as a person repositions, or an extra person enters frame.

Be ready to alter shot size (by zooming or dollying) for wider action or new grouping. Avoid overwidening and then having to correct on shot.

■ 9. *Dolly movement* Check that the mounting is in correct mode to dolly (track) or truck (crab). Be ready to steer in a new direction. (Turn and slightly push while off-shot, to prevent wheel jerk as you move.)

Start dollying with gentle pressure and have your foot ready to push pedestal base. Always be ready to 'ped' up or down, which means that the column must be unlocked. Avoid hitting upper or lower column limits. Be aware of obstructions before dollying. Dolly smoothly and do not overshoot your floor marks.

■ 10. *Intercom* Listen to all instructions during rehearsal, including those for other cameras. In rehearsal visual signals may be used such as 'Yes' (tilt up and down); 'No' (pan left and right); 'I've a problem' (pan and tilt in circles); 'I've focus problems' (rapidly rock focus); and 'I want to speak on the private wire' (quick in and out zooms).

Inform the director of *problems* such as subject masking, insufficient time for dolly moves, composition, focusing, that need his reorganization. But avoid undue intercom chat.

■ *11. Shot sheet* Check your next shot, move to the new position and then be ready.

■ *12. Cable* Ensure that you have sufficient cable and that it is not trapped around objects. Never pull cable tight. Route cable in anticipation of moves. To avoid cable drag noise you may need assistance.

■ *13. Obstacles* Anticipate obstructions such as cables, scenery, people, cameras and booms. Avoid elevating or depressing on nearby obstacles.

■ *14. Problems* Be calm. If you are late on shot, try not to lose it altogether. Do not lose the next! If you misjudge then correct as unobtrusively as possible.

4 The persuasive camera

Anybody can do the obvious, and point a couple of cameras at the subject. One takes an overall *cover(ing) shot*, while the other shows closer detail. Subjects and cameras move around for the sake of variety, and to interlink areas. TV and film cameras can be used at this level. But this is making pedestrian use of a persuasive tool.

The selection of the right techniques at the right time, distinguishes the creative artist from the hack.

1. Good techniques are meaningful and apt. They add persuasively and significantly to the raw program material.
2. Poor techniques are mechanical conveniences, often unmotivated, ambiguous, unstimulating.

Defining the shot

In defining the *length of shot*, we are really considering how much

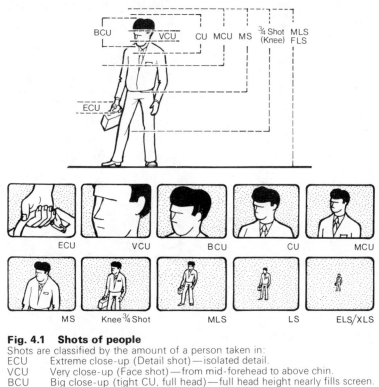

Fig. 4.1   Shots of people
Shots are classified by the amount of a person taken in:
ECU	Extreme close-up (Detail shot)—isolated detail.
VCU	Very close-up (Face shot)—from mid-forehead to above chin.
BCU	Big close-up (tight CU, full head)—full head height nearly fills screen.
CU	Close-up—just above head to upper chest (cuts below necktie knot).
MCU	Medium close-up (bust shot, chest shot)—cuts body at lower chest (breast-pocket, armpit).
MS	Medium shot (mid-shot, close medium shot, CMS, waist shot)—cuts body just below waist.
KNEE; ¾ shot	Knee shot, three-quarter length shot—cuts just below knees.
MLS	Medium long-shot (full-length shot, FLS)—entire body plus short distance above/below.
LS	Long shot—person occupies ¾ to ⅓ screen height.
ELS	Extra long shot (XLS), extreme LS.

of the screen the subject fills—how close it appears to be. Some descriptions are quite general.

1. *Long shot* or *full shot*—a distant view of unspecified coverage.
2. *Wide shot* or *cover shot*—a broad view taking in all action.
3. *Close shot* or *tight shot*—a detailed view, often excluding other nearby subjects.
4. Shooting people, we have the approximations: single, 2-shot, 3-shot, group shot.

A shot is classified according to its *effect*, irrespective of how it is actually obtained. For example, a *close shot*, could result from a close wide-angle lens, or a distant narrow-angle (long-focus) lens; although the effective perspective, distortions, camera handling, etc., will vary between these two extremes (but the depth of field remains similar).

Selecting the length of shot

The TV screen's size limitations do not prevent you from presenting effective variations in shot size. Although a small screen does less than justice to panoramic or spectacular situations, the TV camera can successfully explore the scene and action, from vista shots to microscopic close-ups.

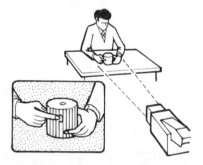

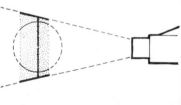

Fig. 4.2 part 1   Restricted framing
In close shots, action must be kept within the frame area. Avoid moving items in/out of frame (it draws attention to limited viewpoint).

Fig. 4.2 part 2   Limited depth of field
Confine movement for big close-ups (use firm, marked position) as depth of field is extremely restricted.

Fig. 4.2 part 3   Combined close-up and other shot
Combined shots can show detail and wider view simultaneously. This can be achieved by *detail insert* (left) or showing foreground detail (right). The latter method requires considerable focused depth.

Longer shots reveal location, establish mood, show inter-relationships, or follow broad action. But sustained long-shots can deprive the viewer of detail he is anxious to see (e.g. individual paintings in a wide view of an art gallery).

Closer shots lay emphasis, dramatize, reveal reactions, demonstrate detail. But sustained close-ups can be very restrictive. They can prevent the viewer from looking around the scene, from seeing the response of other performers, from watching action, or observing other aspects of the subject.

By using the wrong type of shot you may frustrate the viewer, leaving him uninformed, leading him to wrong conclusions, underplaying important information, or simply depriving him of what he wants to see at that moment. Occasionally you can arrange a combination picture to show a wide view of the scene and close detail simultaneously (Fig. 4.2).

The very long shot (vista shot)

Here the environment predominates, and the viewer often develops an impersonal, detached attitude to people within the scene, or a feeling of surveillance. In most studios, such wide views are only achievable by using a distant wide-angle lens and electronic insertion into a pictorial background.

Appropriately used, the very long shot enables you to establish broad location, provide an overall atmospheric impression, to coordinate several small action groups, or to accommodate widespread activity.

The long shot

Often used at the start of a scene, the long shot establishes location and atmosphere, and enables you to follow the pattern and purpose of movement.

As the shot tightens (less of the scene shown) the influence of setting and lighting will lessen. The impact of people within the scene grows. They become more important; their gestures and facial expressions become correspondingly more forceful.

Instead of beginning a sequence with a long *establishing shot*, you can build up impressions gradually, shot-by-shot; satisfying curiosity a little at a time. This encourages a sense of expectation, speculation.

But take care that delayed piecemeal introductions do not confuse; especially where the locale is unfamiliar. There must be sufficient pointers for our audience to be able to interpret the time, place and action correctly. Where action has been seen from closer viewpoints for some time, longer shots may be desirable to remind, or re-establish the locale in the viewer's mind.

Medium shots

These range from full-length to mid-shots; their value lying somewhere between the environmental strength of the long shot and the scrutiny of closer shots. From *full-length* to *three-quarter shots*, large bodily gestures (outflung arms) can be contained. But they are likely to pass out of shot if the shot becomes tighter. Then you have the option of restricting performer movement, taking a longer shot, or tightening to exclude arm movements entirely.

The closeup

An extremely powerful shot, the closeup concentrates interest. With people, it draws attention to their reactions, response and emotions. Closeups can reveal or point out information that might otherwise be overlooked, or only discerned with difficulty. They focus attention, or provide emphasis.

When introducing close-ups, you have to ensure that the viewer wants to look that close, and does not feel that:

1. He has been cheated of the wider view, where something more interesting may be happening.
2. He has been thrust disconcertingly close to the subject—he may become over aware of facial blemishes in overenlarged faces.
3. Already familiar detail is being overemphasized.
4. Information is diminished by reduced clarity. This occurs when depth of field is limited, or no further detail becomes visible, or irrelevant detail becomes prominent.
5. The closeup fragment becomes detached from the whole, so that the viewer loses orientation. He may lose a sense of location, or forget its relationship to the main subject.

Generally speaking, shots should not be held for long enough to entirely satisfy the viewer's curiosity, or to encourage attention to wander. And this is particularly true for close-shots containing limited interest.

The prudent director remembers, and allows for, the camera-man's various operational problems when shooting closeups:

1. Restricted depth of field, which therefore requires critical focusing.
2. Handling difficulties with narrow-angle lenses.
3. Depth compression.
4. The distortions and camera shadowing hazards when using close wide-angle lenses.
5. The difficulties in framing and following close movements.

Sometimes you will need to pan over closeup detail on a large flat surface (map or photograph). Then you must avoid the geometric distortions (keystoning) of a close short-focus (wide-angle) lens, or the restricted focused depth of a long-focus (narrow-angle) lens. It is best to move the lens parallel to the surface by trucking (crabbing) past it; with a long mural this is the only successful solution.

Deep focus techniques

As you saw earlier, the *depth of field* in a scene varies with the lens f stop, lens angle, and focused distance. You can change it by altering any of these parameters.

Stopping the lens down (e.g. f 11), everything from foreground to far distance appears sharply focused. The cameraman has no problems in following focus and there is little danger of subjects becoming soft focused. There is an illusion of spaciousness and depth; enabling shots to be composed with subjects at various distances from the camera.

One weakness of this technique is that where there is little camera movement or few progressively distant planes in the picture, it can appear unattractively flat. Surfaces or subjects at very different distances can merge or become confused. *Too much* may be sharply and distractingly visible.

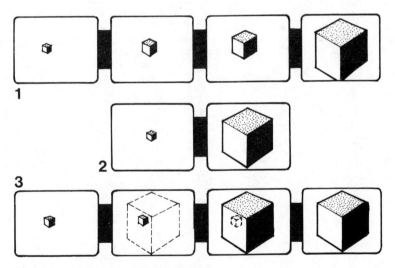

Fig. 4.3 Varying the length of shot
1, Dollying: Forward dollying can offer a gradual build-up of interest or tension, the importance of the subject growing progressively. Dollying back allows interest, tension or importance to be released gradually. The change can be made imperceptibly or fairly rapidly, the size and rate being controlled throughout. (Zooming has a simiar effect for many purposes.) Dollying avoids disruptive editing, but can appear laborious when rapid viewpoint changes are needed. 2, Cutting: Cutting shock-excites, thrusting the second viewpoint forward forcefully. Cutting to closer shots helps to emphasize momentarily the subject's strength and importance, while cutting to longer shots tends to cause sudden drops in tension and subject-importance. 3, Mixing (dissolves): Mixing has something of the transitional smoothness of dollying while taking less time. It enables the viewer to interrelate two viewpoints more readily than cutting, but the double-exposure of slow mixes is highly distracting.

Shallow-focus techniques

Using a wider lens aperture (e.g. $f2$) restricts focused depth. It enables you to *isolate* a subject spatially, keeping it sharp within blurred surroundings; and avoids the distraction of irrelevant subjects. You can display a single sharply focused flower against a detailless background; concentrating attention on the bloom and suppressing the confusion of foliage. Sharply defined detail attracts the eye more readily than defocused areas.

Table 4.1   Why change the lens angle?

To adjust framing	A slight change in lens angle:
	Where you want to exclude (or include) certain foreground objects, and repositioning camera or subject would spoil proportions.
	Where a normal lens would not provide the required shot size or framing, without repositioning the camera or subject.
For otherwise unobtainable shots	Using a narrow-angle (long-focus) lens:
	To shoot remotely situated subjects—separated by uneven ground or inaccessible.
	Where the camera is isolated—on a camera platform (tower); shooting through scenic openings.
	Where the camera cannot be moved—static tripod; obstructions.
	Using a wider angle (shorter focus) lens:
	Where the normal lens does not provide a wide enough shot—space restrictions.
	To maintain a reasonably close camera position (e.g. so talent can read *prompter*) yet still provide wider shots.
To adjust effective perspective	Altering lens angle and changing camera distance to maintain same subject size, alters relative subject/background proportions and effective distances (Table 4.3).
	Using a wider-lens angle (short-focus lens)—enhances spatial impression; increases depth of field.
	Using a narrower-lens angle (long-focus lens)—reduces spatial impression; compresses depth, e.g. bunching together a straggling procession.
Insufficient time to change camera distance	Altering apparent camera distance (shot size) by changing lens angle:
	During fast inteructting sequence.
	When camera repositioning would involve complicated moves.
To provide simpler or more reliable mechanics	Zooming in/out instead of dollying may produce smoother, easier changes in shot size (but perspective and handling effects change).
	Zooming provides rapid changes in image size more safely than fast dollying (for dramatic effect or to suddenly reveal detail).
	Zooming in/out on a flat subject is indistinguishable from dollying, but avoids focus-following problems.
	Lens angle changes avoid close-up cameras coming into picture on wider shots.
To increase production flexibility	Where dollying is undesirable—because it distracts talent or obscures action from audience.
	When using only one camera.

Using a deliberately restricted depth of field, you can 'soften' the artificiality or obtrusiveness of backgrounds. People stand out from their surroundings.

On the other hand, restricted depth can prove embarrassing when essential details of a close object are badly defocused. The cameraman has the problem of continually refocusing to keep a close moving subject in focus. For instance close shots along a piano keyboard can demand considerable dexterity of the camera-

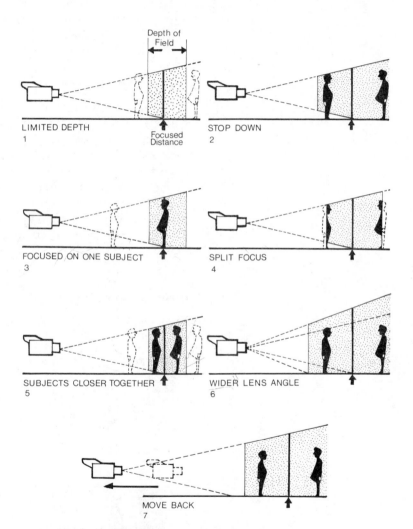

Fig. 4.4   The problem of limited depth of field
Where 1, depth of field proves too *limited*, the solutions are:
2. *Stop down*—depth of field increases, but higher light intensity necessary.
3. *Focus on one subject*—permitting other(s) to soften.
4. *Split focus*—spread available depth between both subjects (both are now softer than optimum).
5. *Move subjects closer together*—making subjects more equidistant from camera.
6. *Use wider lens angle*—depth of field increases but subjects now appear smaller. (Moving closer for larger image, re-introduces limited depth.)
7. *Pull camera back*—depth increases but shot is smaller.

man when following focus on quickly moving fingers. Where there are two or more subjects in shot, he may not be able to maintain sharp overall focus but have to split focus between them, or select one to focus on.

Occasionally, by changing focus between subjects at different distances (*pulling focus*, *throwing focus*), you can move the viewer's attention from one to another. Although, this trick easily becomes disturbing, unless coordinated with action. Blurred color pictures can be frustrating. Whereas in monochrome defocused planes merge, in color the viewer may find himself trying to decipher unsharp detail and the effect can be less acceptable.

Moving the camera head

In everyday life, we respond to situations by making particular gestures or movements. These reactions and actions often become very closely associated. We look around with curiosity; move in to inspect an object; withdraw or avert our eyes from a situation that we find embarrassing, distasteful or boring.

It is not surprising to find, therefore, that certain *camera movements* can evoke associated responses in our audience, causing them to have particular feelings towards what they see on the screen. And these effects underly the impact of persuasive camera techniques.

Panning the camera

Panning shows us the spatial relationship between two subjects

Fig. 4.5   Focus techniques
Deep focus (top illustrations) provides overall clarity. This technique allows the cameraman to compose in depth and also accommodate close shots of three-dimensional subjects. *Shallow focus* (bottom illustrations) isolates subjects from distracting or confusing backgrounds, softens off unwanted detail.

or areas. Cutting between two viewpoints does not provide the same sense of continuity or extent. When panning over a wide arc, the intermediate parts of the scene connect together in our minds, helping us to orientate ourselves. We develop an impression of space. But avoid panning across irrelevant areas—such as the 'dead' space between two widely separated people.

Panning should be smooth; neither jerking into action nor juddering to a halt. Erratic or hesitant panning irritates. With a correctly adjusted panning head (suitable drag/friction) and a properly stanced cameraman, such unevenness usually only occurs when using very narrow lens angles or where a subject makes an unpredicted move.

■ *The following pan*   This is the commonest camera movement. The camera pans as it follows a moving subject.

In longer shots the viewer becomes aware of the interrelationship between the subject and its surroundings. Visual interaction can develop between the subject and its apparently moving background pattern, creating a dynamic mutual impact (*dynamic composition*).

In closer shots, the background becomes incidental; often indecipherably blurred.

■ *The surveying pan*   Here the camera slowly searches the scene (a crowd, a landscape), allowing the audience to look around at choice. It can be a restful anticipatory action; providing there is something worth seeing. It is not enough to pan hopefully.

The move can be dramatic, with high expectancy—the ship-wrecked survivor scans the horizon, sees a ship . . . but will it notice him? But the surveying pan can build to an anticlimax too; the fugitive searches to see if he is being followed . . . and finds that his 'pursuer' is a friendly hound.

■ *The interrupted pan*   This pan is a long smooth movement, that is suddenly stopped (sometimes reversed) to provide visual contrast. It is normally used to link a series of isolated subjects. In a dance spectacle, the camera might follow a solo dancer from one

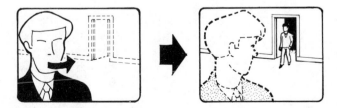

Fig. 4.6   Pulling focus
As the subject turns his head to welcome a newcomer, the camera refocuses.

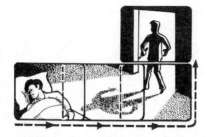

Fig. 4.7   Slow panning
To provide a dramatic development.
Panning slowly from the sleeping
victim . . . along the intruder's
shadow . . . then a rapid upward-tilt,
revealing the intruder's identity.

group to the next, pausing awhile at each as it becomes the new center of interest.

The technique has comic applications: the camera follows a bunch of sailors. They pass a pretty girl. The camera stops in mid-pan to watch her. One sailor comes running back and grimaces to the camera, beckoning it to hurry and catch them up.

In a dramatic application: escaping prisoners slowly stagger over treacherous marshland. One man falls exhausted, but the camera stays with the rest. A moment later it stops and pans back— to see only the last traces of the straggler remain.

■ *Panning speed*   A slow prolonged pan can create opportunities or disappointments, according to how it is used. If the camera pans slowly over a series of objects that are increasingly interesting or significant, the movement will continually hold our attention; even build up to a climax. But a slow pan around without any real points of interest, will quickly pall, as the viewer's initial expectancy falls.

■ *The whip pan*   A *fast* pan may produce nothing but a series of broken-up, half-seen images due to stroboscopic effects. But a *whip pan* (*swish, zip, blur pan*) turns so rapidly from one subject to the next, that the intermediate scene becomes a brief streaking blur.

Whether the effect arouses excitement or annoyance, largely rests on how the preceding and following shots are developed. As our attention is dragged rapidly to the next shot, it gives it transitory importance.

The *whip pan* usually produces a dynamic relational or comparative change:

1. *Joining different viewpoints of the same scene*—The pan *direction* should be compatible with the location of these viewpoints i.e. panning right to a viewpoint located to camera right.
2. *Providing continuity of interest*—Connecting a series of similar subjects or themes.
3. *Changing centers of attention*—From one area of concentration to another e.g. a golfer drives . . . we whip pan to the awaiting hole.
4. *Showing cause and effect*—A whip pan from a cannon . . . to the demolished castle wall.
5. *Comparing or contrasting*—Relating situations e.g. wealth with poverty, old with new.
6. *Transferring in filmic time and space*—An aircraft takes off . . . whip pan to it landing at the destination.
7. *Dramatic change of direction*—Changing the viewpoint with a sudden climax e.g. the camera subjectively scans the distance,

we hear a noise and it whip pans to see an intruder behind, about to attack.

A whip pan continues the pace between two rapidly moving scenes. It can provide a tempo-bridge between a slow scene and a fast one. But at all times it remains something of a stunt.

A whip pan has to be *accurate* as well as appropriate, to be successful; no fumbling, reframing, refocusing at the end of the pan! To avoid such problems, the effect may be cheated by intercutting a brief shot of 'blur' between two static cameras' shots.

Tilting the camera head

Tilting, like panning, enables you to visually connect subjects or areas that are spaced apart, and would otherwise require intercut shots or a more distant viewpoint to encompass them. This technique can be used:

1. To emphasize *height or depth*—tilting up from the mountaineer, . . . to show the steep cliff face to be climbed.
2. To denote *relationships*—as the camera tilts from the rooftop watcher . . . down to his victim in the street below; or from the victim . . . up to the rooftop, revealing that he is not alone.

Fundamentally, *tilting upward* engenders feelings of rising interest and emotion, expectancy, hope, anticipation. Conversely, *tilting downward* is allied to lowering interest and emotion, disappointment, sadness, critical inspection.

Where the viewer is conditioned beforehand by dialogue or action, responses become more complex. The *downward* tilt can become an act of enquiry; the upward tilt, a gesture of despair. Again, where a subject no longer holds our interest, the relief of a downward tilt can produce an anticipatory feeling—although the movement is not itself anticipatory.

In practice, of course, you will not stop to hypothesize about each camera movement. You come to recognize the 'feel' of such treatment. In a dramatic situation, for example: A woman hears of her son's death, and the camera *tilts down* with her, as she sits. This treatment conveys her grief more convincingly than either a static shot or a depressed viewpoint.

Fig. 4.8   The impact of viewpoint
Viewpoint immediately influences our response to the sleeping old man. 1. Drunk? 2. Quiet contemplation? 3. Lonely? Deserted?

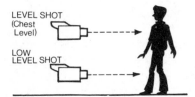

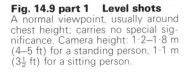

Fig. 14.9 part 1   Level shots
A normal viewpoint, usually around chest height; carries no special significance. Camera height: 1·2–1·8 m (4–5 ft) for a standing person, 1·1 m (3½ ft) for a sitting person.

Camera height

The angle from which you shoot a subject can have a considerable influence on the audience's attitude towards it. Intercut high and low angle shots of a person, demonstrate this immediately. In a drama you may deliberately choose the camera's height to emphasize or diminish a person's dramatic strength, or to control the impact of dialogue. Similarly, if you shoot a piece of sculpture from a low viewpoint, it will appear imposing, forceful and impressive; while shot from above it loses vitality and significance.

How you get that viewpoint is immaterial. Shooting an architect's miniature/model of land development, for example, you could equally well obtain an overhead view from a crane shot, a suspended camera, a mirror shot, or by attaching the subject vertically to the studio wall and shooting straight on to it. The effect is the same, although in some cases camera movement will be very restricted.

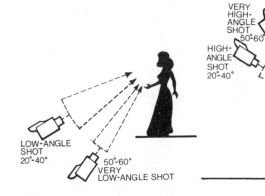

Fig. 4.9 part 2   Low shots
Low angle shots make most subjects appear stronger, more imposing, overpowering, strange, ominous. A person can seem threatening, pompous, authoritative, determined, dignified, benevolent, according to his attitude and environment. Dialogue and movement become significant and dramatic. The closer we are, the stronger these impressions. In a very-low-angle shot, the subject takes on a strangely distorted, even mystical, appearance. At greater distances, it appears remote and unknown.

Fig. 4.9 part 3   High shots
High angle shots give the audience a sense of strength or superiority, an air of tolerance, even condescension, towards the subject; this feeling increases with distance. Therefore high shots can be used to imply unimportance, inferiority, impotence, etc. Very high angle shots give an attitude of peering down, scrutinizing. From a height of 3–6 m (10–20 ft) surveillance gives way to complete detachment. Overhead shots emphasize pattern and movement in formation (as in ballet and dance spectacle). Reveal isolation or congestion.

■ *Extreme camera angles* Extreme angles have an unhappy knack of drawing attention to the abnormality or ingenuity of the camera's position. If our audience is left wondering how we got that shot, mechanics have intruded over artistic purpose. Similarly, during the now familiar swooping and climbing of helicopter shots, the viewer can become overaware of wind-beaten grass or tell-tale shadows of the aircraft on the ground below.

Where extreme angles arise *naturally*, viewers accept them readily: looking down from an upper storey window; looking up from a seated position; even an eavesdropper peering through plank flooring to the room below. But an unexplained extreme usually becomes a visual stunt; for example: a bird's-eye view of a prisoner pacing a ceilingless cell, or shooting up through glass floors.

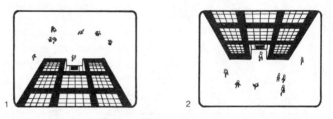

Fig. 4.10   Verticals in overhead shots
In overhead shots, the position of the camera relative to strong verticals in the scene can considerably affect shot impact. 1, Verticals at the bottom of the frame cause tension and instability. 2, This impression is absent where foreground verticals are at the top of the frame.

■ *Emotional impact of extreme viewpoints*   As you have seen, while *high shots* generally create an audience-strength/subject-weakness relationship, low shots reverse this influence. But the associations are often much more subtle than that. Imagine a high-angle shot of an old man trudging through rainswept streets. The audience does not feel 'strength' or 'superiority', but pities him, or feels sorry for his loneliness. But these are, of course, psychological variations on the fundamental theme of 'viewer-strength/subject-weakness' that the high shot provides.

In a very high angle shot, the viewer has an overall view; in fact, more than people on the spot. This underlying response colors their attitude to what they are seeing.

In an *aerial view* we survey the extent of the landscape; we follow the course of a river, seeing houses clustered along its banks, and its toy-like boats. We feel remote, seeing it all at a glance. We see the searchers moving towards the hiding fugitive. We see the magnitude of the invading army. But there is emotional detachment in this inspectional viewpoint; none of the involvement of more normal camera positions. Dynamics are reduced. Some subjects, such as high-wire acts, juggling, balancing feats, can lose their appeal entirely if shot from above, whereas a very low angle position emphasizes their difficulty and suspense.

■ *Portraiture*   The camera's height modifies what people look like, too. *Elevated shots* (especially on wide-angle lenses) em-

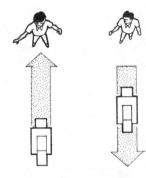

Fig. 4.11 Dollying (tracking)
Moving to or from the subject. *Dollying-in* causes increased interest, build-up of tension, but the closer view may result in disappointment, disillusionment and, consequently, diminished interest. *Dollying back* results in lowered interest, relaxed tension; unless unseen subjects are revealed, or when curiosity, expectation, or hope have been aroused. Attention tends to be directed towards the edges of the picture. Rapid dollying is visually exciting, but space and safe speed restrict fast moves. (Controlled zooming may produce allied effects.)

phasize baldness and plumpness in women; also through foreshortening and distortion, they can produce a dumpy sawn-off effect.

Low-angle shots can emphasize dilated nostrils, upturned or large noses, heavy jawlines, and scrawny necks. Men with high foreheads or receding hair may look entirely bald. Protruding ears may appear very prominent.

Moving the camera

How freely you can move the camera around is determined by the type of mounting used. Whereas a crane boom shot offers considerable flexibility, it may not be able to relocate the viewpoint as rapidly as a highly mobile pedestal. Well-chosen camera moves add visual interest and vitality, as well as engendering certain audience reactions. The intermovement of planes (parallactic changes) that takes place as the camera moves, provides an illusion of solidity and depth; a realness that no static zooming camera can achieve.

But camera movement needs to be motivated, appropriate, smoothly controlled, and at a suitable speed, or it can become restless and disturbing.

■ *Prominence of camera movement* The effects of camera movement are most prominent when shooting a static subject. Movement impact is correspondingly lessened where the camera moves with the subject; e.g. following in *tail-away shot* as the subject moves away; or dollying out as the subject approaches the camera in a head-on shot. And if respective direction and speed match in a *traveling shot*, the dynamic effect of movement will come largely from the passing background. Whenever camera and subject are heading toward or away from each other, the overall impression is considerably heightened.

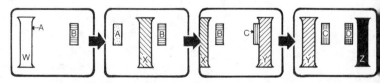

Fig. 4.12 Trucking (crabbing)
Moving across the scene, parallel with it. Trucking becomes associated with an attitude of inspection, critical observation, expectancy, intolerant appraisal. The lateral displacement of planes (parallactic movement) introduces a strong illusion of depth and solidity, but the restriction of the picture's frame becomes over-apparent if trucking stops abruptly.

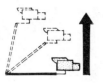 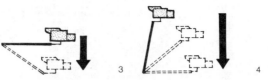

Fig. 4.13 part 1   Craning from a static dolly
1, When craning up from level, audience response is elation, superiority, eventual detachment as height increases; while the subject becomes less significant, even unimportant. 2, When craning down from level, the response is depression, inferiority, even a feeling of being dominated; and the subject becomes more important, impressive, significant. 3, Craning down from an elevated height with superior viewpoint will produce an effect of return to normality, with some sense of depression. 4, Craning up from low-level to level produces the same effect, but with a slight feeling of elation.

■ *Move the camera—or the subject?*   There are intrinsic differences between the effects of moving the *camera*, or moving the *subject*. You can see this by watching a mute TV screen in a darkened room. You will feel your peephole vision moved around, almost physically, as the moving camera *subjectively* examines the scene. Where the camera is static, the performers will seem to move to and from you; an *objective* effect.

Camera movement (or zooming) becomes 'our behavior' by proxy, towards the subject. *We* the audience are moving; going up to look at or meet the subject . . . to satisfy our curiosity or to see detail more clearly. Forward camera movement is strong and exploratory; while dollying back is usually a weak move.

When a performer moves towards or away from the camera, we the audience become recipients of 'his attitude towards us'. Our reactions to his move largely depend upon his demeanour; whether his action is forceful or casual. Circumstances may alter situations—we may experience an 'audience superiority' effect as someone walks the length of a corridor to meet us, but 'inferiority' if they sit while we (the camera) have to move up to meet them.

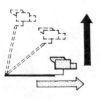 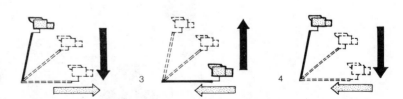

Fig. 4.13 part 2   Craning from a moving dolly
1, Forwards and craning upwards creates an air of freedom, flight, lack of restriction, elevation, joy. 2, Dollying forwards and craning down, implies swooping, power, strength, importance relative to subject when fast; but when done slowly suggests depression, return to normality. 3, Dollying back and craning upwards gives a feeling of complete release; detachment from the subject or action. 4, Dollying back and craning down is a recessive move, often saddening, disillusioning; a return to earth; no longer detached from subject or action.

Table 4.3 Production uses of lens angles

About 50° *(F=17 mm [0·6 in])**	The widest lens angle generally used in the studio. Used for wide shots at close camera distances. Used to exaggerate setting size, space, distance, scale. Exaggerates dolly movements and speed. Exaggerates performer movements to and from the camera. Used to frame fast on erratically moving subjects. Facilitates very smooth camera handling. Provides considerable depth of field. Provides deliberate grotesque geometric (barrel) distortion—e.g. of CU faces, forward arm movements. But camera is liable to *overshoot* (*shoot-off*), and get lens flares.
About 35° *(F=25 mm [1 in])**	Frequently used for close shots of small objects, but camera shadows may arise. Noticeable increase in spatial impressions, without the exaggeration of wider lens angles. Generally emphasizes dolly movements and speed. Provides easy camera handling. Depth of field facilitates easy focusing. Appreciable camera movement may be necessary to correct framing.
About 25° *(F=35 mm [1·5 in])**	The 'normal' lens angle, used for most purposes and providing natural perspective. A generally effective lens angle for good camera handling. Generally convenient working distances for floor title cards and closer views. Focusing and available depth of field suitable for longer shots (deep focus). Yet close-ups provide sufficient depth for most subjects, with background defocusing providing isolation.

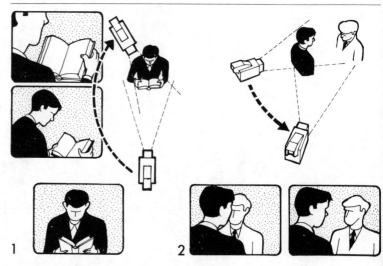

Fig. 4.14 Arcing
1, Circling round a subject to see it from a different viewpoint, tends to be a self-conscious move, drawing attention to the action. Spurious side-slip effects can arise. 2, Slight position changes can help when one performer masks (obscures) another.

About 15° *(F=50 mm [2 in])**	Useful narrow angle for 'close shots' from a distance—whether due to subject inaccessibility, or preventing other cameras coming in shot. Isolates individuals from a group. Noticeable compression of depth and perspective distortion. Camera handling is not easy. The lens angle is only really suitable for slight dolly movements. Depth of field is somewhat restricted.
About 10° *(F=90 mm [3½ in])**	Achieves close-ups from around 3·5 m (11½ ft) but depth compression is considerable. Very 'compressed' perspective. Depth of field is very limited for close shots. Camera handling becomes coarse; needs skilled operation, even for limited movements. Movement of close subjects is often hard to follow accurately.
About 5° *(F=170 mm [6 in])**	Used for static close shots on a distant camera. Considerable depth compression. Mainly suitable for flat subjects, where perspective distortion is absent. Most unsuitable for close-ups of people, because of extreme perspective distortion. Camera handling is extremely coarse. Subject movement is very difficult to follow. Depth of field is extremely limited.

*This is the focal length with standard $1\frac{1}{4}$ in plumbicon tube. Double the focal length for I.O. angles.

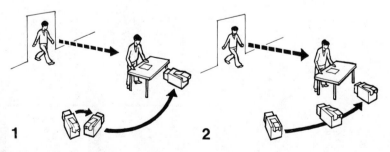

Fig. 4.15 Camera and subject movement
Someone enters a room (in long shot) and crosses to a table (arriving in close-up). There are two methods of single-camera treatment, each providing a different audience impact. 1, The entrance and walk are taken in long shot. When the subject has stopped, the camera dollies in to close-up. 2, The camera pans and trucks throughout the walk, with subject and camera arriving simultaneously in a close-up.

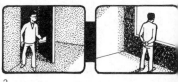

Fig. 4.16 part 1 Following the subject—long shot throughout
Audience impact comes chiefly from the moving environment; subject importance being limited. The pace of action is relatively slow. Varying composition or tone during the pan, can produce emotional changes—e.g. the 'uplift' on panning from low to high key. Sudden mood changes are achieved by, for example having the actor raise window blind to let in daylight, or switching room lights.

Fog. 4.16 part 2 Following the subject—close shot throughout
The subject dominates, its strength depending on camera height. The influence of the environment is limited, and pace varies with dynamic composition. Slightly off-centering the subject in the direction of movement creates a feeling of anticipation, expectancy. 1, Profile—weak against a plain background, but an impression of speed and urgency against a detailed background. 2, Three-quarter frontal—can be dramatically strong. By preventing us from seeing the subject's route or destination, marked curiosity can be encouraged. 3, Elevated frontal—although himself weakened, the subject dominates his environment. 4, Depressed frontal—the subject is especially forceful, dominating his environment.
Rear shots introduce a subjective effect
5, Three-quarter-rear—partially subjective; the viewer moves with the subject, expectancy developing during movement. 6, High rear shot—almost entirely subjective; producing increased anticipation; searching. 7, In level and 8, low shots there is a striking sense of depth. The subject is strongly linked to the setting and other people, yet remains separated from them.

Fig. 4.16 part 3 Following the subject—long shot to close-up
The subject walks up to the camera. Environment predominates at first, but the subject grows increasingly stronger as it approaches. The impact can be modified by dollying or zooming as the subject approaches.

Fig. 4.16 part 4   Following the subject—close-up to long shot
A weak, recessive movement, generally accompanied by lowering tension as the subject's strength falls and that of the environment increases. Used to depict pathos, anticlimax.

Fig. 4.16 part 5   Following the subject—then arriving at his destination before him
1, The camera moves more quickly than the subject, to his destination—the quickening pace providing an exciting introduction to the new position; but it must be sustained to be effective. The shot shows spatial relationships and gives the viewer time to assimilate the destination before the subject arrives. 2, Here the person begins to move, you cut to his destination, and he then arrives. Often mechanically convenient, the cut gives the destination shot impact. Audience-superiority can arise through foreknowledge of the subject's destination, and provide pathos, bathos, stage irony, climatic effects. It avoids lengthy dollying shots, and can imply non-existent spatial relationships. Irrelevant intermediate movement is eliminated and pseudo filmic time effect achieved, but interest and tension can flag while awaiting the subject.

Fig. 4.16 part 6   Following the subject—cutting from the static subject to his destination
A simple but weak treatment, unless dialogue or gesture has indicated that he is going to move. Spatial relationship or significance may not be clear to the viewer. Static action cuts of this kind can make the viewer over-conscious of the time taken for the subject's move. On commenced-action cuts, as in part 5, this is less pronounced.

Subjective camera movement creates a participatory effect for the audience. But if a director moves us when we do not want to do so, or fails to show us something we wish to see, we feel resentful or frustrated. The skilled director persuades his audience to want a change in viewpoint, or a move. The unskilled director thrusts it upon them.

■ *Imitative camera movement*   Apart from the familiar ploy of moving cameras to suggest jogging vehicles or rolling ships,

camera movements can provide subjective comments on th[e]
action itself. Staggering as it dollies after a drunk; copying th[e]
bounce of a dandy's gait; swaying in waltztime as we follow a gi[rl]
home from her first dance.

Drawing attention to itself quite openly, the camera movemer[t]
becomes a tongue-in-cheek observation between the director an[d]
his audience. When it comes off, this device has persuasive appea[l.]

Fig. 4.17   The canted shot
1, Canting is most pronounced with subjects inherently associated with hor[i]
zontal or vertical stability. Left: slight canting (about 10°) is ineffectual. Centr[e]
typical tilt around 20–30° is used to convey instability, abnormality, dynamisr[n]
Right: excessively canted (50–60°) the effect is excessive overbalancin[g]
2, From low-angle (or high-angle) positions, canting causes subjects to le[an]
into, or out of frame.

Using the zoom lens

As you saw in Chapter 3, the zoom lens brings both advantag[es]
and pitfalls for the unwary. It is too easy to use lens angl[e]
indiscriminately, just to change subject image size.

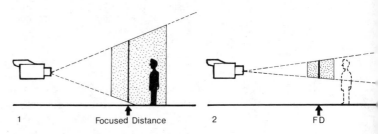

1   Focused Distance   2   FD

Fig. 4.18   The need to pre-focus
1, The considerable depth of field available with a wide lens angle broade[ns]
focusing, so that subjects may appear sharp although not at the focused plan[e]
2, However, on zooming in to a narrow lens angle (without re-focusing) dep[th]
of field becomes restricted, and the subject now falls outside the available dep[th]
of field.

There is a great temptation to stand and zoom, rather than move around with a normal lens angle. It is an easy operation, demands little of the cameraman or director, and avoids problems such as keeping close cameras out of long shots. There is just the need for a prezoom focus check, before zooming in.

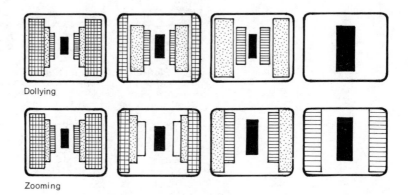

Dollying

Zooming

Fig. 4.19 Comparing dollying and zooming
1, Dollying: When a camera moves through a scene, relative sizes and spacing of subjects change proportionately. This continual parallactic movement causes planes to become visible and disappear with position changes. Perspective remains constant. 2, Zooming: As the field of view alters with lens angle, the shot is simply enlarged or diminished. There is no parallactic movement, no planes appear/disappear, or change relative proportions. So effective perspective appears to change (unless the viewer adjusts his viewing distance to compensate) and visible 'depth-stretching' or compression results.

Zooming is extremely convenient. But it only *simulates* camera movement. There are no natural parallactic changes; and scale, distance, and shape become distorted through zooming. A slow zoom made during panning, tilting or subject movement may disguise these discrepancies. A rapid zoom during an exciting fast-moving ball game, would scarcely disturb even the pedant. Much depends on the occasion.

Zooming can provide a visual bridge from the wide view to the close-up, without the time and effort involved in dollying or the interruption (and possible disorientation) of *cutting*. A *rapid zoom-in* produces a highly dramatic swoop onto the subject. An instantaneous *crash zoom-in* flings subject detail at the audience. Such effects can be fantastically forceful—or plain annoying!

All zooming should be decisive. Avoid the nauseous results of rhythmical or jerky in-out zooms, or quick *slight* angular changes that look like an operational error.

Zooming on *flat surfaces* (maps, illustrations) avoids the need to *follow focus* that dollying would entail. On title cards you can produce carefully controlled size changes (titling growth or shrinkage) by zooming.

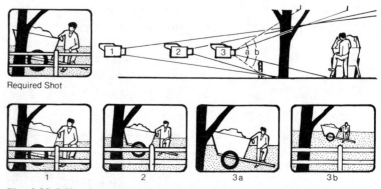

Fig. 4.20 Adjusting proportions
Altering relative subject proportions in the picture can be achieved by changing
to a different lens-angle and readjusting camera distance (see fig. 3.10). 1, A
distant narrow-angle lens provides the required subject/foreground proportions.
2, A closer position using a normal lens angle, shows required foreground
proportions, but the subject appears too small. 3a, Dollying-in will correct the
subject size but loses the foreground fence. 3b, Widening the lens angle now will
bring the fence into shot again, but the subject image appears small and the
subject-foreground distance increased.

Use zoom-ins to direct attention, to increase tension, to give
powerful emphasis, or to restrict the coverage. But the zooming
action itself should be used discriminately for particular occasions
—like wipes, star filters, diffusion discs, and similar productional
tinsel. Certain effects, like that of zooming in and dollying back
simultaneously (keeping the subject-size constant), are bewilder-
ing, to say the least!

Table 4.2   Adjustable picture proportions

Subject size	Background appears	Lens angle correction	Camera position
Too large	Too large*	Use wider lens angle (zoom out)	Keep camera still
Too small	Too small*	Use narrower lens angle (zoom in)	Keep camera still
Satisfactory	Too distant	Use narrower lens angle	Increase camera distance
Satisfactory	Too near	Use wider lens angle	Move camera closer
Too large	Satisfactory	Use narrower lens angle	Increase camera distance
Too small	Satisfactory	Use wider lens angle	Move camera closer

*But relative subject/background proportions satisfactory.

5 Composing the picture

People have very diverse attitudes towards the study of pictorial composition. Some try to follow the 'laws' slavishly, with varying success; others use a few routine rules-of-thumb; others dismiss it with suspicion and arrange shots in a hit or miss fashion. Any meaningful study of pictorial composition must be realistically related to *practical working conditions*, and this is what experienced practitioners do.

A study of pictorial composition helps you to produce attractive significant pictures, that direct audience attention and influence their feelings about the subject and its surroundings. There are really three basic kinds of picture making in photography:

1. Composition by *design*.
2. Composition by *arrangement*.
3. Composition by *selection*.

Composition by design

This occurs where you have an *entirely free hand* in composing your picture. The artist approaching his blank canvas can arrange line, tone and color in any way he chooses, without concern for accuracy or feasibility. The physical limitations are minimal and are inherent in his medium.

Canaletto 'improved' his paintings of London by actually repositioning entire buildings for effect, whatever the resulting topographical errors. Only occasionally does this freedom exist for the camera (e.g. in process work).

Composition by arrangement

Here you *deliberately position subjects* before the camera to produce an appealing meaningful result. A set designer dressing a setting does this by placing furniture, flowers and ornaments to form appropriate arrangements. Sometimes even a single carefully introduced foreground object (a leafy branch perhaps), will help the cameraman to devise appropriate composition.

More usually, a designer creates *potential* rather than specific composition opportunities; an environmental package suiting the mood and mechanics of the occasion, rather than calculated composition arrangements.

There is always the danger when arranging natural situations, that the resultant effect will look deliberately contrived. You can become overaware of how 'cleverly' the picture has been manipulated.

Fig. 5.1 Why compose the picture?
The unguided eye wanders around the scene, finding its own centers of interest.

Composition by selection

This is the situation most cameramen encounter. The camera is

81

positioned at a certain viewpoint chosen by the director (or the cameraman himself) and he composes the shot *using what is there*, to show the subject most effectively.

He can move his camera in three dimensions to select visual components. He adjusts shot size, camera height, moves slightly left/right, arranges framing, and perhaps selectively focusing.

As the camera is moved vertically or laterally, foreground subjects change frame position more markedly than others further away. By adjusting lens angle and distance he can modify proportions. In these ways, the cameraman composes his picture.

The director and composition

The director's attitude to pictorial composition varies between individuals and the type of production. For many shows the director's preoccupation is with what is being said, with performance, continuity and mechanics. His instructions concerning composition are indicative rather than specific. He thinks principally in terms of shot size, of singles, two shots, group shots; and the cameraman composes accordingly.

In other types of production (primarily drama), performers are deliberately grouped to provide particular composition arrangements—for dramatic purpose or to direct audience attention. In some instances the director (or the designer) may have prepared a *storyboard sketch*, showing the detailed composition of certain *key shots*.

Composition principles

Composition principles are not laws. They are indications of how people respond to distribution of line and tone. Whether you put that tone into the scene, or are selecting from whatever is already there, is not important. What *is* important, is that if you do not organize pictures appropriately, your audience may well react by looking at the wrong things, interpret the picture inappropriately, or become bored by unattractive shots. *Composing shots* is not just a matter of 'pictorial packaging', but a method of controlling continuity of thought.

The effect of the picture frame

The camera does much more than 'put a frame' round a segment of the scene. *It inherently modifies whatever it shows*. Because the screen *totally isolates* its subjects (the viewer cannot see whatever else is happening), and because the resultant picture is flat (and cannot reproduce stereoscopic depth) unique relationships develop within it, that are not present in the actual scene.

No shot directly portrays *reality*. In many cases our own experience enables us to rationalize and interpret, so that we make a pretty accurate assessment of what we are seeing, e.g. a tourist photograph composed to make the Statue of Liberty rest on his outstretched hand . . . or is it a model after all? But this does not always happen; particularly where the viewer is not familiar with the situation.

The subjective effects that arise in the flat framed picture take several forms:

1. Although one plane may be some distance behind another, they can appear conjoined or even merge in the shot.
2. Spatially unrelated areas of tone and color are often juxtaposed in the picture; interacting and influencing pictorial blanace. Simultaneous contrast effects result.
3. Seen in perspective on the flat screen, the shapes of objects and composition lines can change with camera angle.
4. A series of quite unrelated subjects at varying distances can combine pictorially to form a composition group.
5. Within the picture, *imaginary* composition relationships can seem to develop between subjects, and visual tensions form between them.
6. As subjects approach the edge of the screen, impressions of tension or compression can develop; particularly in screen-filling shots.
7. The lateral position of subjects influences their composition impact.
8. Various dynamic effects arise (dynamic composition).

Proportions

There are no formulae for beauty. But centuries ago, artists discovered a widely accepted principle guiding harmonic proportions—the Golden Section or Golden Mean. Great painters, sculptors, architects, have used these ratios in many ways (often unwittingly, perhaps) when expressing their concepts of beauty.

If you cut a straight line into unequal parts providing the most pleasing relationship, you will find with incredible regularity that the line lengths are divided according to the Golden Section; in a constant ratio, so that the small part is to the larger, as the larger is to the whole. These proportions work out in practice to about 8:13 (often quoted as 3:5).

Such proportioning has quite far-reaching applications, for it gives us a guide to ratios that most readily please the eye. Be suspicious of such formulae; but see how often they seem to validate themselves.

Fig. 5.2 part 1   Subjective lines
The mind tends to seek pattern and to see relationships, even where none really exist. Individual subjects seem to combine or interact to form patterns.

■ *Rule of thirds*   When composing a shot, you should avoid mechanically dividing the screen into regular sections. A screen cut into halves or quarters can produce a dull, monotonous balance.

You will find the 'rule of thirds' often advocated when selecting proportions. Main subjects are placed at the intersections of lines dividing the screen vertically and horizontally into three equal parts. A useful enough mechanical concept to guide in composing pictures, it should not be allowed overslavishly for it can develop a recognizable sameness.

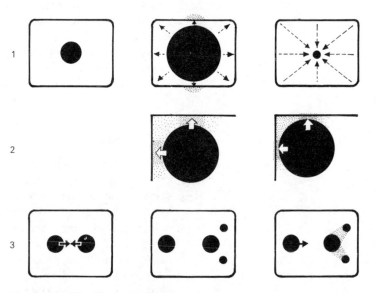

Fig. 5.2 part 2   Illusory attraction
1, A disc in the center of the frame appears at rest, surrounded by empty areas of equal tension. A change in subject size alters this reaction between the subject and its frame. An over-large subject bulges the frame, squeezing-out the surrounding space, while a too-small subject becomes compressed by the large area of empty space surrounding it. 2, This compression effect gives way to a forceful thrusting as the intermediate space lessens. 3, Tension is built up between subjects in a picture. The two close subjects convey a feeling of mutual interlink-ing—illusory attraction. In a more complex example, the isolated subject appears to be attracted towards the more stable supported group.

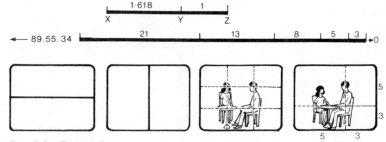

Fig. 5.3 Proportions
The Golden Section: Dividing for ideal proportions tends to produce YZ:XY in the same proportions as XY:XZ. Making several such divisions, a series appears, and this ratio of division, the Golden Section, is about 3:5. Dividing the frame: An equally divided frame allows only formal balance—usually dull and monotonous. Thirds can lead to quickly recognized mechanical proportions. Dividing the screen in a 2:3, or a 3:5 ratio achieves a far more pleasing balance.

Framing

In framing or lining up a shot, you are deciding several important matters:

1. You are choosing *exactly* what is to be included within the picture (*on-frame*, *in-frame*); and what is to be excluded from (*off-frame*, *out-of-frame*). You might be doing this to concentrate attention and avoid distractions, or to enable the viewer to see more subject detail. Or you might be omitting information deliberately, which will be revealed later through camera movement or by changing to a new viewpoint.

2. At the same time you are positioning subjects—and hence line and tone—within the frame. According to how you frame the shot, so the picture impact can be modified.

All parts of the frame do not have equal pictorial value. The implied effect changes, depending on where you place the main subject.

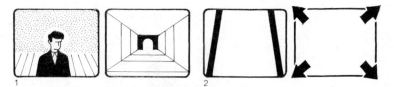

Fig. 5.4 part 1 Framing—frame positions
1, Center screen is generally the weakest area for holding the viewer's attention. Continual or sustained use of picture-center becomes monotonous. 2, Positions near frame edges seldom give good balance (particularly scenic lines running parallel with them). Frame corners exert an outward pull on subjects placed there. Subject strength and importance tends to increase higher in the frame and towards the right.

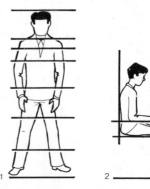

Fig. 5.4 part 3   Framing people
1, Avoid frame-cutting people at natural joints; as shown intermediate cutting—points appear more attractive. Similarly avoid framing so that they touch the screen edge as in 2.

Fig. 5.4 part 2   Safety margins
As TV screen edges are usually masked off, avoid placing important action or titling near the frame edges.

Appropriate framing is a vital aspect of good camerawork. It will not only alter composition balance, but can influence the viewer's interpretation of events. Framed in a certain way, a two-shot might lead him to expect someone is about to enter the room, or that an eavesdropper is outside a door.

Headroom changes proportionally with the length of shot; lessening as the shot tightens. But as there may be differing

Fig. 5.4 part 4   Relevant framing
Do not just frame the subject for an attractive effect, but to suit the program purpose.

Fig. 5.4 part 5 Tight framing
Close shots of people bring problems such as arm gestures, or explanatory items remaining unseen out of shot. Tight framing emphasizes the screen's confines. Head movements can pass out of frame, leaving it empty; thus requiring catch-up panning, which is a fidgety and obtrusive operation.

Fig. 5.4 part 6   Headroom
1, For good vertical balance, people should be framed with appropriate headroom. Avoid the cramping effect of insufficient headroom, or the bottom-heavy effect of excess. 2, Whether compositional elements become an incidental border, or oppressively overhang the action largely depends on how they are framed.

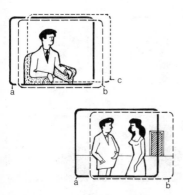

Fig. 5.4 part 7   Offset framing
For side or ¾ views, center-frame positions provide an unattractive balance. Slight offsetting, sometimes referred to as 'looking room' is usually preferable; although this must not be overdone unless you want to draw attention to that side of the frame, or imply isolation. Horizontal framing can affect significance too. Notice when the door is included in the shot, someone's entry is expected.

interpretations between cameramen, it is as well for the director to check that comparable shots match.

Pictorial balance

You should generally aim to produce *balanced* composition. Not the equal balance of formal symmetry, for that is uninteresting, but a picture with *equilibrium*.

Balance stems from:

1. The *size* of a subject within the frame.
2. Its *tone*.
3. Its *position* within the frame.
4. The *interrelationship* of subjects in the shot.

Occasionally, the dynamic restlessness or tension of *deliberately unbalanced* arrangements is refreshing; but this should be used sparingly. Balance is not a fixed composition factor, and you can readjust it by moving a person, altering framing, etc., in order to redirect attention to another subject, or even alter the picture's import.

Pictorial balance is largely an instinctive choice, but we can detect useful guiding principles:

1. A center-frame position is satisfactory; safe but dull to watch.
2. As a subject moves from picture center, the shot feels progressively unbalanced.
3. This effect arises more strongly with bigger and/or darker objects; especially if high in the frame.
4. A subject or tonal mass to one side of the frame, usually requires compensatory counterbalancing in the remainder of the shot. This could be an equal opposite mass (symmetrical balance), or a series of smaller areas that together counterbalance the main offset region.
5. Tone influences visual *weight*: darker tone subjects look heavier and smaller than light-toned ones.
6. A small darker area slightly offset, can balance a larger light-toned one further from picture center.
7. Darker tones towards the top of the frame produce a strong

Fig. 5.4 part 8   Reframing
As subjects leave or enter frame, the cameraman normally readjusts framing unobtrusively for the next situation. For dramatic purposes you may deliberately NOT reframe; the tension from the unbalance composition emphasizing the second subject's departure or absence.

Fig. 5.5 part 1   Adjusting balance
1, Balance pivots about the picture center. 2, Unbalanced shots are visually unstable and unattractive. 3, But formal *symmetrical* arrangements usually prove monotonous and uninteresting. 4, Certain compositional grouping has considerable stability, but must be correctly framed to appear balanced. 5, To reduce symmetry, you may *angle* subjects (avoiding head-on positions); for balance can change with shape. 6, Or you can adjust relative *sizes* of subjects. 7, Altering the subjects relative *distances* from center-frame changes pictorial balance. (Try adjusting positions of cut-out card shapes.) 8, Isolation gives a subject weight. 9, By *grouping* individual subjects, they have greater collective 'weight'. (See unity.) 10, *Size and tone* combine to affect overall balance. Larger areas and darker tones should be framed carefully. Dark tones towards the top of the frame produce a strong downward thrust—top-heaviness, a depressed closed-in feeling. At the bottom of the frame, they provide a firm base for composition, lending it solidity.

Fig. 5.5 part 2   Changing balance
Balance and emphasis can be altered in several ways: 1, By changing the lens-angle (zooming)—size change. 2, By altering the camera-distance—proportions change. 3, Readjusting subject height. 4, Altering subject grouping. 5, Changing camera height. 6, Changing viewpoint—different tones, masses come into shot. 7, Lighting changes—altering tones or colors of areas.

downward thrust—top-heaviness, a depressed closed-in effect. At bottom frame they introduce stability and solidity.

8. Balance is more influenced by vertical elements than horizontals; although the overall horizontal effect determines the final balance.

9. Regularly shaped subjects have greater visual weight than irregular ones.

10. Warmer colors (red, orange) appear heavier than cooler ones (blue, green), bright (saturated) hues looking heavier than desaturated or darker ones.

■ *Unity (order)* Pictures must be cohesive to be effective. You should avoid randomness or scattering items around the frame. Instead, group them with real or imaginary composition lines to emphasize interrelationships. Similarly, between successive pictures in a scene there should be a unity of style, purpose, interest and mood. Where depth of field permits, you can compose *in depth*, unifying two or more planes of action in the foreground and distance.

Fig. 5.6 Unity and order
Unity is cohesion, interrelationship of masses. 1, A picture can be balanced, yet not be unified, or 2, a group unified without overall balance. Real or imaginary compositional lines create visual unification. 3, Lines or tones that divide the screen disrupt unity.

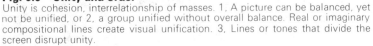

■ *Visual rhythm* In music and poetry the ear prefers a recognizable, but not too elementary, rhythmical beat; the eye also is attracted by a variety of pattern. *Visual rhythm* is a pattern term; a variation of visual emphasis, of outline. While a solemn quiet

Fig. 5.7 Visual rhythm
1, Pictorial arrangements can produce a monotonous repetitive effect, or interesting variety of emphasis in shape or tone; a visual rhythm. 2, While verticals and diagonals give an emphatic rhythm, horizontals and slow curves convey a restful rhythm. 3, Although identical repetition is dull, used as a progressive visual development, repetition can build and strengthen a theme.

mood is suited by slow smooth flowing lines; a rapid spiky staccato would match an exciting dramatic situation.

Emotional influence of tone

The set designer and the lighting director use light and shade to create the picture's emotional key. Where light tones predominate, the effect is cheerful, delicate, airy, open, simple, and weak. Predominantly dark tones create a heavy, somber, sordid, forceful, dignified, or significant impact. When relieved by smaller distinct light areas, the effect tends to become mysterious, solemn, grave or dramatic.

A picture with an overabundance of dark tones does not televise well. Large areas of unrelieved dark or light values, that contain little tonal contrast, can be disappointingly dull and uninteresting. Large, well-marked areas of contrasting tone give a picture strength, vigor, significance. A light tone, relieved by small distinct areas, conveys liveliness and delicacy.

Scale

Scale is concerned with the apparent size relationships a picture conveys. An audience judges how large, how distant, how high,

Fig. 5.8 part 1   Tonal impressions
Sharply-defined tonal contrast (left) is used to isolate and define areas, suggest crispness, hardness, vitality, dynamism. *Graded tone* (right) blends areas together, guiding the eye from darker to lighter regions. It suggests softness, beauty, restfulness, vagueness, lack of vigor, mystery.

Fig. 5.8 part 2   Tonal emphasis
Tonal contrast emphasizes shape and mass. A small dark area becomes subjugated within a large white background, whereas a small light area stands prominently from a dark tone.

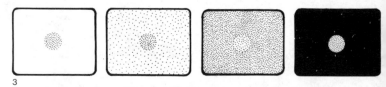

Fig. 5.8 part 3   Simultaneous contrast (spatial induction)
Adjacent tones interact. A light tone can make a nearby one look darker; while a dark area apparently lightens a light tone further. The same central tone here appears different as its background changes. The sharper the contrast and borders, the stronger this effect.

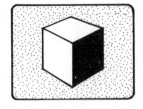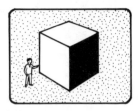

Fig. 5.9 Scale
The viewer judges scale subjectively; by how fully the subject fills the screen and by the presence of other objects of known size. If size-references are omitted (absent or obscured by viewpoint) accurate interpretation of scale may be impossible.

etc., by interpreting a series of visual clues:

1. By comparison of the surroundings with the subject's known real-life size.
2. By comparing the subject shown with adjacent subjects of known size.
3. According to how much of the picture frame is filled by the subject.
4. By relating to perspective clues in the shot.

Subject prominence

Inappropriately presented, a subject can lose strength, importance or vitality. It may even become overlooked or lost within its environment.

The subject's *surroundings* have a considerable influence on its prominence, and on our attitude towards it. Consider the difference between a coin imposingly displayed on velvet . . . or heaped with others in a rusty junk box.

Subject isolation gives emphasis. Some of this emphasis stems from the subject's tone relative to its surroundings, or lighting treatment. But compositionally you can influence subject prominence in various ways: according to your camera height, shot size, framing, background form, differential focusing and spatial separation from other subjects.

Fig. 5.10 Foreground and scale
Foregrounds influence our impression of scale and distance; particularly for isolated and remote subjects. Try to include foreground and perspective clues to define and unify depth; but avoid over-emphasized, stylized, and artificial relationships.

Subject attitude

A performer's associative attitudes can modify his effective strength, whether he looks forceful, cowed, submissive.

His general *posture* is significant, too. *Weak attitudes* include side or rear views, lying down, looking down, bowed, stooping, clasped hands, and slow movements. *Strong attitudes* include frontal view, uptilted head, hands clenched, stamping feet, and fast movement.

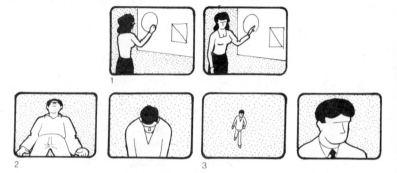

Fig. 5.11 part 1   How subject strength is influenced by viewpoint
1, Unsuitably arranged, a subject's strength is diminished. 2, Low-angle shots emphasize subject strength and importance; while high-angle shots suggest weakness, unimportance. 3, Subject-strength grows as the shot gets closer.

Fig. 5.11 part 2   Subject strength—frame position
Set higher in the frame, the subject gains strength.

Fig. 5.11 part 3   Subject strength—size proportions
Relative proportions of subject size and surroundings will affect the subject's 'power'.

Fig. 5.11 part 4   Surroundings influence subject strength
1, A supported subject appears weaker than an unsupported one. 2, A subject can be dominated by dynamic scenic lines, or by other stronger subjects. 3, Background line and tone can strengthen or weaken the subject.

Camera treatment can enhance or nullify the performance's effect. Shooting a ranting dictator in a high-angle long shot would make his gestures appear futile, weak, and ineffective. However, a low-angle mid-shot would give them a powerfully dramatic force.

Where you use a strong camera treatment for weak performance attitudes, you strengthen them; so that an old woman making weak submissive gestures could seem to have a dignity, and an inner strength against adversity when shot from a depressed viewpoint.

Picture shape

A picture's shape can affect the viewer's feelings towards the scene. A *horizontal* format can give it stability, restfulness, extent . . . A *vertical* format can imply height, balance, hope . . .

The original motion picture screen shape (aspect ratio) had 4:3 (1·33:1) proportions. But subsequently, to enhance big-screen presentation, several others were developed (1·65 to 2·55:1). Television adopted the original 4:3 format; and while this cannot be intrinsically altered, you can change its *effective shape* by introducing mattes (wipes), masks, scenic restriction or by lighting restriction.

Fig. 5.12   Picture shape
The effective picture shape can be modified by 1, electronic wipes, camera mattes (vignettes); or 2, by using a scenic opening as a border; or 3, by restricted lighting.

Unifying interest

You should normally arrange a picture so that the eye can find satisfaction within the frame; concentrating on certain features, while the rest become a subordinate background. Avoid three-ring circus techniques, with several things happening simultaneously. They divide attention, creating ambiguity and confusion. One picture—one focal point.

Unifying the center of interest does not prevent you moving it around freely. This continual flow, linking together parts of the picture, is often termed *transition* or *continuity*. The eye moves naturally from one area to the next.

Speed of composition lines

The eye scans some shapes at a leisurely rate (gently curved lines), while others are appreciated quickly (straight lines, zig-zags). Through suitable selection, you can adjust the picture's vitality and the speed with which the eye examines it. Faster lines move attention quickly and would allow fast intercutting.

Simple lines and shapes direct attention more readily than complex or disjointed ones. Interest is strongly attracted by clear-cut geometrical shapes, particularly when they seem to arise accidentally. An excess of simple line soon palls though, and leads to a stark unsympathetic atmosphere.

Fig. 5.13 Split interest
Aim to localize interest. Avoid split centers of attention, except for dramatic purposes.

Continuity of centers of interest

Although our eyes continually move around the picture, we usually find that interest is localized at any given moment to about one-twelfth of the screen area. You can divide the 4 by 3 frame into twelve equal segments.

Where successive centers of interest are widely spaced, the viewer's concentration must be disrupted to some extent while he seeks each new spot. Fast interlinking composition lines help here. This situation particularly arises, of course, whenever you cut to another shot. If the new picture presents an entirely different interest location, the audience becomes overaware of the transition; so the points should be reasonably matched.

Poorly matched transitions, especially between numerous brief duration shots, can accelerate mild audience resentment into marked hostility. This situation arises all too regularly where editors demonstrate their prowess by rapid cutting to the beat of fast music. Even if synchronism is perfect, inept matching of centers of interest (often with too little time to examine each shot) becomes totally frustrating.

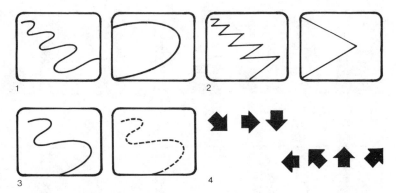

Fig. 5.14   Speed of compositional lines
1, A pattern or direction-of-line over which the eye lingers is termed 'slow'. Predominating curved lines suggest leisure, beauty, deliberation. 2, 'Fast' lines are usually straight or angular, creating an impression of speed and vitality. 3, Unbroken lines are faster than broken ones. 4, Direction of movement will affect 'speed'; the first three directions being generally faster than the remaining four directions.

Composition continuity in multicamera production

A regular problem in multicamera production, is that a composition set-up look great from one viewpoint, may prove quite disappointing from another. Various weaknesses can arise: the new picture may lack cohesion, unity or balance. It may remain attractively composed but inept; laying wrong emphasis, perhaps.

Although these difficulties may only be slight where the set-ups

Fig. 5.15   Related shots
The compositional arrangements and action can move the eye about the picture. On cutting to another shot, a new center of interest presents itself. If the centers are too dissimilar, as in the first pair of illustrations, the viewer becomes over-aware of the change. But where reasonably matched, the transition may be almost unnoticed.

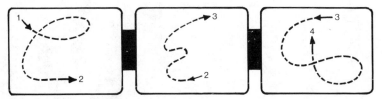

Fig. 5.16   Movement of interest
Good visual continuity between shots does not necessitate static arrangements. Composition and action can introduce unlimited interest-movement, providing successive shots match at the transition.

Fig. 5.17   Cross-cut shots
1, If successive pictures are matched accurately, some visual confusion can arise when inter-switching between cameras. 2, Where alternative foci are continually being intercut, it may be less disturbing to space (offset) centers of interest.

are simple or haphazard, for more elaborate situations some anticipation is necessary.

There are several ways of coping with this situation:

1. You might dwell on the compositionally effective shot, accepting the weakness of others.

2. In *discontinuous* shooting you could rearrange (*cheat*) people's positions, set dressing, lighting, for each new camera set-up. This is time consuming and can cause continuity errors, but provides optimum results.

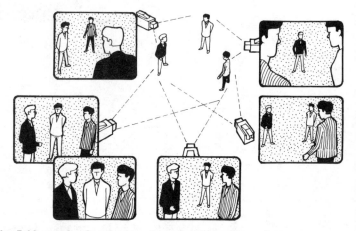

Fig. 5.18 part 1   Continuity of composition—two shot
The subject can be arranged to provide appropriate composition from several viewpoints.

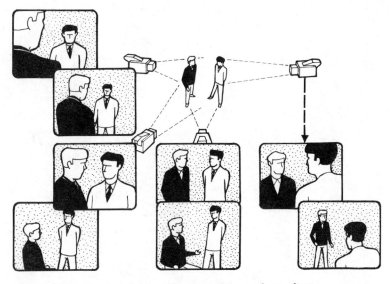

Fig. 5.18 part 2   Continuity of composition—three shot
Although composition continuity becomes more difficult with larger groups, a variety of shots is available with careful grouping.

3. You may arrange the subjects so that they suit all the required angles (as in interviews).

4. Performers can be repositioned at the moment of transition to a new camera; e.g. having delivered a line, the actor turns away . . . on the cut.

5. You can recompose the first shot shortly before the transition, to suit the next shot's composition.

Fig. 5.19   Care in grouping
Cameras shooting actuality, have to select from available grouping. But by attention to people's positions a wider range of shots becomes possible. 1, People must normally work closer together than in everyday life, to avoid widely spaced shots. 2, Where close spacing is inappropriate, cross-shooting groups them. 3, Although parallel face-to-face conversation may seem more natural, slight angling improves shot-opportunity. 4, Similarly, closed groups are better opened up.

Fig. 5.20   The influence of viewpoint
The viewer's feelings about the subjects, their environment and their relationship
to it, can be modified according to the viewpoint selected for the camera.

Color and the picture

Most of us take color for granted. We accept that a surface is a
particular shade of green, for example. But in fact, what we are
interpreting as 'green' can vary considerably: with the color of
the light falling on the surface, its angle, any other nearby color,
and various other factors.

■ *Surface effects*   The color of a smooth surface generally looks
'purer' (more saturated) than a rough-textured one of identical
color; its brightness too, will change more noticeably with the
light direction. The color of a surface can seem to vary with the
quality of the incidental light; appearing brighter and more
saturated under hard light, than under diffused lighting. A color
surface may reflect light onto a nearby subject—so the face of a
person wearing a yellow sweater may be tinged yellow by its
reflected light.

■ *Fine color detail*   Our eyes cannot readily detect fine detail in
color. Where a scene contains tiny areas of color (whether small
patterns or distant subjects), these will seem to get paler as they
grow smaller. The actual color of fine detail can become difficult
to identify. Yellow, for instance, tends to become indistinguishable
from light grey, and blue detail can become confused with dark
gray. Similarly, bright areas of green, blue-green and blue appear
identical when very small. Eventually as detail size diminishes,
even strong reds and blue-greens become indistinguishable, so
that you can only detect differences in brightness between them.

■ *Warm and cold colors*   When looking at colors, you will often
find yourself unconsciously attributing 'temperature' and 'dis-

98

tance' to particular hues. Red, yellow and brown seem 'warm', and areas of these colors tend to look bigger and nearer than the 'cool' hues such as blue and green. You will find too, that darker and more saturated colors seem nearer than lighter or desaturated ones.

Subjectively, the pictorial 'weight' of a color is often affected by its 'warmth': warmer tones seem heavier than cooler ones, saturated colors look heavier than duller desaturated ones. For these reasons, it is desirable when composing masses in a picture, to keep areas of saturated color small, for they can easily balance much larger ones of desaturated color.

■ *Simultaneous color contrast (lateral color adaptation)* The appearance of any color can be considerably affected by the background against which it appears; an effect known as *simultaneous contrast*. A green dress against gray drapes will not only look greener than normal, but the gray background itself may tend to take on the complementary hue (magenta) and appear 'warmer'. If you place a white object in white surroundings, and illuminate it with both white and magenta light, its shadow appears to be green! Clearly, the eye and the brain can play strange tricks!

The same turquoise object may look lighter, darker, bluer, greener, according to its background colors. When a strong color appears against a pale version of the same hue, the background seems grayer. A strong color will appear more vibrant when backed by white, than against black. Conversely, a darker color is more pronounced against a black background. Any color will look brightest and strongest against its complementary color.

Even *neutral tonal values* (white, grays, black) are modified by their background. The darker the background, the lighter will the subject tones appear to be (and vice versa). Light tonal areas look larger. Darker tones look heavier. All these factors can strongly influence our impressions of size, distance and pictorial balance.

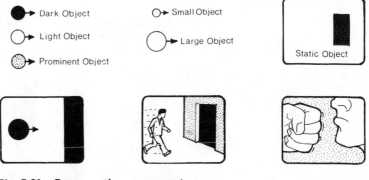

Fig. 5.21 Representing movement
Symbols can represent many different dynamic situations—the subjects, their direction, and their path within the TV frame.

■ *Separation of color*   Although color surfaces may in reality b
some distance away from each other, they can appear adjacen
when seen together within the shot. So there is a mutual interactio
between them.

■ *Bright colors*   Bright colors hold the attention—particularl
scarlets, bright yellows and orange. But their prominence ca
become embarrassing when they are defocused, for the eye turn
to them, yet cannot see detail there. Pastel (desaturated) hues d
not pose this problem—but they on the other hand can lack vigo
and visual appeal if widely used.

Where there are reasonably equal areas of strong color,
disharmonious tension can develop between them, particularl
where they are not complementary. So it is preferable to have on
or more of them desaturated (i.e. grayed-off, less vivid).

■ *Approximate color constancy*   This illusory effect continuall
influences our evaluation of color. We experience it when lookin
at a picture containing a familiar subject of known color—
mailbox, perhaps. Having fixed on this item, we go on to judg
other colors in the picture accordingly. Even where the reproduce
color of this subject is considerably different from that of the rea
thing, the brain adjusts its interpretation, so that we are pre
disposed to see it as 'correct'; perhaps regarding others nearb
as inaccurate, although they are actually satisfactory.

All interpretation is subjective. We regularly misinterpret a
'white', areas that in reality are quite blue or yellow. Blue puddle
on the sidewalk may look unnatural, although they truly reflect
blue sky. A 'black' object in bright sunlight may actually be re
flecting more light than a dimly illuminated 'white'—althoug
our brain, making a comparative judgment, refuses to believe this

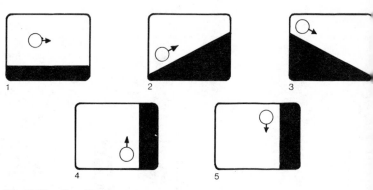

Fig. 5.22   Gravity
There is a tendency to interpret movement within the picture as if gravity wer
always involved. So in these illustrations movement is seen as: 1, Norma
progress. 2, Moving uphill; climbing—3, Moving downhill; descending. 4
Rising. 5, Falling. Such movement is regarded as difficult, easy, forceful, weak
etc.

■ *After images* After looking at a color for some time, you will often experience a brief color 'hang-over' on switching to the next shot. Following exposure to red, you will see a blue-green (cyan) ghostly after-image on a black screen. After orange, a peacock-blue after-image follows. Yellow gives blue; green gives purple (magenta); while after blue an orange-yellow image is seen.

Bearing this in mind, you would expect some interaction between a succession of colors. Such *successive contrast* causes white to appear bluish-green after looking at a red screen. Similarly, yellow appears bright green; blue appears more intense and greenish; while repeated red has a grayed appearance.

Admittedly, these are extreme situations, but color distortion on quick cuts and the visual fatigue that follows overexposure to a strong color are not to be disregarded.

■ *Color associations* Color and emotion are inextricably inter-linked. Color associations are legion. For example:

Red—with warmth, anger, crudity, excitement, power, strength.

Green—with spring, macabre, freshness.

Yellow—with sunlight, the Orient, treachery, brilliance.

White—with snow, delicacy, purity, cold.

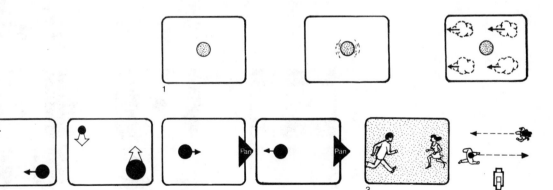

Fig. 5.23 The impression of movement
1, The interpretation largely depends on framing, and the apparent speed of other subjects and background. Firmly framed against a plain background, there is no sense of movement. If the subject's position wavers and it is known to be moving, the movement is assumed to be so fast as to be difficult to follow. Add signs of a background racing past (e.g. aircraft in clouds) and the sense of movement and speed grows considerably. 2, To this there can be added four general axioms. Movement across the screen tends to appear exaggeratedly fast. Movement to or from the lens seems slower for narrow-angle, faster for wide-angle lenses. Camera movement in the direction of subject movement reduces its speed. Camera and subject moving in opposite directions increase their mutual speed and impact. 3, The flat screen can cause the viewer to misinterpret spatial movements. On camera the people seem to be running to meet—but the plan view shows otherwise!

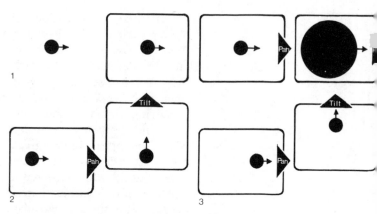

Fig. 5.24 Restriction of movement by the frame
1, Movement in free space is interpreted as having unrestricted freedom. But onc
a frame is placed around the subject, freedom is relative to these limits. There
potential restriction. Where the moving subject is held near center-frame, th
restriction is not excessive. But close shots must always be cramped. 2, Restrictic
becomes less when the subject is kept near the lagging edge of the fram
3, Restriction is emphasized when kept near the leading edge.

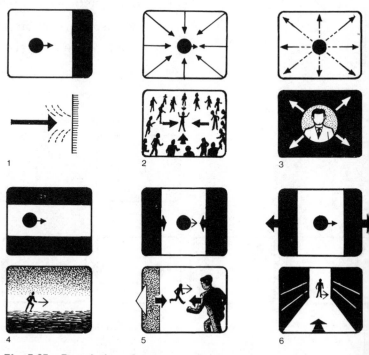

Fig. 5.25 Restriction of movement by masses
The shape and position of masses (moving or still) in the picture can alter th
interpretation of events. 1, The mass in the path of the moving subject sugges
imminent impact or arrest. 2, Converging, encircling, crushing, as when a crow
presses in on a central figure. 3, Expansion, new freedom, isolation, as when usin
an iris-out wipe. Even complicated situations can be arranged to emerge qui
naturally: 4, The restricted feeling aroused by the traveler moving at dus
between darkened sky and ground. 5, The compression effect as the camera par
left (following the pursuer), and the wall moving frame-right create a pince
compression movement towards the distant victim. 6, As the camera dollie
down the alley, the distant light area grows—creating an expansive effect.

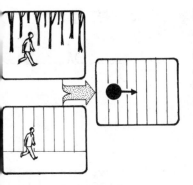

Fig. 5.26 Background patterns
As subjects move past them, background patterns can exert strong effects upon picture impact.

Fig. 5.27 Background and subject movement
Where movement is in the direction of background pattern, the movement will appear unimpeded if slightly restricted. The line emphasizes direction. Where the subject moves across (against) the pattern, effort is emphasized; movement appears stronger. Subject shape may modify these effects.

Fig. 5.28 Moving background and subject 'strength'
Against a plain background, the effect of this low-angle shot remains constant as the subject walks and the camera dollies back. But if a series of horizontals move downwards (crossboards in a ceiling) during the move, it is considerably strengthened.

Dynamic composition

Where you have static or semi-static shots, long-established composition principles clearly apply. But television and film are *dynamic media*. Therefore, it is essential that the distinctions between the still and the moving picture are understood; a study that may be termed *dynamic composition*.

Potentials of the still picture

The *still picture* has the particular advantage that we can study its details at leisure; concentrating on the aspects that attract our interest. A moving picture allows us only a limited time to examine the picture (the shot duration), before another shot replaces it.

Paradoxically, a frozen-action shot may have a stronger more sustained impact than the movement seen in real time, e.g. a boxer falling from a knock-out punch. Realizing this, some directors have introduced brief *freeze frames* or slow-motion sequences during powerful action, to savor and emphasize the dynamics.

This indefinable feeling of 'suspended animation about to recommence' is the underlying appeal of many great paintings and sculptures. A composition balance that is so forceful in a still version, would often be impossible to maintain or even ineffectual in a moving form.

Action that is intriguing when seen in a fragmentary glimpse, can appear commonplace when carried to its natural conclusion. Hence the production techniques of cutting between brief shots of uncompleted action, or fading before a sequence has logically concluded are often used. This method is normally employed at a climax, where a continuing shot would result in an anticlimax.

A shot that is well composed while it is *still*, may become less pleasing when movement rearranges its component parts.

Potentials of the moving picture

The moving (or movable) picture has certain properties that the static shot lacks. It can depict *change*. You can modify the picture as the audience watches; altering subject prominence, redirecting attention, introducing or removing information, modifying the mood. You can convey directly the dynamics of movement, variation, growth, development, etc.

The moving picture offers *continued interest*; unlike the still picture whose attraction usually diminishes quite quickly. Movement attracts. It gives the director the opportunity to maintain interest in the subject aspects *he* intends, by presenting continually changing stimuli.

However, you must not forget that while the still picture allows the viewer freedom to scrutinize and assess, the moving picture

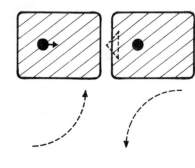

Fig. 5.29   Background and direction
1, When the subject moves across a static patterned background, the viewer feels an undertow of movement. Where the subject is held in frame and the background passes the direction of this undertow changes. 2, As the subject moves over the upward-sweeping background, an upward force seems to be exerted on the subject. In addition, background tones modify its apparent size and importance.

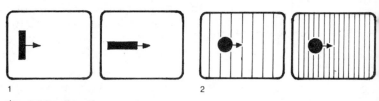

Fig. 5.30   Speed
The subject's shape can affect its apparent speed. 1, Although moving at the same speed the 'tall' object appears to be moving faster than 'flat' object. 2, The apparent speed is also increased as the background becomes more detailed. Shot-size will modify how much of the background is visible and hence effective speed-impact.

often gives him time only to grasp essentials; particularly if action is fast or brief. You have to take care not to build resentment as you snatch one shot away and replace it with the next. Much will depend on the pace and the complexity of information; and a sophisticated or informed audience can actually welcome brief segmentized dynamics.

A theory of dynamic composition

We interpret movement in everyday life, by making a series of comparative assessments. We watch a subject's rate of progress relative to its surroundings and to other subjects. We evaluate relative displacements. This happens when we watch pictures too, but the confining frame and the two-dimensional images impose further influences on our judgment.

Whether we see movement *directly* (the subject traveling within a static scene), or deduce it through secondary effects (passing background, light or shadows passing over it) our interpretation is at best very subjective and conditioned by previous experience.

Our evaluations of picture dynamics are modified by a number of often irrelevant factors:

1. We compare apparent *relative speeds* of subjects within the picture.
2. The amount of *border* round a moving subject modifies our impression of speed.

104

3. Where there are simultaneous movements in the picture, we frequently associate the movement pattern with *compression* (impact, collision) or *expansion* (parting, stretching, strain).
4. We judge a movement's *strength* or *speed* according to the apparent size, power, or force involved.
5. We subconsciously relate subject position and movement in the frame to *gravity effects*; even where there are no actual gravitation relationships.

Fig. 5.32 part 1. Patterns of movement
Some movement patterns have widely accepted associative feelings. 1, Continuity. 2, Vigor; excitement, indecision. 3, Beauty, charm. 4, Expansion. 5, Contraction, collapse. *Smooth* action suggests control, evenness, simplicity; while *jerky* irregular motion suggests the clumsy, erratic, uncontrollable.

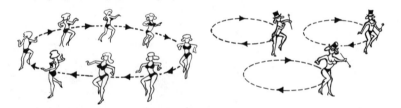

Fig. 5.32 part 2 Patterns of movement
In massed dance and marching, the viewer sees simple pattern formations from group movement, and individual movements.

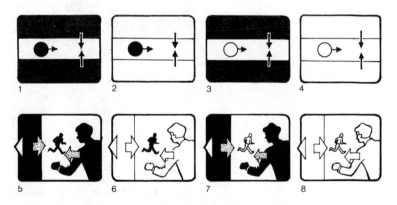

Fig. 5.33 The dynamic effect of tone
The effective strength of static and moving masses varies with tone. 1, When a dark subject moves through converging dark areas, the mutual impact is powerful. 2, The 'crushing' effect is lightweight against a heavy substantial subject. 3, The *apparently* larger but lightweight subject suggests less resistance to approaching forces. There is a hint of destruction. 4, Both subject and approaching forces are lightweight, so the effect is less dynamic than in the first example. 5, Dynamic pursuit. 6, Ineffectual pursuit of strong subject. 7, Annihilation of vulnerable subject. 8, Undramatic. Following rather than pursuing.

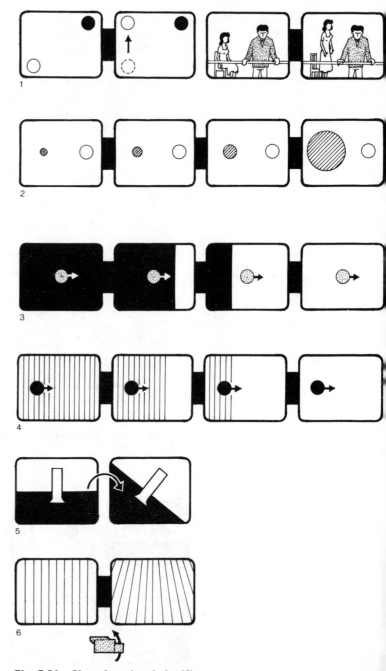

Fig. 5.34 Changing visual significance
The visual significance or strength of a subject can be modified by altering its appearance (e.g. by lighting), position or size in frame, or its background. 1, The upward movement and higher position strengthens the left-hand person. 2, Striped object is approaching. At first it is insignificant, but with increasing size, its importance and interest exceed that of the static subject. 3, As the subject passes from the dark background to the light one, there is an emotional uplift. 4, Passing from a region of restrictive verticals to a plain background produces a sense of freedom, and an effective speed change (fast to slow). 5, As the shot cants off the horizontal, pictorial balance becomes unstable. 6, The camera tilts upwards, and the verticals in this background converge into diagonals.

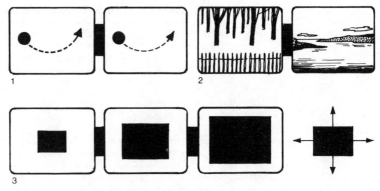

Fig. 5.35 Dynamic effects between pictures
Seeing a rapid succession of shots (still or moving), the viewer subconsciously compares or contrasts them. 1, Cutting between scenes with similar movement-form or shape, emphasizes this effect. 2, The vigorous verticals of one scene contrasts effectively with the restful horizontals of the next. 3, The successive size differences here produce expansive jumps.

6. The visual impact of movement can depend on where we happen to fix our attention in the shot. Looking skyward, we see moving clouds and static buildings—or static clouds and toppling buildings.
7. The sounds accompanying a picture often influence our interpretations of speed.
8. The impact of movement may be greater in slow motion than at normal speed.

Accepted maxims

You see the principles of dynamic composition applied daily; and from experience, some excellent widely accepted working axioms have emerged.

■ *Direction of movement* Like vertical lines, vertical movement is stronger than horizontal, which is the least arresting.

A left-right move is stronger than a right-left action.

A rising action is stronger than a downward one. Hence a rise from a seated position has greater attraction than a downward sitting movement.

An upward move generally looks faster than a horizontal one.

■ *Diagonal movement* Like diagonal lines, this is the most dynamic movement direction.

■ *Movement towards the camera* All forward gestures or movements are more powerful than recessive action away from the camera: a glance, a turned head, a pointing hand. Similarly, a

Fig. 5.36 Area of interest
1, Sloping to the right, the eye tends to favor the distance and so run out of the frame. 2, With the reverse slope, the eye tends to favor the foreground.

Table 5.1   Lateral reversal effects

Pictorial balance	Subjects tend to look heavier (affecting balance more markedly) on frame *right*. So frame left can support more weight (i.e. greater density or mass) than right. A predominant mass on frame right can make the picture unbalanced or lopsided. The space between subject and left frame may appear excessive where masses predominate on frame right.
Pictorial stability	The picture may look more stable one way round than another; dormant in one version—but dynamic when reversed.
Apparent proportions	Large dark foreground areas on frame right may produce a crowded, shut-in, or heavy feeling; while the reverse gives a more open impression.
Subject strength	A composition element that seems unimportant when located on the left, can become obtrusive on frame right (although the reverse seldom holds). A subject overlooked on left frame may become compositionally or dramatically significant on the right.
Area of interest	Our reaction to diagonal lines can vary with their slope direction. So our attention center can be directed in depth within the scene.
Concentration of interest	The eye tends to wander towards frame right (even with a predominant frame left subject). Where strong subjects are positioned frame right, left-hand subject may go almost unregarded.
Picture significance	Part of the scene may take on an implied significance in a mirrored version.
Visual interpretation	Our interpretation of what is happening or what the picture depicts, may be modified by left-right disposition.
Direction of slope	A down-slope to the right tends to be interpreted as downhill; and an up-slope as uphill. This influences how difficult or forceful lateral movements along such slopes appear to be as they move in an 'uphill' or 'downhill' version. Similarly, a vertical object leaning or falling to the right tends to 'fall forward;' rather than backward.

Fig. 5.37   Concentration of interest
1, Placed on the left, the window is just part of the scene; the foreground person dominates. When on the right, the viewer anticipates important action through the window; the person becomes less important. 2, The viewer is very conscious of the watcher on the right, but only half-aware of his victims. Placed on the left, the watcher is nearly overlooked, attention focusing on his victims.

shot moving towards a subject (dolly/zoom in) arouses greater interest than one withdrawing from it (dolly back/zoom out).

Where someone moves in front of another person (or scenery) they attract more attention than when moving behind them. Avoid such cross-moves in shots closer than mid-shot.

■ *Continuity of movement*   While a moving subject attracts attention more readily than a static one, continuous movement at constant speed does not maintain maximum interest. Where action is momentarily interrupted or changes direction, the impact is greater than one carried straight through. Converging movements are usually more forceful than expanding ones.

Fig. 5.38   Direction of slope
Direction of the slope can alter a picture's attractiveness, there being a tendency for version 1 to seem more dynamic than 2.

Mirror images

If you look at any collection of photographs direct and then in a mirror, you invariably find a difference in the appeal of the two versions. Differences may be great or slight; but they do modify the picture's impact.

The various effects that arise regularly through 'right-handedness' in composition treatment can influence which way round you arrange your picture to suit any particular purpose.

Fig. 5.39   Examples of combined effects
Several aspects of right-handedness can appear simultaneously.

109

6 Editing

In television production you will meet editing in three forms:

1. Real-time video switching (vision mixing).
2. Post-production videotape editing (Chapter 12).
3. Film editing (Chapter 11).

Although these processes are quite different, the *artistic effects* of editing in each can be identical.

The nature of editing

In purely *mechanical* terms, editing is concerned with:

1. When and how you are going to change from one shot to another.
2. The instant, method and duration of transitions.
3. The order and durations of shots.
4. The maintenance of good picture and audio continuity.

These decisions can all have a direct bearing on the appeal, interpretation and emotional impact of the program material.

Artistically, the potentials of editing are far reaching:

1. An entire production can be shot on a single camera—yet through skilled editing the audience is given a sense of unlimited spatial and temperal freedom.
2. Editing can juxtapose events occurring at quite different times and places. It can seemingly expand or contract time.
3. Editing can insert or omit information; correcting, excluding or censoring. It can equally well excise the extraneous or the essential. It can introduce the apposite—or the irrelevant.
4. Editing creates interrelationships—that may or may not have existed.
5. Editing is selective, and depending on the choice and arrangement of this selection, you can influence the audience reactions and interpretation of events.

Editing and the TV director

In *film making*, a specialist editor and his assistants are responsible for the mechanics and techniques of the editing process; submitting the product for the director's critical appraisal in review. From the rough-cut, the final fine-cut version is developed.

In *television*, the director is invariably his own editor, particularly where he operates the *production switcher* (*video switching panel, vision mixer*) himself in real time. Even where the *technical director* or a specialist *switcher* (*vision mixer*) takes over the mechanics under his instruction, the bulk of the editing for many TV productions is virtually completed 'on the hoof' *during* the recording. Subsequent videotape editing in such cases is mainly cosmetic (inserting corrected shots, cutting overmatter). Live transmissions, of course, are necessarily a fait accompli!

TV directors may approach a production with a marked-up camera script showing all intershot transitions; or arrange editing during rehearsal; or make spontaneous decisions during recording/transmission.

Watching the preview picture monitors continuously showing his various video sources, the director selects at the required moment, switching the chosen sources onto line (transmission, main channel, studio output). The shots may have been planned precisely, approximately, or offered up by the cameramen and modified as necessary. Prerecorded film or videotape contributions are either totally inserted on cue, or excerpted from prearranged sections.

Because preview monitors *continually* display each source's output, they show not only the contributory material, but we can see cameras off-shot, moving position, film inserts running on to their next cue point, leader cue frames, graphics, titles, contributory remotes standing by . . . and so on. Concurrently, the director will be guiding, instructing, selecting and generally coordinating the production team's work. All this, in addition to keeping on schedule, and coping with various contingencies as they arise. As transitions from one source to another require us just to push a button or slide a fader lever, it is little wonder that under these conditions, 'editing' can degenerate into mechanical switching!

Television editing techniques

With the advent of videotape recording, various production methods became possible that were quite impracticable during *live* transmissions. The elaboration of these techniques largely depends on the sophistication of the videotape editing facilities available. Several production processes are widely used:

■ *Live on tape* The program is recorded continuously in its entirety, as if it were a live transmission. The videotape is used here as a straightforward recording medium to provide more convenient scheduling of facilities and transmission, repeat transmissions, later analysis, etc. All editing is carried out on the studio production switcher during performance, and the show is ready for transmission at the conclusion of the recording session.

■ *Basic retakes* Here the production is recorded continuously as for transmission, but any errors (performers, production, technical) are re-recorded and the corrected sections substituted. *Duration trimming* may require some editing cuts—often covered by introducing *cutaways/nod shots/reaction shots*. As most programs can be improved by such editing, this method is widely used.

■ *Discontinuous recording* The show may be taped in a series of sequences or scenes, each retaken as necessary to correct or improve. These *takes* can be shortened, action removed, or their order changed. Sometimes several versions are shot; the preferred one being selected later. The production switcher is used for most intershot transitions.

During later videotape editing, sections (containing their own intrinsic editing) are dubbed off in their final order to form an edited composite tape. Further audio treatment can be added at this stage (background music, audio effects) from unsynchronized audio disc or tape. Various video effects may be introduced during postproduction editing (wipes, multiimages, color synthesizer treatment, detail insertions, etc.).

■ *Single-source recording* Here each camera's output is separately recorded on its own dedicated videotape recorder (VTR). No production switcher is used; all transitions and selections being made during a postproduction dubbing session to form a composite tape.

Videotape editing is carried out using complex computer-controlled facilities incorporating digital editing techniques, storage (cart, or video disc), all guided by time-code shot identification. As a one-hour program may involve 300-400 individual edits, the operation clearly requires systematic organization of sophisticated postproduction facilities.

Three types of editing are widely used that rely on the audience interpreting the interrelationship of successive shots:

1. Continuity cutting.
2. Relational cutting.
3. Dynamic cutting.

Continuity cutting

In most TV productions editing is simplistic, serving the practical purpose of interrelating viewpoints, creating a feeling of consecutiveness and order, interlinking dialogue and action. At best, the technique provides a clear straightforward narrative, a smooth pictorial flow, and unobtrusive transitions. At worst, such *continuity cutting* becomes a purely functional routine.

Example:
Mr. Brown speaks—cut to CU of Brown.
Mr. Smith speaks—cut to CU of Smith.
Anchorman speaks—cut to three-shot of group.

Relational cutting

Here through intercutting, shots that have no direct connection in reality, are deliberately given an implied relationship.

112

Examples:

Woman seen waking up to house, *cut* to woman in a room. We assume she entered house, and is now in a room there. (Actually shot with different women in similar costume, at different locations.)

Shot of aircraft in flight, cut to pilot at controls.

(A library model shot, has been intercut with a studio mock-up. Neither is real.)

Man approaches foot of stairs, *cut* to him coming upstairs.

(Action may have been shot on different occasions.)

Types of editing

Whenever you join two shots together by a *cut*, you immediately establish a relationship between them. This juxtaposition is both *physical* and *intellectual*.

Physically, the viewer's eyes become aware of the change, and begin tracing a fresh interest pattern in the new picture.

Intellectually, the viewer has to interpret the new picture. (Where are we now? What is this? What is happening?).

These responses are, of course, interrelated. Where the shots are compositionally matched (matched cuts) and the significance of the second shot obvious, it will be a *smooth cut*.

Where the shots are pictorially unmatched (jump cut, mismatched cut), the effect is physically disturbing, even though intellectually the viewer instantly appreciates what the second shot means. Even perfect physical matching of shots is insufficient to prevent disruption, if the viewer cannot immediately understand the significance of the second shot and its intended relationship to the first. These are not simply academic considerations, but are essential factors whenever you intercut shots.

Dynamic cutting

Sophisticated intercutting can create dramatic emphasis, convey moods or abstract ideas that are not readily expressed in more direct terms. The ideas themselves may not be implicit in the component shots. Cause-effect relationships are often interpreted.

Examples:

Shot of broken window, *cut* to small boy crying.

(A lost ball? Punished after breaking the window?)

Intercut shots symbolically suggesting 'Spring', 'Progress', 'Terror', etc.

The cut

The cut is the simplest transition. It is dynamic, instantly associating two situations. Sudden change has a more powerful

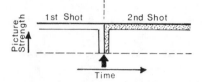

Fig. 6.1 The cut
The mechanics of picture transitions can be shown graphically. Here the *cut* (straight cut, flat cut) is shown— from one shot or scene to another.

audience impact than a gradual one, and here lies the cut's strength.

We *cut to* or 'take' another shot for several reasons:

1. To emphasize.
2. To redirect attention to another aspect of a subject.
3. To prevent it passing out of shot.
4. To maintain a particular view when the subject moves.
5. To show the subject's position relative to others.

Cutting, like all production treatment, should be purposeful. An unmotivated cut interrupts continuity and can create false relationships between shots. Cutting is not akin to repositioning the eyes as we glance around a scene, for we move our eyes with a full knowledge of our surroundings and always remain correctly orientated. On the screen, we know only what the camera shows us; although guesses or previous knowledge may fill out the environment in our minds.

Cutting jumps attention between new viewpoints or locations, so the viewer has continually to relocate and interpret each new shot. This is a problem during fast intercutting between different locations.

Where quite dissimilar shots are intercut, your audience may have some difficulty in appreciating continuity or relationships. But you can establish an immediate connection through:

1. *Dialogue*—introducing or implying the next shot's substance.
2. *Action*—establishing a cause–effect relationship.
3. *Common reference points* (a person) in both shots.
4. *Audio continuity.*

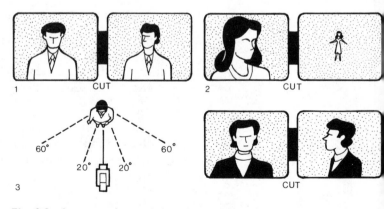

Fig. 6.2 Intercut shots of the same subject
1, Avoid cutting between similar or near-similar shots. 2, Extreme changes in shot size (e.g. over 5:1) are distracting, unless explained by dialogue or story-line. (See fig. 15.1). 3, Viewpoint angle cuts of less than 20° or over 60° can be visually disturbing.

Fig. 6.3 Visual continuity
1, The viewer can lose sense of direction where there are dissimilar backgrounds. 2, This is unlikely to happen where there is background continuity, or 3, compositional links between pictures.

Fig. 6.4 Cutting between moving pictures
Cuts during movement create strong subjective effects. 1, *Directional* cuts between subjects moving in the same direction, suggest continuity of action or direction, pursuit, similar purpose. 2, *Reverse cuts* between opposite directions can suggest converging forces, impending meeting or collision, or 3, diverging forces, parting, expansion.

Occasionally, and *only* occasionally, you may want to mislead the viewer's continuity deliberately for a comic or dramatic purpose.

Where a quick viewpoint change seems obtrusive, it has usually not been clearly *related*, and has not fulfilled the *purposes* outlined earlier. It is a change that the viewer has no reason to want or expect. Unmotivated changes rarely satisfy.

It is generally inadvisable to intercut between shots of widely differing size (LS to BCU), or over a wide angle unless the attention is tightly localized.

■ *The moment for the cut* Cuts should preferably be made on an *action* or *reaction*—turning head, gasp of astonishment, a rise. However, there are two philosophies about the moment to make the cut. Some editors contend that the cutting moment should be *just before* or *after* an action. A transition during a movement can disrupt it and create a spurious cutting rhythm. Others maintain that by cutting *during* a movement you avoid the illusion of its being jerked into action or suddenly halted (from an in-motion shot, to a static one). Where the aim is to transfer attention rather than create a visual shock, the cut can be hidden within action, as the audience is preoccupied with the movement itself.

■ *The delayed cut* A *late* cut is frustrating. Badly timed, it has missed the optimum moment for change. But the *delayed cut* has exciting possibilities. It deliberately withholds the new shot until after the expected moment—in order to create audience suspense, interest, and anticipation. Example:

1. *Cutting on a reaction*:
   Knock at door. Writer looks up—
   CUT—Door opens. Woman enters. 'May I come in?'

2. *Delaying the cut*:
   Knock at door. Writer looks up. Door heard opening. Woman's voice, 'May I come in?'
   CUT—To woman at door, who enters.

■ *Cutting between moving pictures* When moving shots are intercut, you invariably create spatial interactions between them, even where none exists. Intercutting such shots suggests:

1. Moving together (meeting, collision).

2. Moving apart (parting, expansion).

3. Moving in the same direction (following, chasing, similar destination).

To avoid these effects, cutaway shots or intermediate movement directions (to or from camera) can be introduced. Where the viewer is watching a single moving subject, any direction changes should be obvious or deducible. Otherwise he will see distracting direction-jumps or may lose orientation.

■ *Cutting between static and moving pictures* Cutting from a static scene to a moving one, suddenly accelerates audience interest. The new shot springs into action; its speed, energy or violence being accentuated by its sudden appearance.

Cutting from a moving scene to a static shot, causes a sudden collapse in tension; which, if unsatisfied, can lead to a rapid fall-off in interest. Or it can produce a highly dramatic interruption. For example panning with a searcher around an empty room . . . and then CUT—to a static shot of someone watching his actions.

■ *Cut to begin or end action* These devices must be used with caution. They can too easily look like operational errors!

The cut in: a shock introduction to vigorous action, catastrophy, chaos.

The cut out : a severe, definite conclusive action as you cut to a blank screen at the end of a shot. An emphatic gesture or comment, of finality.

The fade

■ *The fade-in* This provides a quiet introduction to action. A slow fade-in suggests the forming of an idea. A fast fade-in has rather less vitality and shock value than the cut.

■ *The fade-out* A quick fade-out has rather less finality and suspense than the cut-out. A slow fade-out is a peaceful cessation of action.

Fig. 6.5 Avoiding directional reversal
Where a subject appears in successive shots showing movement in opposite directions, you can avoid accidental subjective effects by introducing an inter-mediate buffer shot showing a head-on, view (perhaps including the actual change in direction).

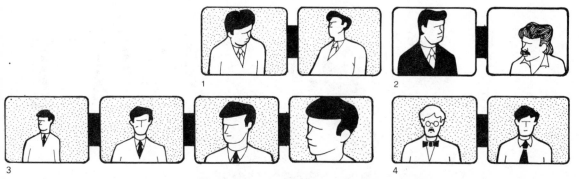

Fig. 6.6 part 1   Matched and mismatched cutting
Mismatching on cutting can take several forms, but in each case results in visual
disruption. 1, Height changes—mismatched lens-heights (for no dramatic
purpose). 2, Headroom changes—mismatched headroom. 3, Size jumps—
instantaneous growth or shrinkage occurs if subject size changes only slightly
on the cut (e.g. a–b, or b–c). A difference of at least a–c (4:3 previous size) is
desirable. 4, Transformations—if too closely matched, the viewer may see
'magical transformations' on the cut!

Fig. 6.6 part 2   Matched and mismatched cutting
1, Jump cuts—if the same subject is included in successive shots, it may jump
frame-position on the cut—a common dilemma in table shots. 2, It can arise
also in group shots. Cameras 1–2 or 1–3 intercut, but not 2–3. 3, Twists—a
slight difference in the viewpoint, can produce a turn or twist on cutting. Hopefully, the disruption may be disguised by 'motivating' the change (e.g. a singer
turns to face the new camera), or by a slow dissolve. 4, Continuity matching—
when shooting *discontinuously*, you must match shot detail and action for each
cut, during editing; e.g. the moment of turning a knob, opening a door, lifting a
glass. Otherwise continuity is disrupted.

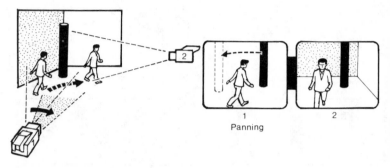

Fig. 6.7   Rediscovery on a cut
Camera 1 pans, following the action. The black pillar passes out of shot (frame left). *Cut* to Camera 2 (to prevent shooting off) and the pillar is rediscovered on frame right again! A similar situation arises when someone who has just entered a shot is seen, after a cut, to re-enter the next.

■ *The fade out-in*   Linking two sequences, the fade out-in introduces a pause in the flow of action. Mood and pace vary with their relative speeds and the pause time between them. You can use this transition to connect slow-tempo sequences, where a change in time or place is involved. Between two fast-moving scenes it may act as a momentary pause, emphasizing the activity of the second shot.

A cut-out/cut-in with a pause between shots is valueless. It always looks like a lost shot—or a mistake.

The mix (dissolve, lap-dissolve)

The *mix* is produced by fading out one picture, while fading in the next. The two images are momentarily superimposed; the first gradually disappears being replaced by the second.

Mixing between shots provides a smooth restful transition, with minimum interruption of the visual flow (except when a confusing intermixture is used).

A *quick-mix* usually implies that their action is concurrent (parallel action).

A *slow-mix* suggests differences in time or place.

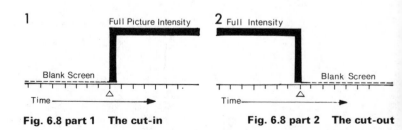

Fig. 6.8 part 1   The cut-in   **Fig. 6.8 part 2   The cut-out**

Mixes are generally comparative:

1. Pointing similarities or differences.

2. Comparing time (especially time passing).
3. Comparing space or position (mixing a series of shots showing progress).

4. Helping to relate areas visually (when transferring attention from the whole subject to a localized part).

Mixes are widely used as 'softened-off' cuts, to provide an unobtrusive transition for slow-tempo occasions where the violence of a cut would be disruptive. Unfortunately, they are used also to hide an absence of motivation when changing to a new shot!

A very slow mix produces sustained intermingled images that can be tediously confusing. Between different viewpoints of the same subject, they result in twin images.

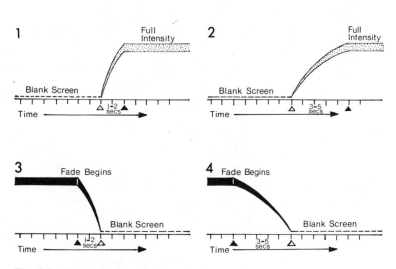

Fig. 6.9   The fade-in and fade-out
1, Fast fade-in. 2, Slow fade-in. 3, Fast fade-out. 4, Slow fade-out. The fade-out and fade-in may be combined in permutation (slow-out, fast-in, etc.) to link two shots producing a fade out-in.

■ *Matched dissolves* By mixing between carefully matched shots with similarly framed subjects, you obtain a transformation effect. Its main applications are to suggest time changes (boy to man, a seasonal change), or *flashbacks* (reversed time), or magical changes (pumpkin to Cinderella's coach).

■ *Mixes on movement*   Mixes during movement are usually only completely satisfactory when their relative directions are similar. Mixing opposite directions can arouse feelings of confusion, impact or expansion, without necessarily indicating that these subjects are involved. But it is a transition to be used cautiously.

■ *Defocus dissolve*   The first camera gradually defocuses, while you mix to a second completely defocused shot which then sharpens. Normally used for *flashbacks*, but sometimes introduced for transformation effects, or decorative interscene transitions.

■ *Ripple dissolve*   Here the first shot becomes increasingly broken up by horizontal rippling (weave), during which you mix to a second rippling shot . . . which steadies to a normal image. Generally introduced for flashback effects.

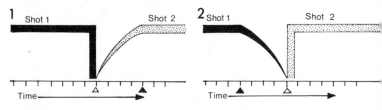

Fig. 6.10   Combinations of fade and cut
1, The cut/fade-in results in a surge, a visual crescendo, strongly introducing the new shot. 2, The fade-out/cut-in produces a visual punch; used between a succession of static shots (e.g. of captions, paintings), provides a dynamic introduction to each.

The wipe

The wipe is a visual effect, mainly used for decorative transitions. It can be produced by several methods, but most production switchers include electronic wipe facilities. It had a particular vogue in early movies, and is now mostly found in trailers and commercials. There the visual impact of the wipes tends to disguise the raw edges of compositionally unrelated fragments.

The effect of the wipe is to uncover, reveal, conceal, fragment; according to how it is applied. In all forms it draws attention to the flat nature of the screen, destroying any three-dimensional illusion.

The wipe direction can aid or oppose the subject's movement, modifying its vigor accordingly. The edge may be sharp (hard-edged/hard wipe) or diffused (soft-edged/soft wipe), the latter being much less distracting. Broad soft-edged versions can have the unobtrusiveness of a mix.

Wipes take many geometric forms with a variety of applications. A circular (*iris*) wipe, for example, may be used as a transition between close-up detail (a soloist musician) and a wider viewpoint (the full orchestra).

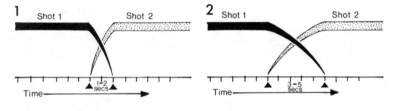

Fig. 6.11   The mix
The mix usually compares successive shots. 1, Fast mix. 2, Slow mix.

■ *The split screen*   If a *wipe* is stopped during its travel, the screen remains divided; showing part of both shots. In this way you can produce a localized *inset*, showing a small part of a second shot, or where the proportions are more comparable, a *split screen*.

The split screen can show us simultaneously:

1. Concurrent events.
2. The interaction of events in separate locations (phone conversations).
3. A comparison of appearance, behavior, etc., of two or·more subjects.
4. A comparison of before and after (developments, growth, etc.).
5. A comparison of different versions (comparing a relief map with its aerial photograph).

Superimpositions (half-lap dissolves, supers)

In film, superimposition necessitates exposing each frame twice or more (double exposure) in the camera, or more usually in the laboratory's optical printer. In television, you simply fade up two or more picture channels simultaneously.

Because the tones of superimposed pictures are additive, the light areas of any picture generally break through the darker tones of other shots with which it is superimposed.

Whether superimposed subjects appear solid or transparent will, therefore, depend on relative picture tones. Where a lighter-toned subject is superimposed on a *black background*, it will appear

Fig. 6.12 part 1   Uncover wipe
Second image is progressively uncovered.

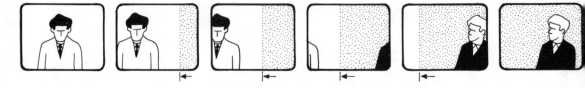

Fig. 6.12 part 2   Pushover wipe
Second image pushes over, displacing original one.

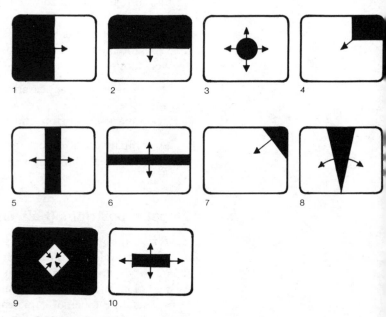

Fig. 6.12 part 3   Common wipe patterns
1, Horizontal wipe/line-split. (When static, this provides a split-screen.) 2, Downwards vertical wipe. 3, Iris/circle wipe (out or in, expanding or contracting). 4, Corner wipe—top right. (When static, this provides a *corner inset*) 5, Barndoors/line bar/horizontal split wipe (expanding or converging). 6, Vertical split wipe/field bar (wipes out or in). 7, Diagonal (top right). 8, Fan wipe (in or out). 9, Diamond wipe-in. 10, Box wipe-out.

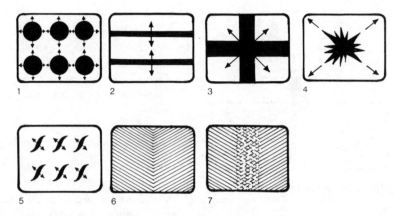

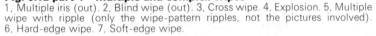

Fig. 6.12 part 4   Multiple and complex wipes
1, Multiple iris (out). 2, Blind wipe (out). 3, Cross wipe. 4, Explosion. 5, Multiple wipe with ripple (only the wipe-pattern ripples, not the pictures involved). 6, Hard-edge wipe. 7, Soft-edge wipe.

solid. On lighter-toned backgrounds, it will be proportionately diluted. Superimposed over multitones, the result appears 'transparent'.

But you must always guard against messy confused images, particularly where multicolored or multitone shots are combined (color mixtures and lightness variations result). You should also avoid any movements (or zooms) of superimposed cameras, where supers are being used to provide subject inserts, unless you want growth, shrinkage or side-slip effects.

Superimposition can be used productionally, to convey several ideas:

1. *Spatial montage*: suggesting that two or more events are occurring concurrently in different places.
2. *Comparison*: showing the similarity or difference between the subjects juxtaposed. (Bringing together action, events, locale, having parallel significance.)
3. *Development*: displaying stages in a process. (A half-built house with architectural sketches superimposed.)
4. *Relationship*: showing subjects' interrelationships, how components are positioned. (A mechanism superimposed on a faint image of the entire machine.)
5. *Thoughts*: a close shot of a person, with images of his thoughts superimposed.

Superimposition has a number of practical mechanical applications:

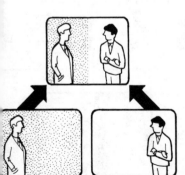

Fig. 6.13   Split screen
The screen is split into two or more segments, showing a series of pictures or respective portions of these components.

1. To obtain transparent-looking images (ghosts).
2. To achieve larger or smaller proportions (giants, dwarfs), by 'solid' superimposition.

Fig. 6.14 Superimposition
Two or more pictures faded up at the same time; their relative strengths being controlled by video faders.

3. To permit appearances and disappearances.
4. To combine titling with another picture.
5. To insert subjects (solid or transparent) into a separate background shot.
6. To add 'surface texture' (e.g. canvas), or a decorative border.
7. To emphasize or identify detail (map route, selected area, pointer).

■ *Montage sequences*   These can take two forms:

1. *A rapid succession of shots* that together imply some total, often abstract, concept (progress, age).
2. A *multiimage presentation* (superimposed, quad-split, multi-split) showing a series of related situations; for example combining different shots of the same scene, similar subjects, or similar action.

Order of shots

It is a characteristic of the human mind that we continually seek meaningful relationships in what we see and hear—even where none exists. In any rapid succession of shots, we subconsciously try to establish a connection between them (slow mixes between them reduce this effect).

We can interpret a succession of shots in several ways:

1. Shot A may cause (or lead to) Shot B's situation.
2. Shot A may be explained by Shot B.
3. Shots A and B together may give rise to Shot C.
4. Shots A and B together may be explained by Shot C.
5. Shot A juxtaposed with Shot B may imply an idea 'X' that is not implicit in either A or B alone.

Fig. 6.15 Montage
1. A rapid succession of related images, or 2. juxtaposed images or 3. superimposed images developing a common theme.

In various circumstances you may find that the order in which you present a series of shots will influence your audience's interpretation of them.

Even a simple example shows the nuances that easily arise. *A burning building—a violent explosion—men running towards an automobile.*

Altering the order of these shots, can modify what seems to be happening:

1. Fire—automobile—explosion
   Men killed while trying to escape from fire.
2. Fire—explosion—automobile
   Running from fire, men escaped despite explosion.
3. Automobile—explosion—fire
   Running men caused explosion, firing the building.

Not only is the imagination stimulated more effectively by *implication* rather than direct statements, but indirect techniques overcome many practical difficulties.

Supposing you join two shots: *A boy looking upward . . . tree falling toward camera.* One's impression is that a boy is watching a tree being felled. Reverse the shots and the viewer could assume that the tree is falling towards the boy, who sensing danger looks up. The actual pictures might be totally unrelated; a couple of library shots.

■ *Cause–effect relationships* Sometimes pictures convey practically the same idea, whichever way they are combined: *woman screaming* and *lion leaping*. But there is usually some distinction, especially where any cause–effect relationship is suggestible.

Cause–effect or effect–cause relationships are a common link between successive shots. Someone turns his head—the director cuts to show the reason. The viewer has become accustomed to this concept. Occasionally, you may deliberately show an unexpected outcome:

1. A bore and his victim are walking along a street.
2. Close-up of bore . . . who eventually turns to his companion.
3. *Cut to* shot of companion far behind, window gazing.

The result here is an amusingly sardonic comment.

But sometimes the viewer expects an outcome that does not develop and he feels frustrated or mystified, having jumped to wrong conclusions:

1. A lecturer in long-shot beside a wall-map.
2. *Cut to* a close-up of map.
3. Lecturer now in an entirely different setting.

The director used the map-shot to relocate the speaker for the next sequence, but the viewer expected to find the lecturer beside the map, and became disorientated.

Even more disturbing, are situations where there is no visua
continuity, although action has implied one:

1. Hearing a knock at the door, the girl turns . . .
2. *Cut to* a shot of a train speeding through the night.

The director thought that he would create tension by with
holding who was outside the door, but he inadvertently created
false relationship instead. Even where dialogue or action explain
the second shot, this is an unsuitable transition. A mix or fade out
in would have prevented misunderstandings.

Quite often, in an attempt to shorten scene duration and t
create filmic time, directors cut from one to another in which th
same person appears in different costume at a different time an
place. Whether the technique is acceptable or confusing, i
challengable.

Duration of shots

If a shot is too brief, the viewer will have insufficient time t
appreciate its intended information. Held too long and attentio
wanders. Thoughts dwell on the sound, eventually giving way t
rumination—or channel switching! The limit for most subjects i
probably around 15–30 seconds; and for a static shot, much less
For a mute shot, it could be as little as 5–10 seconds. The belie
that a close shot can sustain interest longer, is arguable.

The 'correct' duration for a shot depends on its purpose. W
may show a hand holding a coin for half a minute as its feature

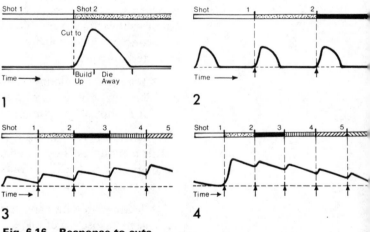

Fig. 6.16 Response to cuts
1, The viewer's response to a cut is not instantaneous. When the cut is made
the audience-impact is slightly delayed; building to a peak and dying awa
gradually. 2, A series of cuts may achieve individual impacts; there bein
sufficient time for each reaction to subside. 3, Faster cutting may produce
cumulative build-up. 4, Through repetition, surprise may diminish with each cu
resulting in declining tension.

are described by a lecturer; whereas in a drama, a one-second shot would tell us that the thief has successfully stolen it from the owner's pocket. Given the right occasion, even a single subliminal frame ($1/30$ or $1/25$ second) can be seen and responded to!

Many factors influence how long you can usefully hold a shot:

1. The amount of information you want the viewer to assimilate (general impression, minute detail).
2. How obvious and easily discerned the information is.
3. Subject familiarity (its appearance, viewpoint, associations, etc.).
4. How much action, change or movement the shot contains.
5. Picture quality (marked contrast, detail, strong composition, all hold most interest).

Audience attention is normally keyed to production pace. A shot flash of information during a slow-tempo sequence may pass unnoticed, while in a fast-moving sequence it would have been fully comprehended.

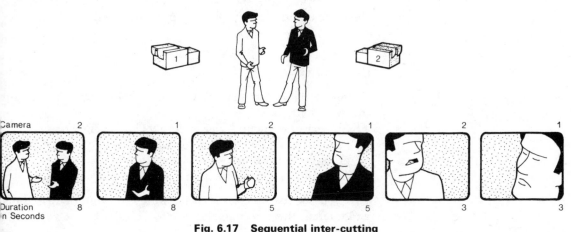

Fig. 6.17 Sequential inter-cutting
Each subject accuses the other of treachery, tension is rising. The camera dollies closer and closer, shots become larger and larger, and of increasingly shorter duration.

Cutting rate

The frequency with which you can cut within a series of shots depends initially on successful shot matching, continuity, whether the transitions are motivated, and their purpose. Cuts may be used relatively unobtrusively; or used as deliberate interruptions to shock, surprise, emphasize, accentuate. With the latter, you have to judge carefully the borderline between pleasurable shock and distracting annoyance.

Where the picture contains fast action, the cutting rate can be high; whereas in a slow-moving scene, fast cutting would usually be intrusive—unless, perhaps, the accompanying sound has a rapid tempo.

Although some production switchers provide an ultrarapid interswitching facility, such an effect soon becomes visually disturbing, so that its use is quite limited. Even using more orthodox techniques, unless we are intercutting similar pictures, fast switching between video sources soon exhausts shots available.

A fast average cutting rate during continuous action imposes considerable strain on the entire TV production crew, as the switcher cuts to cue and cameras rapidly compose their next shots. Unless the production treatment is straightforward, camera moves limited, and action reliable; operation standards are liable to deteriorate. Typical live telecasts have contained some 200 shot changes in a half-hour show; but statistics as such reveal little without details of the opposite complexities involved.

Videotape recording, like filming, permits any cutting rate we may choose. But clearly the less editing carried out on the switcher and the briefer the shots to be inserted into the program tape, the more exacting and time consuming will be the VT editing.

Cutting rhythm

Within a picture sequence, we can become aware of certain intrinsic rhythms:
1. Subject movement— rhythm of moving machinery.
2. Composition changes—the dynamic effects of passing backgrounds (a line of trees).
3. Superimposed movement—regularly flashing street sign.

If you take two pictures (each with its own internal rhythms) and intercut them, a comparative effect results.

A *cutting rhythm* can be set up according to the durations of successive shots. Shot durations might, for example, be progressively shortened as tension rises—an increasing cutting rate. Long and short durations might be contrasted to compare leisurely and urgent paces. Similarly, you can use cutting rhythm to emphasize audio rhythm, or vice versa.

The production switcher

At first sight, the *production switcher* (*video switching panel, vision mixer*) can be extremely daunting. This is not helped by the

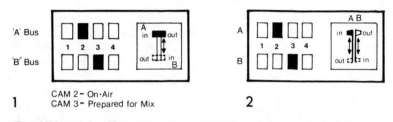

CAM 2 – On-Air
CAM 3 – Prepared for Mix

Fig. 6.18 part 1   The production switcher—principles of the A-B switches

There are two identical rows of push buttons called *buses* or *banks*, labelled A and B (or BUS-1, BUS-2) respectively. Each pair of A and B buttons selects a particular video source (typically 6 to 20 sources per bus—cameras, film, VTR, slide scanner, etc.). The *Bus fader-lever(s)* position determines which bus is in use. 1, A single fader may be fitted (mixing or superimposing between the A and B buses). 2, If twin fader-levers are included, they may be operated together, or split to fade A and B buses separately. With the fader moved to the top, A-bus is active. Switching buttons on A-bus, cuts the chosen source (e.g. Camera 1) to the studio's line-output (main channel, 'on-the-air'). (B-bus switching produces no visible result, unless connected to a pre-view monitor. It becomes a preparatory bus.)

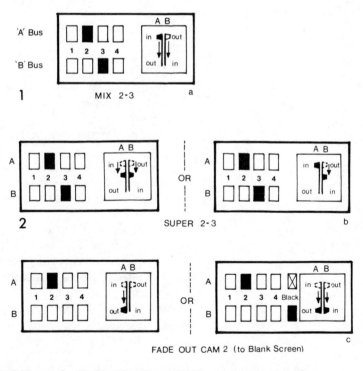

Fig. 6.18 part 2   The production switches—basic operations

1, Mix (dissolve) between Cam 2 and Cam 3: Select 2 on A-bus. It is now on iine (on-air) as bus fader is at A. Prepare for mix, by pressing 3 on B-bus. Push down bus fader(s) from A to B position: output mixes from 2 to 3. Move fader to A-bus: output mixes back from 3 to 2. 2, Super cameras 2 and 3: Either as mix; stopping fader when superimposition strength as required. Or split faders, moving appropriate lever towards but to be added. 3, Fade out Cam 2 (to black screen): Either split faders, pushing A-bus lever to out (B-bus is already faded out). Or where the system would lose chroma using this method, select special *black-level* button in the other bus, and mix to black.

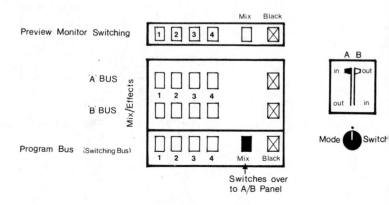

Fig. 6.18 part 3   The 'A-B-C' switcher
A separate program bus is used only for *cutting* (switching to line). The A/B
buses are used for setting up *mixes and effects* (supers, split-screen, wipes,
chroma-key, inserts). The A/B fader lever can be used in either of two ways
(a) *Mix mode*, to provide mixes/dissolves/superimpositions between buses
(b) *Mask key mode*, in which it fades in a keying signal, or makes a pattern
transition, or operates a wipe-pattern *movement*. A *take* button or bar may
interswitch between A and B buses. Preview buttons switch chosen channels
onto preview monitor for checking before use. (Do not affect output to line.)

variety of layouts and designs available. If you divide the panel
into its separate parts though, the operation becomes much
clearer.

The production switcher includes:

1. Arrangements for enabling you to *switch* or *mix* between
   selected video sources (inputs), presenting your choise to *line*
   for videotaping or transmission.

2. A bank of buttons for simply switching sources to *preview
   monitors*. This does not affect the output to line. It is just a
   means of checking combined picture arrangements, special
   effects, source availability, etc.

3. A special *video effects system* (wipes, pattern, key and matte
   electronic insertion); enabling you to manipulate and combine
   shots in various ways.

4. Perhaps *remote control* of slide scanner, film, VTR channels
   (remote start/stop, change).

5. Facilities for *title insertion, synthesized color*, etc.

6. *Indicators* showing which channels are selected, and illuminating camera and picture monitor tally lights (cue lights).

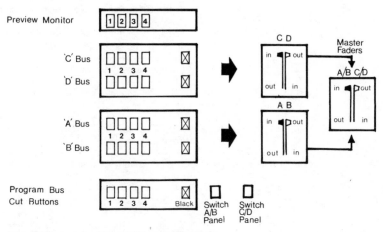

Fig. 6.18 part 4   The 'A-B-C-D' switcher
An elaboration, providing two duplicate pairs of buses—A/B and C/D. Now a
pre-arranged combination of cameras can be set up on one pair of buses (e.g.
for titling or effects), while the other pair is used for cuts and/or mixes. Master
fader levers select the entire A/B and/or C/D combinations as required. A pair of
buses (A/B or C/D) is called a MIX/EFFECTS (M/E) BANK.

Basic production switcher functions

The main production switcher enables you to:

1. Cut (switch); fade up/down; mix (dissolve); superimpose any
    sources.

2. Add/subtract from *superimposed (mixed)* sources by switching/
    fading.

3. Cutting (switching) between *combinations* of sources.

Switchers (vision mixers) are available with various degrees of
complexity and flexibility. Fundamentally, *cut buttons* provide
interswitching between sources; while *fader levers* enable you
to fade/mix between them—selectively or combining outputs.

The almost universal form of production switcher uses an
arrangement known as an *AB*, *ABC*, or *ABCD* format with dupli-
cated switching buses and interbus faders. Another design
approach uses a 'fader-per-channel' format.

In the most advanced equipment, microprocessors enable the
operator to set up transitions or complex multi-source effects at
any time, and memorize (file) them for the moment they are
required. These intersource changes can be selected and introduced simply by pushing a *take bar*.

Table 6.1 Production switcher terms *(See also Table 19.3)*

Audio-follow-video/flip-flop auto take	Facility providing mike-channel switching to correspond with video switching. (Intercutting between both ends of a phone conversation.)
Auto-switching	Provides automatic interchannel switching at an adjustable rate.
Black input/black buttons	Selection prevents loss of color (chroma) during fade to black. (Due to loss of 'color-burst' when fading out a composite video signal.)
Bus/bank	A series of selector cut buttons (one for each source) associated with a particular function.
Composite/coded signal	The complete video signal in NTSC, PAL or SECAM form (including sync pulses).
Cut buttons	Pushbuttons used to intercut sources (*Program bus*).
Decoded signal/RGB	Separate circuits carrying the red, green, and blue video signals of the total TV picture.
Effects bus/mix-effects bus	A bus used for intersource mixes (dissolves), or for combined-source effects.
Channel fader	Individual video fader for each picture source.
Group fader	Fader controlling a group of sources (bank, bus).
Inputs	Switches routing selected video sources to switcher channels.
Non-composite signal	Composite video signal with sync pulses removed (station syncs being re-added after processing).
Preview bus	Switch buttons enabling picture sources to be checked (especially effects: supers split-screen, insertions).
Program bus	Cut buttons interswitching program sources. The result is seen on the line output (studio output, transmission).
Sync/non-sync mode	Adjusts switcher to accept video-sources that are *in-sync* or *out-of-sync* with local synchronizing pulses.
Take-bar	A facility in some switchers that enables you to record a program's transitions and composite effects in a memory/file/store, and play them out to line as required by pressing a *take-bar*. (Similarly AUTO-TAKE, AUTO-DISSOLVE, AUTO-WIPE).

7 Lighting

Looking around the solid three-dimensional world, stereoscopic vision enables you to interpret space with considerable accuracy. Moreover, you are free to move your viewpoint and to choose where you want to concentrate your attention.

The TV screen's flat image on the other hand, only conveys a limited two-dimensional impression of the scene. The picture's viewpoint is selected by the director, and changes frequently. All this can make it difficult to interpret correctly; to distinguish between different planes, to assess size, shape and distance.

If a subject is hidden, or is in shadow or unlit, it is invisible to the viewer. Unless given clues, he may not even realize it is there! Similarly, he is normally unaware of the presence or absence of anything that is out of shot. One surface can easily appear to merge with another far behind it. Even its shape may not be obvious. Sometimes you may deliberately lay false clues or create such ambiguities for dramatic effect, or as a staging technique.

Influence of lighting

Lighting directly influences the viewer's interpretation and re-actions to the flat picture. It can modify his ideas about size, shape and distance; quite apart from the pictorial and environmental impact.

The lighting may enhance form, or suppress it. Lighting may draw attention to texture, or hide its existence. An environment lit in an unfamiliar way may appear excitingly mysterious, or boringly dull and characterless. So lighting not only enables the camera to see, but is a major contributory factor to your audience's responses.

Why is lighting necessary?

■ *The technical reasons* The TV camera can only handle relatively *limited tonal contrasts*, and requires carefully *controlled exposure* to produce the highest picture quality. So you need to adjust the contrast and intensity of the illumination for optimum results, and to minimize various video defects. If you want to select a lens f-stop to give a particular depth of field (shallow focus, deep focus), the light intensity or *level* must be adjusted to provide the correct exposure.

On location, you may need to augment or replace *natural lighting* that does not meet your requirements. For example, the intensity or direction of the available light may be unsuitable, causing high contrast or unwanted shadows. It may be too diffuse, so that pictures lack sparkle. The light's color quality may be unsatisfactory. Quite often, though, the location lighting proves acceptable providing you position the subject and camera to suit prevailing conditions.

■ *The artistic reasons* Lighting helps us to create a three-dimensional illusion—enhancing the impressions of distance, space, solidity and structural form. It can build up a particular atmospheric effect, mood or style. Lighting can simulate or imitate a certain environment; even suggesting the time of day, or weather conditions. You can use light selectively, to emphasize some aspects of the scene, while subduing others; avoiding or reducing distracting features. Light can enhance pictorial beauty, or deliberately create an unattractive sordid atmosphere.

Table 7.1   Light dispersion

	Soft light: This is diffused, shadowless illumination and is obtained from large area soft-light sources, broadly reflected light, or overcast sky.
Advantages	Provides delicate half-tone values, subtle gradations, and gradual ('slow') shading. Avoids emphasis of modeling and texture. Can illuminate shadows without casting additional shadows.
Disadvantages	Light spreads overall and is not readily restricted.
	Can create flat unmoded illumination, suppressing textural beauty and form. Light intensity falls off rapidly with distance and approximately follows an inverse square law', i.e. the level at 2 meters is 0.25 ($\frac{1}{4}$) that at 1 meter; at 4 meters $= 0.06$ ($\frac{1}{16}$); at 8 meters $= 0.016$ ($\frac{1}{64}$).
	Hard light: This is highly directional illumination casting pronounced shadows; it is obtained from spotlights and direct sunlight.
Advantages	Clear-cut vigorous effect, bold and well defined.
	Readily localized to specific areas. Shading (brightness fall-off) on surfaces is abrupt ('fast'). Light intensity is maintained over some distance (because the light rays are parallel).
Disadvantages	Can overemphasize texture and surface modeling.
	Can produce harsh, crude, high-contrast illumination.
	Can create unattractive or spurious shadows.
	Multiple shadows arise when subject is lit by more than one hard source.

Light quality—dispersion

For effective treatment you need a suitable blend of *hard* directional light and *soft* diffused light. Usually the hard light reveals the subject's contours and texture, while the soft reduces undue contrast or harshness and makes shadow detail visible. You may deliberately emphasize contour and texture; or equally deliberately minimize them (suppressing wrinkles and bumps—whether in backcloths or faces).

134

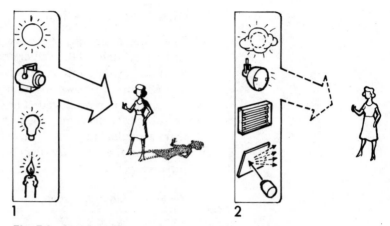

Fig. 7.1   Light quality

1, *Hard light* casts shadows, emphasizes surface modeling (texture, contours). It is emitted by many natural and artificial sources. The smaller the source, the harder its light quality. (Brightness does not affect light quality.) Source-distance increases effective hardness (even for 'soft light' units). 2, *Soft light* is diffused and produces shadowless lighting from scattered illumination. Only large-area sources produce really diffuse light.

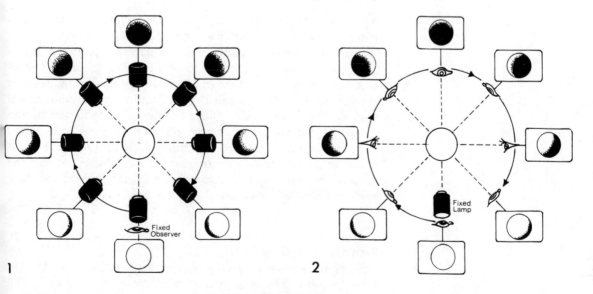

Fig. 7.2   Light direction

1, From a fixed viewpoint, the effect of the lighting changes as *the source* moves round the subject (shading develops on the side opposite the illumination). 2, Using a fixed lamp, appearance alters as *our viewpoint* changes.

Direction of the light

The effects of lighting changes with the angle at which it strikes the subject, relative to the camera's viewpoint. The light's direction will not only determine which parts of the subject are illuminated and which are thrown in shadow, but also how surface contour and texture is reproduced. The lighting direction will emphasize or subdue surface markings and undulations that characterize that particular subject. If you change the light direction or camera viewpoint, these features may become more or less apparent in the picture.

Table 7.2   Light direction

Direct-frontal (from camera position)	*Effect:* Direct frontal lighting along the *lens axis*, reduces texture and modeling to minimum. Valuable for disguising wrinkles or avoiding confusing shadows.
Edge lighting (side lighting)	.Light skimming along a surface (from above, side or beneath) emphasizes texture and surface contours. Ideal for revealing shallow relief such as in coins, embossing, wood grain, fabric and stonework. Produces poor portraiture on full-face shots (ugly facial contouring).
Back lighting	Light directly behind the subject and along the camera lens axis, is largely ineffectual unless the subject is translucent or has edge-detail (fur, hair, feathers).
	Slightly offset back light illuminates the subject's outline, modeling its borders and helping to distinguish it tonally from its background.

Light intensity

The amount of light needed to suitably illuminate a subject, is partly a technical and partly an artistic decision.

Camera tubes are usually quoted as having a particular sensitivity—requiring a certain light level (intensity) for a given f-stop. But this is only a general guide. Much depends on the lightness of surroundings (a dark paneled room requires more illumination than another with white walls); on the color, tone, finish, and complexity of surfaces (whether plain or strongly contoured); and the type and direction of illumination. So ultimately, the light intensity from various directions must be adjusted to suit the subject and the required atmospheric effect (high or low key, environmental effects, etc.). This is the *lighting balance*.

Pictorial quality

Our impressions of pictorial quality are strongly influenced by the tones present. The highest quality normally comes from a full tonal range—from clean whites to deep blacks. Burned-out whites and muddy, detailless gray shadows are unattractive in most pictures.

A picture without darker tones can look 'thin', lacking body (an ethereal effect, perhaps). A picture without lighter tones can look dull and lack 'snap', vigor or sparkle (but may be suitable to simulate a misty locale). A picture with only a few mid-tones normally looks harsh and crude, but may be excellent for highly dramatic or sordid scenes.

Color quality of light

Although our brain can adapt remarkably to wide variations in illumination and still regard it as *white light*, color systems cannot do this. If they detect a different spectral balance from the one to which they are adjusted, the resultant pictures will show a strong color cast. Hence the orange-yellow results if we use 'daylight-balanced' film under tungsten lighting without filtering; or the overblue results when using 'tugsten-balanced' film for exteriors.

The color of light—its *color temperature*—is measured in *Kelvins* (K). Light of a warm yellowish-red quality (candles, dimmed tungsten lamps) is of a *low* color temperature; while cold bluish light sources (north sky light, carbon arcs) have a *high* color temperature.

Unlike color film stock, which is designed to be used with a certain luminant color (daylight—6000 K; artificial light—3200 to 3400 K), a color TV system can be readjusted easily, by altering RGB proportions to suit any balance as required.

In film or TV, a compensatory color filter can be used over the lens or over the light source, to adapt a system to another color temperature. As prevailing conditions can range from *clear sky* (10 000–20 000 K) to tungsten lighting (3000 K), appropriate compensation is essential for good color fidelity. In TV studios, the color quality is standardized around 2850–3000 K to improve lamp life (by being slightly underrun). Variations of 200 K due to dimming, generally result in unnoticed color changes; but even 50–100 K differences may be detectable in critical applications such as a series of lamps lighting a white cyclorama.

Lighting flat surface

Even a simple flat surface requires appropriate lighting. *Hard light* enables you to localize illumination and thus prevent spill onto nearby areas. But it can emphasize surface irregularities such as bumps, wrinkles and creases. It may also not be even overall (fall-off, shading). By using two opposing 45° sources such variations can normally be eliminated. *Soft light* is less liable to produce such spurious effects, but the wide light-spread can spill around uncontrollably.

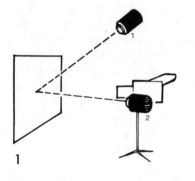
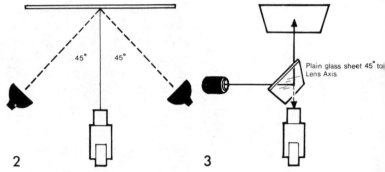

Fig. 7.3 Lighting a flat surface
1, Light from a frontal position near the camera (2), may reflect to produce
hotspots, flares. Offset at 1 will overcome reflections, but may cause uneven
illumination. 2, For critical work, dual lighting from either side (30–45° offset)
is preferable. Soft light sources reduce spurious surface texture or irregularities
(blisters, wrinkles). 3, Shadowless lighting (along the lens axis) is necessary for
multi-layer graphics (Pepper's ghost).

Lighting an object

Most solid subjects require three basic light-source directions—
referred to as *three-point lighting*. The exact angles depend on the
subject and the aspects you want to emphasize.

■ *Key light*   This is a predominant hard light, normally in a
cross-frontal position—for example up to around 45° vertical,
45° horizontal from the lens axis. This lamp creates the principal
shadows; reveals form, surface formation and texture; and largely
determines the exposure. Only one key light should normally be
used for a subject.

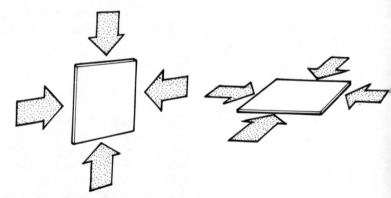

Fig. 7.4 Edge-lighting
Whatever the position of a surface, if light skims along it at an acute angle, any
texture or unevenness will be emphasized.

Table 7.3 Additional lighting functions

Base light or foundation light	Diffuse light flooding the entire setting uniformly—used to prevent underexposure of shadows or excessive contrast.
Rim light or rimming	Illumination of subject edges by backlight.
Modeling light or accent light	Loose term for any hard light revealing texture and form.
Cross light, counter key or balance light	An occasional second frontal key—positioned 30–60° horizontally from lens axis.
Kicker, cross-back light or three-quarter back light	Back light some 30° off lens axis.
Effects light	Light producing specific highlight areas on backgrounds—e.g. around a light fitting.
Bounce light	Diffuse illumination obtained by reflection from a strongly-lit surface such as a ceiling or a *reflector board*.
Set light or background light	Light illuminating the background alone.
Eye light or catch light	Eye reflection (preferably one only) of a light source, giving lively expression. Sometimes from a low-powered camera light.
Camera light, basher, headlamp, camera fill light or spot bar	Small light source mounted on camera to reduce contrast for closer shots, improve/correct modeling, and for localized illumination (e.g. title card).
Hair light	Lamp localized to reveal hair detail.
Clothes light	Hard light revealing texture and form in clothing.
Top light	Vertical overhead lighting (edge lighting from above). Undesirable for portraiture.
Under lighting	Lamp below lens axis used to illuminate, relieve or disguise downward shadows. Also used to create uncanny effects.
Side light	Light located at right-angles to the lens axis. Reveals subject's contours. Creates ugly portraiture for full-face shots, but effective for profiles.
Contrast control light	Soft fill light from camera position illuminating shadows and reducing lighting contrast.
Edge light	Light skimming along a surface, revealing its texture and contours.

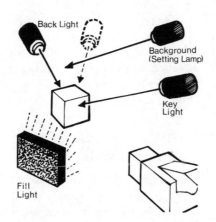

Fig. 7.5   Lighting an object
Effective lighting treatment usually involves four basic lighting functions: *Key light*—Usually one lamp (spotlight) in a cross-frontal position. *Fill light*—Usually a soft light source illuminating shadows and reducing lighting contrast. But may not be required, or alternatively a reflected hard light is used (rarely a direct spotlight). *Back light*—Usually a spotlight (or two) behind the subject, pointing towards the camera. (Occasionally soft light is used.) *Background (setting) light*—Backgrounds are preferably lit by specific lighting, but they may be illuminated by the key or fill-light spill.

■ *Fill light (fill-in, filler)* This soft shadowless illumination reduces tonal contrast between highlights and shadows, revealing shadow detail. It should not modify the exposure nor create additional shadows or modeling. The fill light is usually located 0–30° from the lens axis and on the opposite side from the key light.

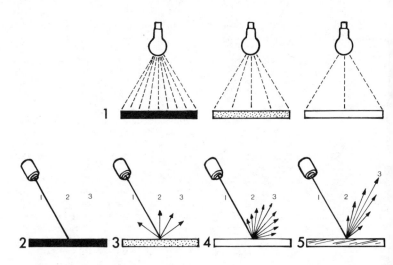

Fig. 7.6   Surface brightness
1, Apparent surface brightness depends on both surface tone and incident light intensity. Each of these examples would appear equally bright. The surface texture and finish, affect how brightness changes with light direction and the viewing angle. 2, Complete absorption (e.g. black velvet). Little or no reflection. Surface appears dark from all viewpoints. 3, Diffuse reflection (rough irregular surface). Light scatters in all directions, fairly bright from all viewpoints. 4, Spread reflection (glossy surface). Fairly dark at 1; fairly bright at 2; bright at 3. 5, Specular reflection (shiny surface). Fairly dark at 1, 2. Very bright at 3.

■ *Back light* This is a hard rimming light behind the subject, separating it from the background. It helps reveal edge contours, transparency and tracery. A single lamp or two spaced back lights may be used.

There is nothing mandatory about these arrangements. The key light is positioned to suit the object's particular features. If a single *side key* is sufficient, neither back light nor fill light may be needed. If the object is transparent or translucent you may only back light it, or even rest it on an illuminated panel for the best effect (e.g. engraved glass).

Lighting people

Although facial details vary between individuals, the basic head formation remains similar enough, so that the unattractive effects produced by poor lighting are much the same for most people.

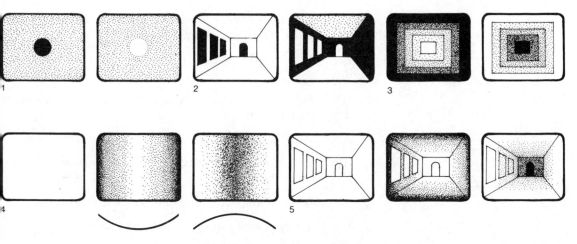

Fig. 7.7   The effects of surface tones
1, A subject's tones affect its prominence (relative to background and other adjacent tones), its apparent size, and its apparent distance. Lighter tones generally appear more prominent, larger, and more distant than darker ones. 2, While our eye stops at the black openings, the white ones suggest space beyond. 3, Progressively lighter planes recede. Progressively darker ones advance. 4, Tonal gradation (shading) affects our impression of form. If a flat plain area is shaded, it appears to have shape. 5, Pictorial perspective conveys an impression of space, which is enhanced by appropriate tonal change—and largely destroyed by unsuitable treatment.

Fig. 7.8   Pitfalls in portrait lighting
Various unattractive effects can arise from unsuitable portrait lighting. 1, *Steep frontal key light*: Black eyes, harsh modeling, long nose shadow. 2, *Two frontal keys*: Twin nose-shadows and shoulder shadows (latter can also arise from two backlights). 3, *Steep backlight*: Hot head-top; bright nose-tip, and eartips; long bib-shadow on chest. 4, *Oblique frontal key*: Talking profile seen on shoulder. Bisected face. 5, *Side key*: In full face bisects the face. Half of it crudely edge-lit, half unlit. 6, *Dual side-keys*: Create central 'badger' shadow effect, and coarse modeling.

■ *General maxims*   There are a number of generally accepted principles when lighting people:

1. Place the key light within about 10–30° of a person's nose direction.

2. Have them look *toward* rather than away from the key light, if reangling their position.

3. Avoid *steep* lighting (above 40–45°) or oblique angles.

4. Avoid a very wide horizontal or vertical angle between the fill light and the key light.

5. Do not have more than one key light for each viewpoint.

6. Use properly placed soft light to fill shadows.

7. Avoid side light on full-face shots.

8. For profiles it is generally best to place the key light on the far side of the face (upstage). This gives improved solidity and avoids casting shadows of the sound boom.

Table 7.4   Lamp positions and portrait lighting

	Back light	Front light
Lamp's vertical angle too steep	Nose becomes lit while face in shadow ('white nose'). Effect most marked when head tilted back. 'Hot top' to forehead and shoulders.	Harsh modeling which gives a haggard and ageing appearance. Gives black eyes, black neck and long vertical nose shadow. *Emphasizes*: forehead size, baldness and deep eyes. Figure appears busty. Casts hair shadows on to forehead. Hat brims and spectacles produce large shadows.
Lamp's vertical angle too shallow	Lens flares or lamp actually in shot.	Picture flat and subject lacks modeling. Produces shadows on background.
Underlighting	Largely ineffectual but can be used for back lighting women's hair. Shadows from ears and shoulders cast on face.	Inverted facial modeling. Shadows of movement beneath head level (e.g. of hands) appear on faces. Shadows may be cast up over background. 'Mysterious' atmosphere when underlighting used alone. Useful to soften harsh modeling from steep lighting. Reduces age lines in face and neck.
Light too far off camera axis	Ear and hair shadows are cast on cheek. One side of nose is 'hot'. Eye on same side as back light appears black—being left in shadow while temple is lit.	In full face, long nose shadow across opposite cheek. An asymmetric face can be further unbalanced if lit on wider side. One ear lighter than the other.

Note: All lamp positions are relative to camera viewpoint.

Table 7.5   Lighting balance

	Too bright	*Too dim*
Frontal light	Back light less effective. Skin tones are high and facial modeling lost. Lightest tones tend to be overexposed. Gives a harsh pictorial effect.	Back light predominates and often becomes excessive. Darker tones are underexposed. Can lead to muddy, lifeless pictorial effect.
Back light	Excessive rim light. Hot shoulders and tops of heads. Exposing for areas lit by excess back light causes frontal light to appear inadequate.	Two-dimensional picture which lacks solidity. Subject and background tend to merge. Picture appears undynamic.
Filler	Modeling from key light reduced and flattened.	Produces excessive contrast and subject too harshly modeled.

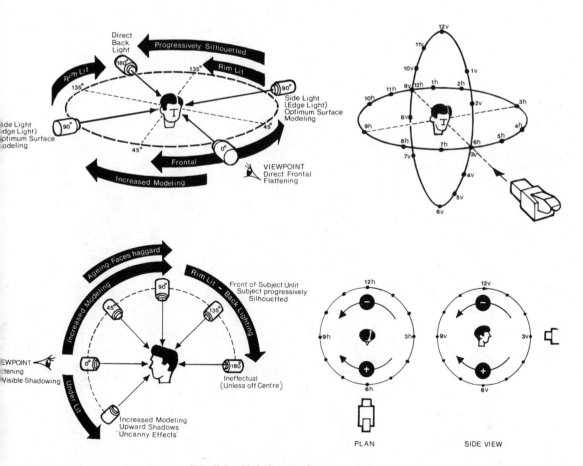

Fig. 7.9   Lighting angles
1, The lighting angle chosen depends on the particular features you want to emphasize or suppress. 2, Using two imaginary clock faces, lamps' positions are easily designated. The camera is always shown as at 3 o'clock V (3V), 6 o'clock H (6H). (Hours are 30° steps, minutes are 6° each.) Intermediate positions between hours are shown by + (clockwise) or − (anticlockwise) signs.

■ *Lighting groups of people* Ideally, you should consider each person in turn. Note the directions in which they will be looking and arrange their key/back/fill lights accordingly. One lamp from the appropriate angle, can often cover two or three people. When people share lamps, problems may arise if they have markedly different tonal values in costume, hair, skin tones. But a compromise intensity setting is usually possible, especially when coupled with suitable diffusers or barndoor adjustments.

Shared lights often have a dual purpose where space or facilities preclude completely individual treatment; for example the key light for one person, may serve as a back light for someone facing him.

Where there are groups of people (audience, orchestra) you can either light them as a whole or in subdivided sections, still using three-point lighting principles and keeping overlaps to a minimum.

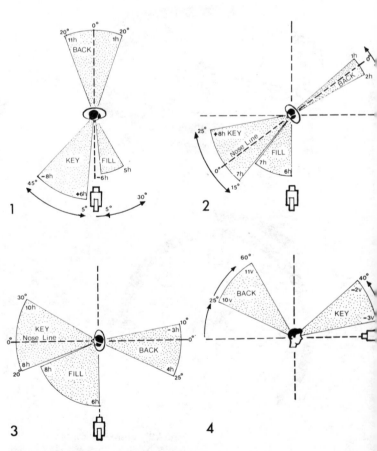

Fig. 7.10   Lighting a single person
The plans and elevations show typical range of lighting angles within which good portraiture is achieved. 1, Full face. 2, Offset head (¾ front, 45° turn). 3, Profile. 4, Vertical angles for each are shown.

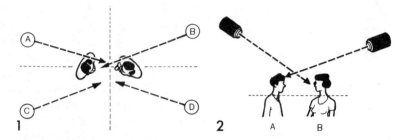

Fig. 7.11   Lighting two people
1, Several arrangements are effective (each has particular advantages) with keys at A and B, C and D, A and D, or C and B. 2, The lamps here may be shared, as Key A/back B, and Key B/back A. Or used specifically as keys, with extra backlights.

Lighting areas

Do not flood areas with light. Particularly avoid the once-common techniques of spreading high-intensity *base* or *foundation light* overall, and superimposing spotlights for modeling and intensity variations. This approach was born with earlier IO TV cameras, and was due to their high gamma, low contrast range and various spurious effects.

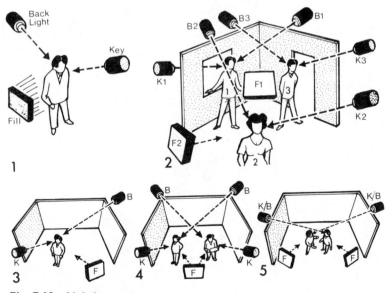

Fig. 7.12   Lighting areas
Where action is localized, the three-point lighting technique (1) can be applied at specific positions throughout the setting (2). One lamp may be used for several positions. Overall three-point treatment can be used for general action (3) or the area can be sectionalized (4). 5, For cross-shooting cameras, lamps over the set's side-walls can be used as key/backlights, with suitably positioned soft fill-light.

Except for high-key pictures, the visual appeal of intense soft light is limited—although it does make key light positioning less critical, as ugly modeling is strongly diluted (*lit out*). Such treatments flatten modeling and produce overlit backgrounds.

Dynamic lighting requires appropriately positioned spotlights, relieved by suitably placed and carefully controlled fill light. (The amount of fill light due to random reflections from floor and scenery, is usually minimal.)

The most attractive picture quality usually comes from analyzing the performance area into a series of *locating points* (by the

Table 7.6   Basic lighting approach

Camera sensitivity	Check typical light levels required at your *f*-stop. For example 1600–2700 lux/ 150–250 fc at 3000 K, aperture *f*4–8.
Subject	Scrutinize the subject and consider the particular features you want to emphasize or suppress.
	How critical will the shots be? (Revealingly close or the background predominating?)
	Is the subject static or moving/moved around?
	Is it being shot from several angles? (Which?)
	Any shadowing problems? Shadows falling onto subject, subject's shadows onto background.
Background	Normally light the subjects (people) before dealing with their background.
	Will subject lighting suitably illuminate the background also?
	Does the background require separate specific lighting?
	Will subject and background lighting mutually interfere (e.g. set lighting spuriously illuminating people)?
Facilities	What facilities are available? (Types and numbers of lamps, sufficient cabling and power?)
	Check methods of lamp suspension available (including safety considerations).
	Work within the available facilities, simplifying treatment if necessary.
Key light	Arrange an offset key light, suitable for camera position, subject direction, and movement.
	Does the key light cover more than one subject? (Check coverage, angle and intensity suitability.)
	Must this key light satisfy more than one subject or camera angle? (Compromise the angle or include a second lamp?)
	Check/adjust key light intensity. (Are diffusers needed?)
	Does light coverage need restricting—unwanted spill onto other areas? (Barndoors, flag.)
	Is key light likely to cause boom or camera shadows, ugly background shadows or hot spots?
	Will key light dazzle performer or prevent their reading prompter? Is key light too steep? (Long nose/neck shadows, dark eyes?) Is key light too offset? (Head profile on shoulder; half-lit face).

Fill light	Position to illuminate shadows, not adding to key light.
	Avoid steep or widely angled filler.
	Fill light must be diffused (place diffuser over sources if necessary). *Exceptionally*, use dim diffused spotlight (despite the extra shadow it casts) to provide localized filler in dark surroundings.
	Check that fill light does not overilluminate other subjects or background.
	Avoid excess fill light (little or none required for very frontal keys). Intensity typically half key light level.
Back light	Is backlight desirable ! Not needed for some subjects, or where edge lit.
	Avoid steep back light because it becomes ugly top light—flattening head, hitting nose tip.
	Avoid very shallow back light because lamps may come into shot and lens flares may arise.
	Check that back light for one camera is not providing ugly side light from another viewpoint.
	Avoid excess back light. (Unnatural hot borders to subjects.) Intensity typically $1-1\frac{1}{2}$ key light level.
Background lighting (set light[	When flat background lighting is appropriate, avoid spurious shadows, patches or hot spots, or emphasizing creases or undulations.
	For interiors, relate lighting to natural environmental effects (windows, light fittings).
	Generally shade off walls to about shoulder height.
	Avoid bright areas near top of shot. Avoid distracting contrasts.
	Avoid light scraping along walls (emphasizing surface blemishes, causing long shadows).

table, at the door, looking out the window) and tailoring the three-point lighting at each to suit the action. One lighting arrangement will often suit other shots or action in that area (perhaps with slight lighting rebalance).

Given sufficient lighting facilities (enough lamps, dimmers, etc.) and adequate time to readjust lamps, you could light to suit each individual shot. But such elaboration is not normally feasible.

Where action is more general or widespread, you must cover the area in a systematic pattern of lamps.

Pictorial treatment

There are three broad styles of pictorial lighting style, through which we can depict the three-dimensional world on a flat screen: *notan, silhouette* and *chiaroscuro*.

■ *Notan* Emphasis is on surface detail, outline and surface tones. Pattern rather than form predominates. The impression is flat (two-dimensional). Notan effects come from high-key, low-contrast treatments, an absence of modeling light, reduced back light and widespread diffused light.

Table 7.7   Approaches to pictorial treatment

Illumination—overall (flat)	*Method* Main consideration is visibility. Lighting is almost entirely frontal and flat. Tends to *notan*. At its best, subjects distinguishable from each other, and from their background. At worst: flat, characterless pictures; ambiguously merging planes; low pictorial appeal.
Ilumination—solid	By a careful balance of frontal and back light, subjects are made to appear solid. This and clarity are the principal considerations of this chiaroscuro approach. The lighting suggests no particular atmosphere.
Realism—direct imitation	*Direct* imitation of effect seen in real life. e.g. sunlight through a window imitated by a lamp shining through it at a similar angle.
Realism—indirect imitation	Imitation of a natural effect, but achieved by a *contrived* method. e.g. simulating sunlight by lighting a backing beyond a window and projecting a window shadow on to an adjacent wall, from a more convenient position.
Realism—simulated realism	An imitation of a natural effect, where there is no direct justification for it within the visible scene. e.g. a window shadow on a far wall, that comes from an unseen (probably non-existent) window. (This may be a projected slide, cast from a cut-out stencil, or a real off-stage window.)
Atmspheric—'natural'	A lighting treatment in which natural effects are not accurately reproduced, but *suggested* by discreet lighting. A pattern of light that highlights and suppresse pictorial detail selectively, to create an appealing effect. Suggesting realism in most instances, but seldom permitting dogmatic justification for each light source and direction. A typical motion picture approach.
Atmospheric—decorative	Associative light patterns. e.g. of leafy branches on a plain background.
Atmospheric—abstract	Light patterns that have no direct imitative associations, but create a visual appeal. e.g. a silhouetted unknown person, lit only by a rectangular slit of light across his eyes. A flickerwheel pattern cast over an exciting dance sequence.

■ *Chiaroscuro*   In this approach, light and shade conjure remarkable illusion of solidity, depth and space. The style is s familiar that we are liable to overlook that it is actually contrived When well handled, this treatment possesses arresting, vita qualities, as the Dutch Master painters found and depicted s convincingly.

Where subjects are set up before a totally black backgroun (cameo), or a white background (limbo), the pictorial effect i more strictly a scenic design treatment than a lighting styl although commonly referred to as such. In *cameo* situations sid light usually predominates (for maximum modeling); whil *limbo* situations are often a notan treatment, avoiding shadows on white background.

Unfortunately, terminology has too often become confused, s that *limbo* becomes used to indicate 'neutral backgrounds' of an tone. Similarly, for some people 'Rembrandt lighting' define selectively localized lighting treatment, while for others it implie strongly modeled portraiture.

■ *Silhouette* Here we concentrate on subject outline, suppressing all surface details. Apart from true silhouettes (black against a light background) used for dramatic, mysterious and decorative effects; *semi-silhouettes* arise in contre-jour (against the sun) shots where back light predominates and frontal light levels are low.

Light sources (luminants)

Several kinds of luminant are used in TV lighting: regular tungsten lamps, tungsten-halogen lamps ('quartz lamps'), gas discharge lamps (metal halide, HMI, CSI, linear light sources), and fluorescent tubes. Carbon arcs are now only used on some remotes (OB's) and location filming—to a decreasing extent.

■ *Regular tungsten lamps* Confusingly also called 'incandescent' (which *all* TV luminants inherently are!) or 'tungsten' (used in quartz lamps too!). Still extensively used, the regular tungsten filament lamp is relatively cheap, has a reasonably long life, exists in a wide range of intensities (power ratings), is generally reliable, and can be mounted in many types of fittings.

Against this, tungsten lamps waste much electrical energy as heat. Their relatively large filaments prevent them being true point sources and providing crisp, clear-cut shadows. Their color temperature is normally relatively low and decreases as they get older. The light output also decreases due to the bulb (globe) envelope blackening.

■ *Overrun lamps* These tungsten lamps are deliberately designed for a supply voltage that slightly exceeds the filament's normal rating. By this means, you achieve a much greater light output and a higher color temperature—but at the expense of a considerably shortened life. *Photoflood* and *Nitraphot* are widely used versions.

Such lamps can be used in *practical fittings* (wall brackets, table lamps) on sets, or location interiors.

Portable lightweight lighting systems make use of overrun lamps in traditional tungsten or quartz form. For example, a *Colortran* unit using only 3 kW of power, can provide the light output of a regular 10 kW tungsten rig. So too, *Sunguns* run from belt-batteries can offer intense hand-held illumination on location.

Fig. 7.13 Internal reflector lamp (sealed beam)
This lamp has a mirror-silvered internal wall and a clear ribbed or frosted front. Available in regular tungsten and quartz (tungsten halogen) forms. Used individually in lamps or grouped.

■ *Tungsten-halogen (quartz lamps, halogen lamps, quartz iodine)* Here the tungsten filament is enclosed within a quartz or silica envelope filled with a halogen gas. This restricts the normal filament evaporation and bulb blackening, so providing a longer lamp life and/or a higher, more constant light output of increased color temperature (e.g. 3200 K), that is ideal for color television.

The bulb must not be handled since body acid attacks the sur face, and the lamp becomes brittle and extremely hot in use Tubular strip and compact designs are available, making lighte and much smaller fittings possible. Twin-filament lamps ar produced which provide three intensities of similar color tempera ture by switching.

■ *Gas discharge lamps* Compact, extremely efficient source with high light output, these lamps employ a mercury arc ignite within argon gas. Special 'rare-earth additives' result in near daylight illumination (54–6000 K). This luminant is increasingl used for effects/pattern projectors, follow spotlights—CSI (com pact source iodide) type, and exterior lighting (sports stadia, etc.— HMI types).

Discharge lamps necessarily require auxiliary circuitry (ignitor and ballast units), and there is usually a $1\frac{1}{2}$–3 minutes build-u time from switch-on (striking) to full light. Dimming methods ar restricted and a lamp cannot usually be restruck quickly afte switch-off. Film shot with this light source may exhibit flicke problems (visible beat, dark-frame).

■ *Fluorescent tubes* The ubiquitous fluorescent tube has bee used in TV lighting as a soft light source for base light, cycloramas etc. Economical but bulky and fragile, this low-intensity non directional source has some use in small installations, but it applications in color TV are limited.

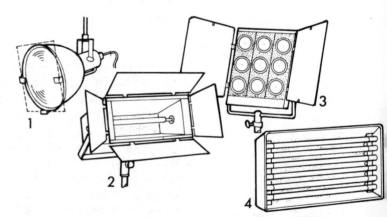

Fig. 7.14 part 1 Soft-light sources—open-lamp types
1, *Scoop* (1–1$\frac{1}{2}$ kW) has a spun aluminium bowl of diameter 0·3–0·46 n (1–1$\frac{1}{2}$ ft). Emergent light is not particularly soft (or controllable) where an oper lamp is used (tugsten or quartz). Spun-glass scrim fits into supports, aidin diffusion. The lamp is obtainable in fixed-focus and adjustable (ring-focus forms; the latter has adjustable light-spread. (Versions with lamp-shields ar available to prevent direct light emergence.) 2, *Small broad* ($\frac{1}{2}$–1 kW) is a tubula quartz light. Flaps or barndoors restrict light. Fairly hard source. (Lamp-shiel versions available.) 3, *Floodlight bank (cluster, nest)* consists of groups o internal reflector lamps (650 watt) that create soft-light through multi-sourc light overlap. Its sections may be switchable and adjustable (e.g. 3×3 to 3×! lamp complement)′ 4, *Fluorescent bank* is a group of fluorescent tubes providin diffuse illumination.

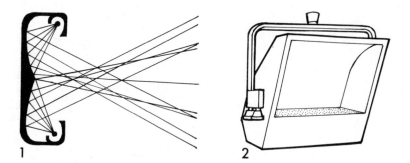

Fig. 7.14 part 2   Soft-light sources—internal reflection types
1, Internal reflection from shielded tubular quartz-lights, produces diffused high-intensity illumination. 2, *Large broad* (1–6 kW typical) has a reflected internal lamp (tubular quartz or fluorescent). A diffuser may also be attached.

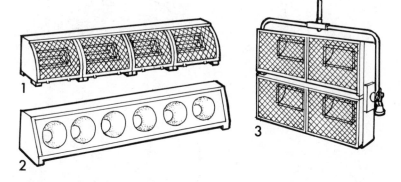

Fig. 7.14 part 3   Soft-light sources—cyclorama lighting
1, *Ground-row* (*trough*) containing tubular quartz units (625 W–1500 W).
2, *Strip-light* (*border light*) consists of a row of internal-reflector lamps. 3, *Suspended cyc light* is a single or multi-unit fitting (for color mixing) with strip quartz-lights.

Types of lamp housing—luminaires

Four basic lighting units are used in TV:

■ *Soft light fittings*   These include *scoops*, *broads*, internal-reflector fittings, multilamp assemblies and strip lights. They are used for fill light (usually slung), and for lighting broad backgrounds (back-drops, cycloramas) from slung, stand and floor positions.

■ *Spotlights/Fresnel spots*   Focused spotlights with Fresnel lens and internal reflector are used as key lights, back lights, background/set lighting and for effects (sunlight, broad decorative patterns, dapples). Types relying on a mirror-reflector alone are less readily adjusted or controlled (peripheral spill-light, uneven beam).

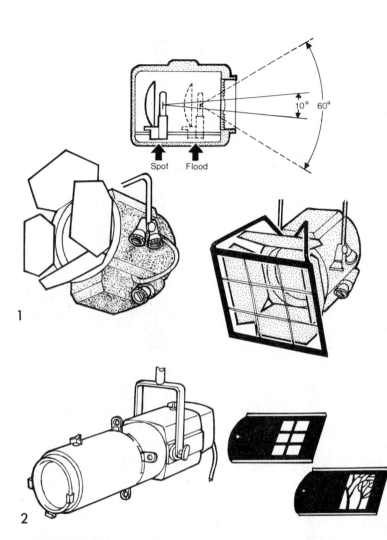

Fig. 7.15   Hard-light sources

1, *Fresnel spotlight* (100 W–10 kW, 8–38 cm/3–15 in lens diameter). Sliding the lamp assembly to/from fresnel lens adjusts light-spread—spot to flood. Adjusted by hand crank (crank or sweep lever), by pole (loop, ring, or T-fitting), or remote electronics. *Controls*: Focus, tilt, pan, filament-switch (where dual-filament lamps used). Barndoor may be permanently attached. Color-frame, french flags, etc., can be attached. 2, *Ellipsoidal spotlight* (profile spot) provides precisely-shaped hard-edged beams which are controlled by shutters, iris diaphragm, or metal plates (profiling blades). Shadow patterns can be projected using metal stencils (masks), wire mesh, etc.

■ *Effects/pattern projectors*   These include effects spots, projection spotlights, profile spots, elipsoidal spotlights and pattern projectors. They are used to project precisely shaped areas of light (soft or sharp-edged cut-off) for localized effects lighting, or to project patterns of stencils or slides (windows, tracery, leafy branches, decorative motifs).

■ *Follow spots*   Used for isolating static subjects or following moving performers (singers, skaters, dancers) in a confined pool of

Lighting

light. These large spotlights (e.g. 5 kW) are carefully balanced for continuous accurate handling and are usually stand or scaffolding mounted. The beam intensity is often shutter-controlled, and its circular spot area (hard- or soft-edged) is adjustable to suit the subject.

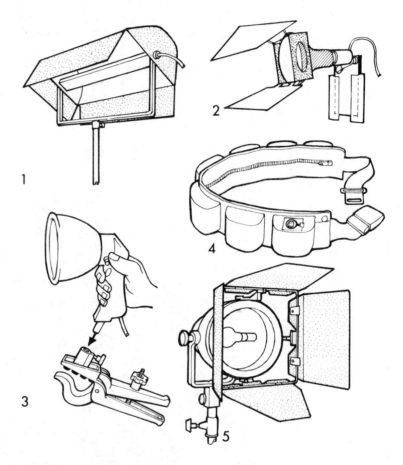

Fig. 7.16 Portable lighting
Typical portable lighting units include: 1, Small broad (600–1000 W). 2, Internal reflector lamp on plate (gaffer-taped to wall, hung, hooked on bar) or clip-light. 3, Sun gun (250 W) powered from 30 volt battery belt (4), hand-held or clamped. 5, External reflector spotlight, in lightweight stand, gaffer-grip, or alligator clip.

Lamp supports

Several methods have become standardized for supporting lamps.

■ *Ground lamps* Lamp fittings may rest on the ground for certain purposes. *Ground rows, cyc lights, strip lights, troughs,* are used to light backgrounds; *spotlights* can provide underlighting effects or be used to reach otherwise inaccessible areas.

153

■ *Floor stands* Floor lights are mounted on telescopic three-castered stands. They have the disadvantages of occupying valuable floor space (perhaps impeding camera or sound-boom movement), casting shadows onto backgrounds, having trailing cables and being vulnerable. But they do permit easily adjusted precision lighting.

In fact, most TV lighting is suspended, to leave floor space uncluttered by lamps or cables.

■ *Pipe grids* Smaller studios frequently use a pipework lattice or ladder beams fixed below the ceiling. This piping (37–50 mm/ $1\frac{1}{2}$–2 in diameter, at $1\cdot2$–$1\cdot8$ m/4–6 ft apart) enables lamps to be clamped or suspended as required. Catwalks may allow overgrid access.

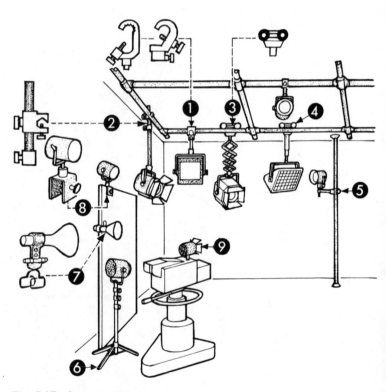

Fig. 7.17 Lamp supports
Studio lighting is mainly suspended (from a pipe-grid in this example). 1, Clamped directly to the grid (with a C-clamp fitting). 2, Lowered from an extendable hanger (sliding rod, drop-arm). 3, A movable trolley holding a vertically adjustable pantograph (extension $0\cdot05$–$3\cdot6$ m/2–12 ft). 4, Telescopic hanger (skyhook, telescope, monopole). 5, In confined space, a counter-balanced spring-loaded support bar (barricuda, polecat) wedged between walls or floor/ceiling. 6, Telescopic floor stand ($0\cdot45$–$2\cdot7$ m/$1\frac{1}{2}$–9 ft). 7, Clip-lamp (spring clamp) attached to scenic flat. 8, Scenic bracket screws to top of flat. 9, Camera light (headlamp, basher). Low-power lamp for eye-light or local illumination. Safety bonds (wire or chain) are fitted to all hung fittings and accessories.

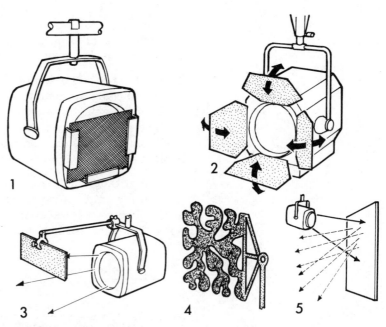

Fig. 7.18 Lamp accessories
Various attachments control the light: 1, *Diffuser* consists of a translucent
material (wire mesh, frosted plastic or spun-glass sheet). Diffuses light (overall or
locally) and reduces intensity. 2, *Barndoor* has independently adjustable flaps
(2 or 4) on rotatable frame; cuts off light beam selectively. 3, *Flag* is a small
metal sheet (gobo) preventing light spill or cutting off light beam. 4, *Cookie*
(*cucaloris, cuke*) is a perforated opaque or transparent sheet that creats dappling,
shadows, light break-up, patterns. 5, *Reflector* is a flat sheet (white or silver
surface) reflecting light to re-direct or diffuse it.

■ *Battens/barrels/bars* These are arranged in a regular parallel
pattern over the staging area and individual battens may be height-
adjusted (counterbalanced by wall weights or motor-winched)
to allow lamps to be rigged, or related to staging requirements.

■ *Ceiling tracks* Crossbars that can be slid along parallel
ceiling tracks offer flexibility in lamp-rigging density. Power rails
in single or dual-track forms may support telescopic hangers or
pantographs for individual or twin-lamp arrangements.

■ *Lamp adjustment* When working from treads (stepladder)
lamp adjustments are precise but time consuming. So many lamps
are now designed with loop or 'T' controls to be operated from
below by a long extendable pole. Lamp adjustments by remote
electronic control are quite practicable (focus, pan, tilt, barndoors)
and invaluable for isolated lamp positions, but add to expense and
maintenance complications.

Light control

To achieve a pleasing lighting balance, you must control the relative brightness of your lamps. You can do this in several ways:

1. By the type and power of lamp you choose.
2. By adjusting the lamp to subject distance.
3. By introducing diffusers.
4. By altering light concentration (flooding/spotting).
5. By selective switching in multilamp fittings.
6. By using a *dimmer* to control light intensity.

Even without dimmers, very effective lighting treatment is possible for simpler productions. But for maximum technical and artistic effect you need comprehensive light-control facilities to adjust exposure, balance, cuing and atmospheric treatment.

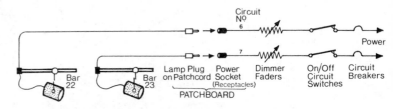

Fig. 7.19 part 1   Patching and control—basic patching
Lamp on Bar 22 has been plugged into nearby batten outlet, connecting it to the patchboard. There a similarly numbered patch cord can be plugged (patched) into any numbered power-circuit outlet (channel), incorporating its own control dimmer/switch (circuit 6 in this case). Similarly, Bar 23 is fed from power Channel 7. They can be dimmed/switched independently. (Avoid exceeding power loading limit e.g. 2 kW. Switch off circuit before patching.) Some installations use patch cords; others have miniature peg-boards, automatically switched or permanent connections.

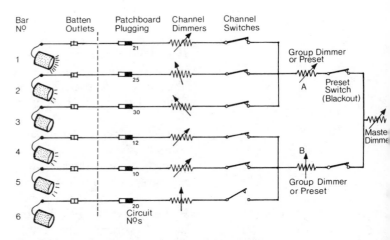

Fig. 7.19 part 2   Patching and control—group presets
A series of individually adjusted circuits can be grouped, to be dimmed/switched communally e.g. fading cyclorama lighting up/down. Typical situation is shown here: *Preset A* (fully faded up). Lamp 1—Switched on, fully faded up (bright). Lamp 2—On, faded down (dim). Lamp 3—On, but faded out (out). *Preset B* Faded down, reducing supply to all group. Lamps 4 and 5—On, individually faded up (but dim due to preset dimmer). Lamp 6—Channel switched off.

Lighting control equipment

Thyristor (SCR—silicon controlled rectifier) dimmers are almost universally used to control tungsten and halide ('quartz') lamps. Thyristors are relatively economical and permit lightweight compact control boards. Efficiency is high and control is smooth; the dimmers producing reasonably even brightness (intensity) changes over their range, even with varying loads.

The thyristor is a semi-conductor device, which controls lamp current according to the timing of a series of electrical 'gating pulses' applied to it. Earlier problems causing 'lamp sing' (filament vibrating at audible rates) or audio induction (mike cables picking up interference) can now be overcome by suitable suppression.

Methods of electrical dimming using resistance, transformer, reactance dimmers and magnetic amplifiers, have been largely superseded.

■ *Dimmer board design* Basically, you want to switch lamps (on/off) singly or in groups, and to adjust lamp intensities individually and/or in groups (to a chosen or variable level). Dimmers (faders) are usually scaled (10 = full, 0 = out) and normally worked around '7', to extend lamp life and permit 'above normal' intensities when needed.

Smaller dimmer boards are usually transportable—often as wheeled installations on the studio floor. These include facilities for combining lamps in *preset* groups for communal control.

Larger sophisticated consoles use digital techniques and can store (file) any pattern of information, including relative intensities and lamp switching. This data can then be recalled manually by push-button, or automatically using punched paper tape or cassette audio tape.

TV lighting problems

Lighting treatment affects and is affected by the work of most other studio activities. Consequently the lighting director needs to cooperate closely with others to achieve maximum pictorial standards and minimum mutual frustration. He has therefore to be aware of a number of potential problems.

Electrical problems

The technical needs of the camera tube can frustrate your artistic aims. If a situation requires us to switch all lights off, or to provide very dim illumination, or to shoot against a very bright background, or have a scene lit by a single (usually unsuitable) source such as a candle, the camera would normally respond by producing inferior picture quality. So instead, the lighting treatment must necessarily be a compromise; a 'cheated treatment' aided by sympathetic video-controlled adjustment (Chapter 21).

(In film making a similar subterfuge is used in conjunction with processing laboratories who are able to adjust processing and the printer light, to introduce exposure and color balance adjustments.)

Sound boom shadows

Shot anticipation and coordination are necessary to prevent boom shadows falling across people and backgrounds. The normal trick is to throw the inevitable shadow out of shot by careful key-light positioning. Obviously, difficulties arise when this lamp position is artistically incompatible.

Table 7.8   Methods of eliminating boom shadows

	Method	Result
By removing the shadow altogether	Switching off the offending light.	This interferes with lighting treatment.
	Shading off that area with a barndoor or gobo.	Normally satisfactory, providing the subject remains lit.
		Shading walls above shoulder height is customary, to provide better prominence to subject.
By throwing the shadow out of shot	By placing the key light at a large horizontal angle relative to the boom arm.	A good working principle in all set lighting.
	By throwing the shadow on to a surface not seen on camera when the microphone is in position.	The normal lighting procedure.
By hiding the shadow	Arranging for it to coincide with a dark, broken up, or unseen angle of background.	Effective where possible, to augment other methods.
	By keeping the shadow still, and hoping that it will be overlooked.	Only suitable when inconspicuous, and when sound source is static.
By diluting the shadow with more light		Liable to overlight the surface, reduce surface modeling, or create multi-shadows. Inadvisable.
By using soft light instead of hard in that area of setting		Occasionally successful, but liable to lead to flat, characterless lighting.
By using floor stand lamps instead of suspended lamps		Light creeps under the boom arm, avoiding shadows. But floor lamps occupy floor space; can impede camera and boom movements; can cast shadows of performers flat-on to walls.
By altering the position of the sound boom relative to the lighting treatment		May interfere with continuous sound pick-up or impede camera moves.
By changing method of sound pick-up	By using slung, concealed, personal mikes, prerecording, etc.	These solutions may result in less flexible sound pick-up.

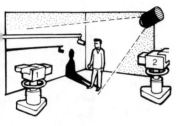 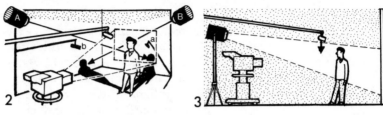

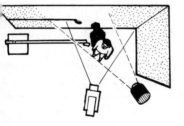 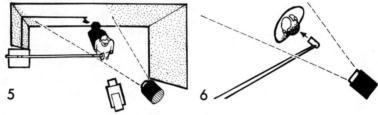

Fig. 7.20 Sound boom shadows
1, From Camera 1, mike shadow is out of shot and lamp provides backlight, but from Camera 2 lamp is now frontal throwing shadow on background behind performer. 2, Mike shadow (a) is caused by frontal key light being in line with boom-arm at A, whilst by placing frontal key at B, at a wide horizontal angle to the boom-arm, shadow (b) is thrown out of shot. 3, Floor lighting avoids mike shadows unless the boom-arm dips into the lamp's beam. 4, Boom shadows can arise when the boom-arm is parallel with a wall; 5, When the boom operator cannot see his boom shadows; and 6, When the performer is playing away from the boom. Where two booms pick up near and far action, the latter can shadow the closer source.

How settings affect lighting treatment

The layout, shape and finish of settings necessarily affect how you can light them.

■ *Size* Usually, the larger the setting, the broader will be the production treatment. Size alone is not significant since a series of small sets often require more lamps than one large set. Also, there are difficulties in providing precision lighting (e.g. for close-ups) over a wide area, without overlaps.

■ *Height of settings* Low-angle shooting usually necessitates higher sets (studio height permitting). This involves higher, steeper, back and side lighting to ensure that lamps do not come into shot.

■ *Overhangs* Appropriate lighting can be quite difficult for people or scenery underneath any overhanging feature such as ceilings, canopies, beams, branches and chandeliers. This is particularly true when using slung lighting.

■ *Shapes of settings* Deep narrow sets (e.g. corridors) and wide shallow sets restrict lighting angles, and can therefore frustrate good overall coverage.

■ *Surface tones* Very dark backgrounds and furniture easily
lose modeling on camera, causing foreground faces to appear
unduly light. Extra lighting may improve reproduced tones, but
can spill onto performers. Overlight surfaces appear glaringly hot
on camera, often making faces look unduly dark. You can try to
keep light off them, but even the illumination from distant soft
light sources may leave them overbright. Darkening by spraying
down with a lower tone (impractical over large areas) can inadvertently cause patchiness, ageing or dirtying.

The illusion of depth requires good subject/background tonal
separation. Contrast ratios of $1\frac{1}{2}:1$ to $2:1$ are typical. Backgrounds
should rarely exceed $1\frac{1}{2}$ times face brightness.

Shiny surfaces facing the camera can be particularly embarrassing, as lamps reflect in window glass, pictures, glossy walls, etc.

Overbright surfaces are a regular problem. Newspapers,
scripts, table tops, light costume, always require care; often
having to be sprayed down (water-soluble color), dipped, or
dulled down (wax dulling spray; anti-flare) to prevent excessive
light bounce. A 'hot' back light can aggravate the situation. Shiny
glossy surfaces inevitably produce bright specular reflections or
glare, and each problem has to be treated as it arises (dull down,
reangle, cover-up, relight).

Other factors affecting lighting

■ *Studio facilities* The elaborateness of lighting is directly
affected by the time and labor available, the number and type of
lamps, as well as suspension methods, light control and the power/
patching arrangements. Where many lamps are prerigged
(*blanket rig*) for immediate selection, time is saved; particularly
with fast turnrounds between productions. Clearly, problems
become more acute as the number of sets, changes, cues and
diversity of camera angles are increased.

Fig. 7.21 Performer's positions
1, Mutual shadowing is often more easily remedied by repositioning people than
by re-lighting. 2, Where people work too close to backgrounds, good backlight
is impracticable. Side lighting or top-lighting (as shown) should be avoided.
3, Where action is in restricted or over-hung positions, lighting can be a compromise—or even impracticable.

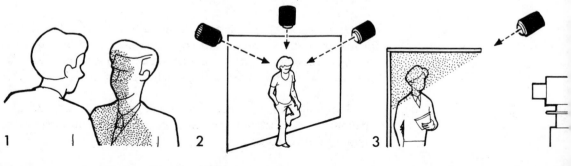

■ *Space limitations* Most staging representing interiors has to be built in reasonably naturalistic proportions. So the working space available *within* the set for cameras, sound booms, and lamps remains pretty restricted. When action or treatment is complicated or continuous, lighting treatment becomes that much more of a compromise. Further lighting difficulties develop when, to conserve space, staging is located near a cyclorama (spurious shadows on cyclorama), or backings are positioned too close to windows for appropriate lighting.

■ *Multi-camera production* Multi-camera shooting does not necessarily frustrate good lighting. Much depends on shot duration, shot variations and duration of takes. Often one set-up can light several viewpoints successfully. Otherwise you will have to introduce lighting changes (unobtrusive or direct) to solve the difficulties.

Lighting for color

Color adds another artistic dimension to the TV picture, enabling you to distinguish between planes more easily (for now there is *hue* in addition to *tone*), and to perceive visual effects that are totally lost in monochrome.

■ *Lighting methods and color reproduction* Many people associate color systems with 'flat lighting' and the extensive use of colored light. Neither is true. Flat overall lighting is as undesirable in color as it is with black-and-white systems. But reasonably *even* lighting is important for controlled results—with neither *hot spots* (paling out and desaturating colors), nor *underlit* (insufficiently illuminated) areas darkening and emphasizing hues. Remember too, that under soft light, colors tend to appear paler; while under hard light, they can look brighter and purer (more saturated).

■ *Exposure and color* Exposure and lighting balance are more critical in color. If you stop the lens down to correct the exposure of a brightly lit subject, its surroundings will reproduce darker as a result. Conversely, if you open up to expose an insufficiently lit subject correctly, other parts of the picture will look much brighter. So exposure compensations for unbalanced lighting can lead to brightness and saturation changes when pictures are inter-cut.

Face tones and color can become considerably modified by changes in exposure or color temperature variations of the light. Caucasian faces look pallid or ruddy when over- or underexposed respectively. Brown skin-tones tend to become yellowish when overlit, and brown-black or terracotta in shadow. So a high

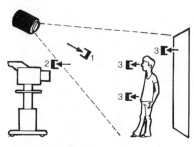

Fig. 7.22   Light meaurement
Light measurements avoid gross over or under exposure. Basic methods: 1, Incident light falling on subject. 2, Average light reflected from scene at camera. 3, Measuring surface brightnesses (tonal values and contrast).

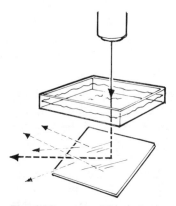

Fig. 7.23 part 1   Water ripple-tray
A spotlight shines through a water-filled glass tray, the rippling light being reflected on to the scene.

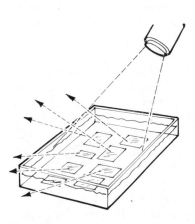

Fig. 7.23 part 2   Water ripple-trough
Light reflects from mirror fragments in water-filled trough.

lighting contrast (key-to-fill ratio) can create color bisection effects. This situation is further exaggerated when lamps are being run at high and low color temperatures. In color too, you become more aware of the blocked-off white highlights from skin shine, or perspiration.

■ *Colored light* In fantasy or highly decorative situations anything goes. Faces may be lit with outrageous color mixtures in the most startling combinations. But for most productions colored lighting (if any) is confined to backgrounds.

Sometimes even the subtlest changes of color temperature will be sufficient to tint light; for instance, low color temperature simulating an oil-lamp or candle-lit atmosphere. Otherwise, you can use plastic color-media sheets in front of selected lamps or occasionally modify the camera channel's color balance. Certain color treatments, such as the blue used to suggest 'moonlight' or red-orange for fire, have become stylized to the point of cliché.

Colored light also enables you to ring changes on neutrally toned scenery and cycloramas. When combined with projected patterns, impressive effects can be obtained quite economically. However, take care that white lighting (from key lights or fill lights) does not dilute the treatment.

Atmospheric lighting

Persuasive lighting treatment, can considerably change the appearance and mood of a scene:

1. Concentrating attention—Emphasizing particular features.
2. Revealing facts—Showing form, texture, surface design, etc.
3. Concealing facts—Preventing our discerning an object or plane; or seeing form, texture, detail, design, etc.
4. Associations of light—By light direction, intensity, distribution, recalling a particular atmosphere (firelight, sunset).
5. Associations of shade—Shadow formations recalling certain subjects, environment, or mood (tree branches, prison-cell bars).

Mood lighting comes from selectively emphasizing and subduing. It is not enough to just imitate natural lighting, for under the camera's scrutiny true environmental illumination often produces ineffectual or ugly results; primarily hot tops, black eyes, half-lit faces and steep harsh modeling. Instead, we must stimulate the imagination with light and shade—intriguing, hinting, evocative effects.

Table 7.9   Methods of light measurement

Incident light method	Reflected light method	Surface brightness method
Meter positioned beside the subject, pointing at light sources.	*Meter positioned* beside the camera, pointing at subject.	*Meter positioned* beside the camera, pointing at the subject.
Measures light intensity falling upon subject from each lamp direction in turn.	*Measures* average amount of light reflected from scene and received at camera lens.	*Measures* brightness of surface at which the instrument is directed.
Providing for (average) subjects of fairly restricted tonal range, typical incident light intensities and balance suitable to camera can be assessed. Base light, key light, fill light, and bak light measured in turn.	*Providing* average reflected light levels suitable to cameras sensitivity. Measure lightest and (darkest tones separately, and use midway reading for guide exposure.)	*Provides* readings by measuring surfaces of known reflectance skin, (standard white, and black). You can then deduce the suitability of light intensities falling upon them. *Also* allows scenic tonal contrasts to be measured to prevent over-contrasty lighting, over lit highlights, under lit shadows.
Ease of operation: method is simple and consistent. Does not require experienced interpolation. Widely used in motion picture lighting.	*Ease of operation*: readings vary with meter angling, and experience is needed to make allowance for subject tones, and contrast. Large dark areas cause readings to be falsely low, encouraging overexposure of highlights. Large light areas give high readings, which may cause underexposed shadows.	*Ease of operation*: method requires some experience in judging the *importance* of individual surfaces' brightness relative to overall exposure.
Advantages: when a show is to be repeated original levels can be duplicated readily. Balance between various light directions readily checked.	*Advantages*: method provides a quick rough check of average light levels. Can facilitate evenness of lighting.	*Advantages*: method is capable of assessing surface brightness and contrast very accurately.
Disadvantages: arbitrary allowance has to be made for subject tones. The amount of light required depends upon the subject—which this method cannot assess. Method only directly useful for 'average' subject-tones. Does not take into account tonal values, proportion of tones and tonal contrast.	*Disadvantages*: meter readings are only of an 'average' nature; which varies considerably with tonal values and proportions. Method does not indicate contrast range of subject or lighting. Meter's 'angle-of-view' seldom identical with the camera-lens'. Where a single surface (e.g. a face), is to be equally exposed in a variety of settings, measured exposure *should* be constant; but will vary as adjacent tones change.	*Disadvantages*: several separate readings are necessary to check evenness of lighting, and contrast. Method measures scenic tones, but does not distinguish their relative importance; and hence the desired exposure. Tonal contrast measurements may not signify: If the tones measured do not appear together in picture; if their proportions are small and unimportant; if they *may* be acceptably 'crushed out' without injuring pictorial quality.

Note: Where meter is held close to subject, measuring individual surface brightness, method becomes as for surface brightness method.

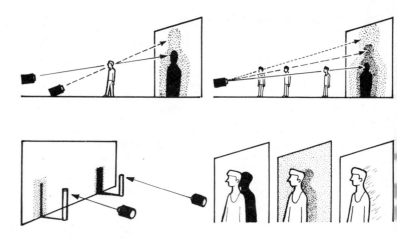

Fig. 7.24 part 1   Shadow size, sharpness, and intensity
Shadow size is always larger than the subject increasing as lamp distance lessens. Close lamps give greater size changes, and exaggerated proportions. Shadow size increases with subject/background distance. *Shadow sharpness* decreases with increased light source area, the lamp's closeness to subject, the subject's distance from background. *Shadow intensity* depends on background tones and finish, and spill-light diluting shadow.

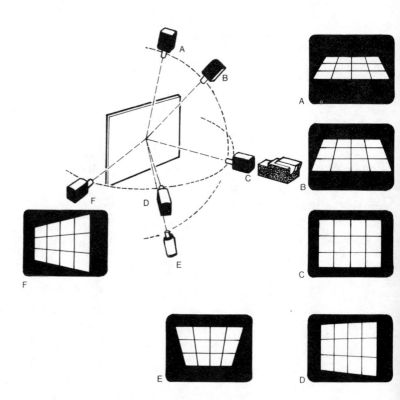

Fig. 7.24 part 2   Shadow distortion
Distortion arises when camera and lamp are not aligned, and not at right angles to background. Tilting background or angling lamp also creates distortion.

■ *Animated lighting* You can introduce life and action into a scene by lighting. The movement of light may be used as a dominating effect, or as part of an environmental illusion. You will often meet examples of this technique:

Rain running down a skylight throws streaking shadows into the room below.
Fluctuating light and shade patterning the interior of a moving automobile.
A room lit by a rhythmically flashing street sign.

Lighting has the particular merit of *flexibility*, for you can change mood or significance in an instant, or imperceptibly alter contrast and balance. For example:

A prowler is suddenly illuminated by a passing automobile . . . and we see that it is not the person we expected.

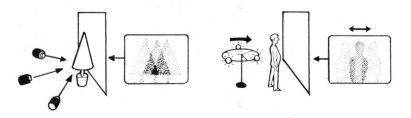

Fig. 7.24 part 3   Multiple shadow effects
These are produced when several lamps combine to light the same subject (particularly successful in color). If they rotate, the multi-shadows weave from side to side.

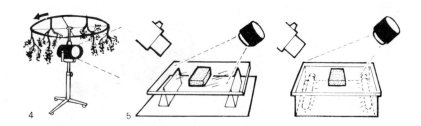

Fig. 7.24 part 4   Passing shadows
A rotating framework fitted over a lamp, and hung with a stencil cut-out, branches, etc., to create passing shadows. Patterns must not be obviously repetitive.

Fig. 7.24 part 5   To lose background shadows
This is achieved by displacing the subject from its background (e.g. on a glass panel), or by using a translucent illuminated panel.

During a brawl, the room light is broken . . . tension rises, as we no longer know who is winning.

Opening a window blind, the room's appearance is transformed by sunshine.

A happy atmosphere . . . becomes sinister . . . and finally horrific. The high key lighting gradually becomes contrasty . . . until finally underlighting predominates.

Lighting effects

■ *Reflected shapes* Light can be reflected from metal foil sheeting or silvered plastic to throw decorative patterns. As the surface is moved or flexed, the pattern shapes and position change. Even stenciled symbols or lettering can be reflected to animate or distort titles.

Countless visual effects can be produced in this way (including nebulae, abstract and magical illusions), reflected directly onto backgrounds, front or rear projection screens, or superimposed.

■ *Firelight* To simulate firelight, move a stick with close linen strips gently before a floor lamp. Smoke or gas jets are generally less convenient and most automatic devices provide mechanical-looking results.

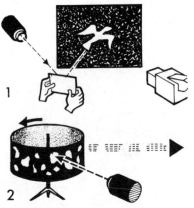

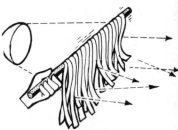

Fig. 7.25 Reflected shapes
Reflective material (metal foil, plastic mirror-sheet) held in a sharply-focused light-beam, projects its image on to nearby surfaces. Bending the reflector distorts the reflected shape. 2. Spotlight focused onto *mirror-drum* faced with plastic mirror, produces 'passing-lights' effect for vehicle interior shots.

Fig. 7.26 Fire flicker
Narrow strips of rag attached to a stick are gently waved in front of a lamp to simulate firelight flicker.

8 Scenery

Television settings—known variously as staging, sets, scenery—have to satisfy several requirements:

1. *Artistically*, settings must be appropriate to the occasion—the subject and the production's purpose.
2. *Staging mechanics* must be practical for the studio—its dimensions, facilities and the production budget.
3. *Design* should provide suitable shot opportunities—operation freedom for sound, cameras, lighting, etc.
4. *The TV camera's characteristics* will influence the tones, colors, contrasts, and finish of settings.

Basic organization

Staging begins with the demands of the script and the aspirations of the director. Much depends on how effectively these can be related to the facilities, time and budget available. As with all craftsmanship in television, optimum results come from a blend of imaginative perception and down-to-earth practical planning. TV set designers achieve minor miracles in making ingenious use and reuse of materials.

Planning begins with discussions between the director and the set designer. Using sketches, scale plans and elevations, production concepts are transformed into the practicalities of man hours, costing, materials . . . For larger productions, there is close collaboration with various specialists, investigating:

Shot opportunities for cameras, performer action and moves, the various considerations of lighting, audio pick-up, camera treatment, costumes, make-up and technical requirements.

In such an interdependent venture, teamwork is essential.

■ *The studio plan* The basis for much of the organization is the standard printed *studio plan*, showing the studio's permanent staging area with such features and facilities as exits, technical supplies, cycloramas, service and storage areas, etc. A typical metric scale of 1:50 i.e. 2 cm = 1 m, is replacing the widely used $\frac{1}{4}$ in = 1 ft scale.

■ *The staging plan (floor plan, ground plan, setting plan)* A rough plan of the staging layout usually begins with drawing potential scale outlines of settings, including their main features—windows, doors, stairways. By moving small tracing overlays around the studio plan, you can make best use of available space, while ensuring maximum opportunities for cameras, sound booms and lighting.

Standardized symbols representing scenic units, furniture, etc., are added until you have a mutually accepted 'bird's-eye view' of the entire staging area. Copies of a final plot are distributed to the production director, technical and lighting directors scenic construction workshops, and anyone else who needs one.

■ *Camera plan (production plan)*   The importance of the staging plan becomes clearer, when you bear in mind that in a closely planned complex production, it is used by the director to evaluate or calculate his shots and devise a *camera plan* showing principal camera positions (Chapter 16).

Fig. 8.1 part 1   Staging—studio plan

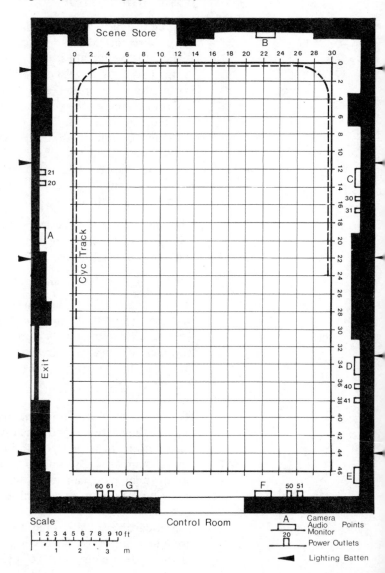

■ *Lighting plot* The lighting director overlays his *studio lighting transparency* (showing battens/barrels, lighting suspension points, lighting grid, rails, supply points, etc.), onto the staging plan. He then assesses the arrangements and types of lamps required for the effects he is seeking, bearing in mind the various production mechanics (Chapter 7).

■ *Furniture/props plot* A series of larger-scale plans may show the disposition of all *set-dressing* (furniture, drapes, etc.) and the main *props* listed on a props list.

■ *Elevations* Devised concurrently with the staging plan and to the same scale, *elevations* provide a scale side-view of surface detail, treatment and dimensions of all vertical scenic planes (walls, doors, windows, pillars, etc.). These not only guide construction and decoration, but help us to imagine the three-dimensional staging in the studio.

If cut out from thin card, these elevations can be attached to a staging plan to create a simple *scale model*. This is particularly useful in complicated situations, to aid visualization of the final scene (for director, performers, specialists, staging crew). A miniature viewfinder will even show shots obtainable from particular camera positions.

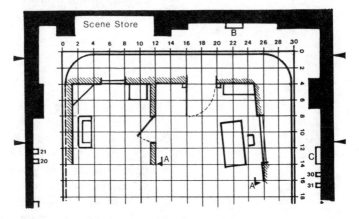

Fig. 8.1 part 2   Staging—staging plan

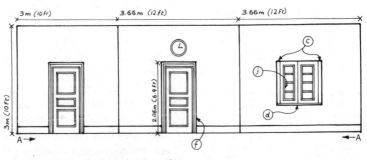

Fig. 8.1 part 3   Staging—elevations

Basic scenic forms

Most studio settings are built from an assembly of prefabricated scenic units. Carefully positioned to the staging plan (using wall ceiling/floor 'footage' marks), they are fastened together and supported where necessary and subsequently dressed or decorated to create the total scenic effect.

Articulated units facilitate transport, *setting-up* (erection) *striking* (dismantling) and storage. Moreover, most of the component parts have been designed to be redecorated and reused over again, in different combinations, for many productions—as *stock units*.

Scenic arrangements can incorporate several basic generic forms:

1. Flats.
2. Set pieces—Built pieces, solid pieces, rigid units (architectural and free-standing).
3. Profile pieces—Cut-outs, ground row.
4. Cycloramas—Cycs.
5. Backgrounds—Drapes, backdrops, backcloths, scenic cloths canvas drops, photomurals, photo blow-ups.
6. Projected backgrounds—Rear (back) projection, reflex projection.
7. Electronically inserted backgrounds—Chroma key insertion color separation overlay.

The flat

Most scenic flats are made from fireproof burlap (hessian) plywood or prepared boarding (fiberboard, hardboard) on wooden frames. Although canvas flats are lightweight, they are flimsy and lack rigidity. While 3·0–3·6 m (10–12 ft) high units are universal in larger studios, 2·5–3·0 m (8–10 ft) are more convenient in smaller studios where the camera is less likely to *shoot off* (*over shoot*) in longer shots. Widths from 0·15–3·6 m (6 in–12 ft) are used, the commonest being 0·6–1·8 m (2–6 ft).

Flattage can rise above 3·6 m (12 ft) but handling and safety problems increase. Tall flats are avoided where possible, for as lamps are forced upward to keep them out of shot. Their steeper angles cause coarser portraiture.

Surface treatments are legion. Cheapest are distemper or casein prepared paints. Wall coverings (paper, vinyl) provide an extensive selection of patterns, tones, and textures. Finally, you can attach such materials as fabric or carpeting; or build up surfaces in relief with stuck-on forms or thin plastic-sheet moldings

Most flats are single sided (*single-clad*), but are made *double-clad* when you need to shoot them from either side. Curved flats are used for some applications.

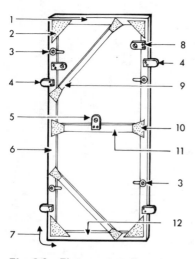

Fig. 8.2 Flat construction
1, Top rail, 2, Corner block. 3, Lash cleat. 4, Stop cleat (to align flats). 5, Brace cleat (brace eye). 6, Stile. 7, Frame (1 × 3 in pine). 8, Lash eye. 9, Diagonal brace (corner brace). 10, Keystone. 11, Toggle (stretcher). 12, Bottom rail.

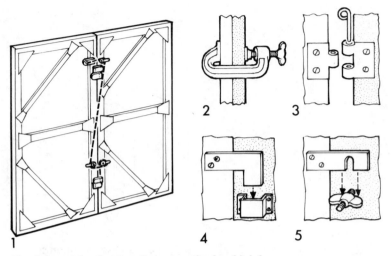

Fig. 8.3 part 1   Setting flats—methods of joining
1, Scenic flats lashed together by line (sash cord). 2, Adjustable cramp/clamp butts flats together firmly. 3, Pin hinge. 4, L-plate clip. 5, Securing nut (wing-nut).

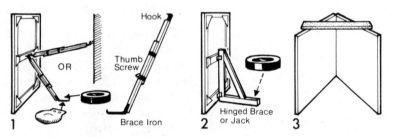

Fig. 8.3 part 2   Setting flats—supports
1, The flat can be supported by extendable stage-brace (held firm with a stage-weight or sandbag), or braced to studio wall plate. 2, A triangular wooden jack can be hinged or hooked onto flat rail (preferably bottom weighted). 3, Top bracing struts may secure unsupported settings. Suspension lines or bottom weighting are alternative arrangements.

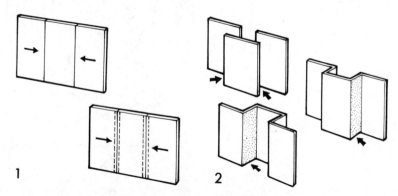

Fig. 8.3 part 3   Setting flats—disguising joins
1, When flats are set edge-to-edge, their joins show. Joins may be covered with pasted-on paper or fabric (stripping); although this prevents later re-angling. Joins still remain visible in plain light surfaces. 2, Where possible breaks or returns are introduced, disguising joins, providing a more interesting effect, and improving acoustics.

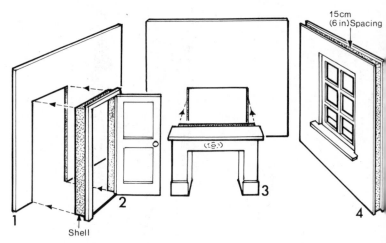

Fig. 8.4 Architectural units
Stock architectural features can be conjoined. 1, Contoured flat or frame (an example of a single-sided unit—viewable on one face only). 2, Door plug. 3, Fireplace plug. 4, Window unit (an example of a double-clad unit—viewable on both faces).

Set pieces—(built pieces/solid pieces/rigid units)

■ *Architectural units* Certain details (doors, windows, fireplaces, niches) can be fitted as required into *profile-flats* (*frames*) in the form of bolt-on *plugs* (*shells*). Alternatively, they may be incorporated as complete units; but these are heavier, bulky and less adaptable. Sometimes instead, dummy inoperative features are attached to the surface of a standard flat (e.g. room doors, closet/cupboard doors).

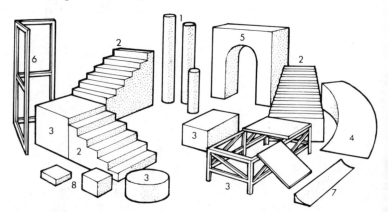

Fig. 8.5 Solid pieces/built pieces
1, Pillars: cylinders or half-shells of 0·15–0·6 m (0·5–2 ft) diameter and up to 4·5 m (15 ft) high. 2, Staircases (stair units): groups of two or more treads, usually 0·15 m (6 in) risers, matching heights of stock parallels (rostra). 3, Parallels (platforms, rostra) which are variously shaped level platforms on folding/dismantling frames with boarded-in sides. 4, Ramp: sloping plane surfaces. 5, Arch. 6, Drape frame: light framework in single or hinged units carrying draperies (otherwise suspended from batten or bar). 7, Cove: shallow sloping surface used to merge horizontal/vertical planes and hide cyc units. 8, Step-box (riser block): wooden shells from e.g. 0·15 m (6 in) to 0·6 m (2 ft) square. All-purpose unit that provides half-steps, display tables, for raising furniture, etc.

■ *Free-standing units*   A second, more extensive group of solid pieces provides not only such architectural features as stairways and arches, but structures for general production applications such as *parallels* (*rostra*) to create platforms or raised areas.

Profile pieces—(cut-outs)

These vertical profiles of plywood, prepared board, etc., may be arranged either free-standing, or attached to stock flats to modify their outline. Typical applications include:

1. Isolated decorative pieces—Flat cut-out representations.
2. Wings—Masking off the edges of an acting area.
3. Ground rows—Concealing background-to-floor joins.
4. Scenic planes—Representing a skyline, and the intermediate terrain.

Although these units seldom bear close scrutiny, at a distance the result can appear very realistic—especially if derived from a profiled photo blow-up. However, when viewed at an angle, the illusion fails completely. For *decorative* planes, this restriction is unimportant.

Profile pieces can provide cheap, very effective substitutes for elaborate scenic arrangements; for as the camera moves, parallactic movement between these planes can create a surprisingly convincing three-dimensional effect.

Fig. 8.6   Profile pieces/cut-outs

1, Profiled flats painted in perspective, create a spatial illusion. 2, Profiled ground-rows suggest progressive planes. 3, Profiled flat supporting plastic-shell rock-face.

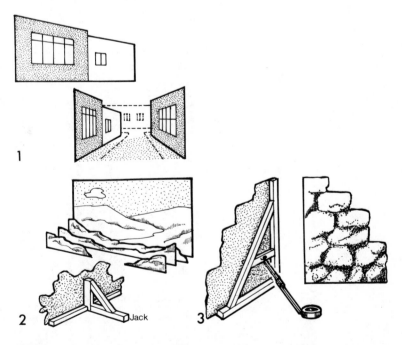

1

2   Jack   3

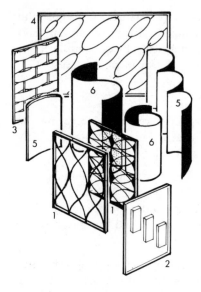

Fig. 8.7 Decorative screens
1, Decorated screens of expanded metal, wire mesh, perforated board. 2, Contoured flats of plastic shell or attached forms. 3, Woven screens. 4, Open screens of stretched wire or cord, poles, slats, etc., supporting motifs or profiles. 5, Curved flats, small solid cycs. 6, Flexible screens.

Cyclorama

Even the smallest studio makes full use of the cyclorama as a general-purpose background. The *cyc* (pronounced 'sike') is a suspended plain cloth; usually stretched taut by weights, battens or tubular piping along its bottom edge where a totally wrinkle-free surface is required.

■ *Cyc track* The cyc is sometimes hung on straight or curved battens or scaffold tubing. But the most adaptable method is by gliders (runners) affixed to a permanent *cyc-track* (*cyc-rail*) round the staging area. Using this, the cyc can be repositioned, changed, and drawn back when not required. Sometimes a range of cyc tones is available for selection on parallel tracks.

■ *Materials* Various materials are used—including duck, canvas, filled gauze, sharks-tooth scrim, and occasionally velours. Usual colors and tones include 'white' (60% reflectance), black, light gray, mid gray, light blue, dark blue, and chroma key blue. Although permanent (*board*) cycloramas are now seldom used, as they restrict staging areas and cause sound coloration, small shallow C-curved flats sometimes provide backings for windows.

■ *Adaptability* The plain light-toned cyc is extremely adaptable, for its smooth unbroken appearance can be considerably modified by lighting treatment. It can be lit to various brightnesses or colors; it can be shaded, or patterned, with projected shadows. With care, you can even attach very lightweight motifs to the cyc (double-sided adhesive tape, or pinned).

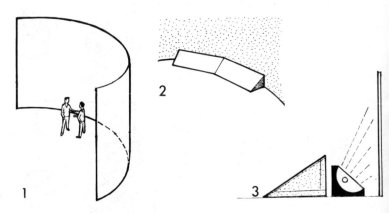

Fig. 8.8. Cyclorama
1, The cyc can be blended into the floor plane by 2, merging cove/ground cove. 3, Cyc lamps may be hidden to illuminate the lower cyc.

g. 8.9 **Drapes**
rapes may be: 1, Suspended from
rs (battens). 2, Attached to a flat
line across a flat. 3, Hung on
bular drape-frame. 4, Hung from
mber arm (gallows arm) hinged to
ge of flat.

Backgrounds

Strictly speaking, any surface seen behind a subject is its *back-ground*. Some people use the term *backing* synonymously, but that really refers to the planes beyond windows and other voids in the setting. These backings imply space and distance and prevent cameras from shooting off the set.

■ *Neutral backgrounds*   Neutral backgrounds are used wherever you want non-specific environments e.g. for talks, *public (current) affairs* programs.

Neutrality does not necessarily involve plainness; although to avoid distraction, any decoration is usually restrained. Surface texturing, uneven lighting (shading, dappling), decorative screens, all have applications. Any drapes should avoid distinct patterning, for it easily becomes obtrusive and too recognizable for reuse.

Limbo (white) or *cameo* (black) backgrounds provide neutrality. Sustained viewing of either can be tiring, and similar subject or costume tones can inadvertently merge with them. The resultant effect easily vacillates between naturalism, fantasy and abstraction. Essential furnishings, props and set pieces (doorways, windows) can be set up in front of the background—either symbolically or to form 'open settings'.

The somber gloom of black backgrounds can be inappropriate for certain subjects; and on many TV receivers may reproduce as gray. Foreground subjects may look overlight against these tones. On the other hand, white backgrounds can appear ethereal and infinite; although subject tones may appear darkened.

■ *Pictorial backgrounds*   Because it is difficult to distinguish in the two-dimensional TV picture between distant objects that are flat and those that are solid, you can simulate three-dimensional effects convincingly using a flat pictorial background. Ideally, it would need to be free from blemishes, evenly lit, and show no spurious shadows; its perspective, proportions, tones, etc., should match the foreground scene; and it should be shot straight-on. In practice, surprising deviations from the ideal will still work very well.

■ *Painted cloths (backdrops, backcloths, scenic cloths, canvas drops)*
Ranging from pure vaudeville to near-photographic masterpieces of scenic art, these large painted sheets are used primarily as window backings. Of canvas or twill, painted cloths are normally hung on battens or pipes, or on wooden frames. In storage, a cloth may be rolled around the pole that weights its lower edge, or *flown* (hung high above the ground).

When used as backings, you may reduce a 'painted look' and suggest distance (the detail-loss of *aerial perspective*) by introducing a light overall spatter or spray. Alternatively, you can stretch a black or white scrim over windows to soften the effect.

175

■ *Photographic enlargements* (*photomurals, photo blow-ups*) Although expensive, enlarged photographs represent the ultimat
realism obtainable from studio pictorial backgrounds. Enlarge
ments are made on sections of photosensitized paper stuck onto
flat, or a canvas support. Color is normally underplayed, dyeretouched monochrome versions being much cheaper and mor
convincing than even high quality color photographs.

Photoenlargements can be adapted and reused by retouching
stick-over alterations, or simply obscuring too recognizabl
details (by a strategically positioned bush or pillar).

Photoenlargements can be used too, as pictorial or decorativ
panels standing or hung within a setting.

■ *Projected backgrounds* Two systems are used. *Rear projectio*
(*back projection*—BP) in which a still or moving backgroun
picture is projected onto the rear of a translucent screen situate
behind the action. *Reflex* or *front projection* in which the image i
projected along the lens axis, onto a special beaded screen (se
Chapter 19).

■ *Electronically inserted backgrounds* (Chroma key, CSO). B
special electronic switching circuits, foreground subjects can b
inserted into the background picture from any other vide
source (photograph, drawing, miniature, TV camera, etc.). (Se
Chapter 19.)

1

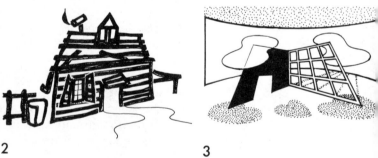

2

3

Fig. 8.10 Skeletal staging
1, May use isolated realistic scenic elements. 2, Decorative symbolism; o
3, Abstract pattern.

Surface detail and contouring

■ *Painted detail* Because you can suggest solidity by suitabl
drawing and shading, it is possible to create an illusion of brickwork, paneling, molding, etc., by brush alone. Realism depend
on the scenic artist's skill, how closely the camera sees the effect
and whether oblique viewpoints or cast shadows reveal the tru
contours.

176

Table 8.1   Surface finish

	Method	Purpose
Flat lay-in	Even-toned surface painting.	Textureless finish.
Dry-brush work	Overpainting a flat tone with a nearly dry brush. Sparse brush marks across the ground color.	Suggests metal, wood, stone fabric.
Stippling	Series of small close dots of color on a different ground tone; using coarse brush, sponge, cloth, wrinkled paper.	Suggests stone, cement, earth.
Puddling	Wet colors flowed together, intermixed for random variations.	Suggests ageing plaster, earth.
Daubing	Dabs of color with rolled rag, patted irregularly over a surface.	Varying density of tone.
Scumbling	Translucent coating, usually of darker tone over a lighter one.	Suggests surface undulations.
Glaze	Transparent dry brush application of lighter tone.	Suggests highlight sheen.
Scuffing (Dragging)	Skimming surface with brush, leaving textural depressions untouched.	Texturing a plain surface.
Wash	Thin coating of lighter or darker tone over background body color.	Suggests highlights or shadows.
Roughcast	Sprinkling material (sawdust, sand) irregularly onto freshly painted surface.	Random changes in tone or texture.
Spattering (Dottling)	Mottling with random brush-thrown splashes.	Textural effect.
Stenciling	Applications of decorative motifs by stencils or rollers.	Decoration.

■ *Wallpapers*   Various photographic wallpapers of brickwork, stonework, wood graining, are easily stuck to flattage—retouched or sprayed over where necessary.

■ *Molded pieces*   Ingenious lightweight imitations now exist for a diversity of subjects. Statuary, architectural features (beams, moldings), tree trunks, brickworks, roofing tiles, cobbled paving . . . are recreated with such conviction that they bear scrutiny— even up to fingernail-scratching distance.

Earlier techniques with papier mâché and plaster, or canvas on wire mesh have been largely superseded by glass fiber, molded rubber and plastic forms. Solid items such as statues and rocks may be hand-carved by hot-wire cutters from lightweight styrofoam (expanded polystyrene foam blocks).

Most outstanding of all, are the molded surfaces of thin plastic sheets (PVC). These are vacuum formed, up to about $2 \cdot 5 \times 1 \cdot 25$ m $(8 \times 4$ ft$)$ to produce realistic contouring shells depicting carved panels, rock-face, stonework, brickwork, tiling, mullioned windows, as well as various wall-furnishings (escutcheons, pistols, fish). These versatile panels are simply cut out, stapled onto stock scenic units, and painted. They have the merits of being low-cost, featherweight, adaptable. But they must be handled with care, to avoid compression damage.

Floor treatment

The studio floor's matte mid-tone surface serves as a neutral ground for most productions. But it can also provide some interesting staging opportunities.

■ *Painting* Temporary floor treatment can be applied, using special water-soluble paints, either to change the overall tone or to decorate the floor with realistic effects (floorboards, paving) or ornamental designs. Whether stenciled, hand-painted, or roller-printed, the treatment can be washed off after use. This means strictly avoiding spilled liquids, for they destroy the treatment and make it treacherously slippery and sticky. Temporary adhesive floor tapes can simplify or augment painted designs.

Painted floors do not frustrate dolly movements as carpets and floor cloths can; but they are liable to become dirtied. Black floors all too readily show wheel and foot marks.

Where facilities allow, lighting may provide projected floor patterns; but these are very easily diluted by other lighting treatment.

Fig. 8.11 Floor treatment
1, The studio floor may be treated decoratively (painted, stenciled, stuck-on designs). 2, Painting simulates surfaces economically, leaving them flat for camera moves. a, crazy paving; b, paving slabs; c, pavé bricks; d, cobbles. 3, The floor surface can be covered by: a, sawdust; b, scattered peat; c, dead leaves; d, sheets of rubber/plastic cobbles, brickwork, etc; e, grass matting (surface contours changed by f, sawdust sacks, or g, sack-filled platforms).

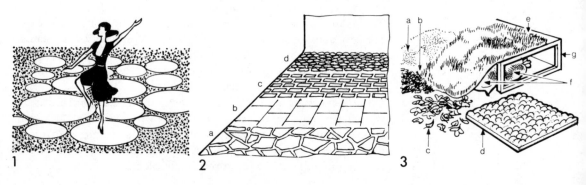

■ *Scattering* Scattering innocuous materials such as peat, sawdust, leaves, cork chips, bark, transforms a surface rapidly for naturalistic effects. However, material tends to stray around the studio (particularly when using wind machines) and must be confined. Avoid salt or sand—they foul up equipment!

■ *Floor coverings* The floor can be decorated by covering it with a *floor cloth* made of heavy duck, canvas, or tarpaulin (plain or prepainted); photographic or patterned papers; or panels of prepared board. Thin materials may tear or ruck-up under action or camera movements. Bulky materials such as turf or carpets can impede cameras.

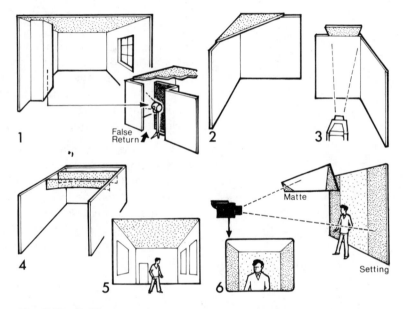

Fig. 8.12 Ceilings
1, Complete ceilings over settings are feasible, providing appropriate lighting can be introduced (through windows, doors, lamps behind furniture, false returns, etc.). Translucent ceilings can help but light may be hot or patchy. 2, Partial ceilings in corners or 3, localized areas may prevent shoot-off. 4, Cutting pieces suggesting beams, vaulting; and suspended lamps (chandeliers) suggest a ceiling. 5, A scenic background (cloth, photo, chroma-key insert) may contain a ceiling. 6, A foreground camera matte simulates a ceiling on the studio setting (Chapter 19).

Ceilings

Ceilings are introduced both the prevent cameras from shooting over sets into the studio beyond, and to create environmental effects. However, unless carefully confined, a ceiling can frustrate the lighting treatment and produce boxy sound quality. Usually, the viewer simply assumes the ceiling, just as he does the non-existent fourth wall of a setting. Where ceilings are essential (in very low or long shots) and where high walls or a void (black drapes) would be unrealistic, various subterfuges are used.

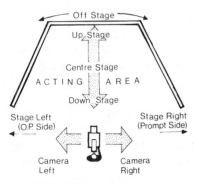

Fig. 8.13　The staging area
Locations are normally described in
relation to a given camera's viewpoint.

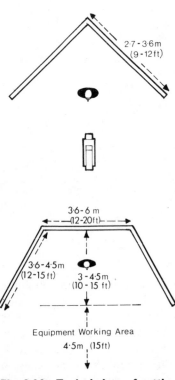

Fig. 8.14　Typical sizes of settings
1. Two-fold set. 2, Three-fold set.

Size and shape of sets

Space and budget restrictions usually prevent you from building
sets much larger than they need be. Instead, you can *suggest* space
by scenic treatment. Where spacious effects are required, wide-
angle lenses can exaggerate distance, so that even quite sawn-off
versions appear impressive on camera. The kind and extent of
action anticipated will influence set sizes. The smaller the setting,
the more restricted will the camera and lighting treatments
normally be.

Shooting off (overshoot)

Wherever you inadvertently shoot past the confines of the setting,
the solution is self-evident: readjust the shot, or add *masking*
(intermediate planes, a screen, border, wall-extension, drapes,
etc.). Anticipation during planning can prevent needless last-
minute rearrangements.

Shoot-off (overshoot) mostly arises when:

1. Cross-shooting on a shallow set.
2. Shooting downstage from an upstage viewpoint.
3. Inadequately backed windows or doors.
4. Reverse-shooting into a set through a window or door.
5. Using low-angle shots.
6. Reflections in mirrors or reflective surfaces.
7. Seeing through communicating openings (windows, arches)
   into adjacent sets.

Height and depth in floors

Vari-height platforms (parallels, rostra) are used where you need
elevated floor areas. Any irregularities are built up with blocks,
additional framework, tightly packed sawdust or sandbags.

To create 'holes in the ground' (graves, bomb craters, trapdoors,
wells) it is best to build up the overall floor level and leave the
required 'hole'. This flooring must be suitably reinforced, and
where heavy loads of scenery, people or equipment are involved,
tubular scaffold structures are essential. Where cameras are to
move around on the elevated areas, surfaces must be level, flat
and non-skid.

To prevent unnaturally resonant flooring, platforms are best
surfaced with felt, flexible urethane foam (foam-plastic sheeting),
carpeting or similar sound absorbants, plus internal packing
where necessary.

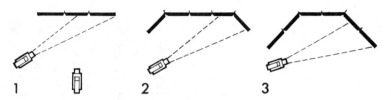

Fig. 8.15 part 1   Proportions of settings—flat backgrounds
On flat backgrounds, cross shots shoot-off easily. 1, The set lacks depth and solidity, but occupies little room (can be piled or nested). 2, Slightly improved, but setting looks disproportionately wide in cross-shots. 3, Very satisfactory proportions for general use.

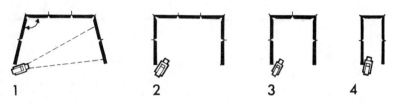

Fig. 8.15 part 2   Proportions of settings—walls
1, Walls may be splayed slightly to improve camera access. 2, Right angled walls may restrict camera movement, but prevent exaggerated room-size in cross-shots. 3, 4, As sets narrow, little deviation from straight-on shooting is possible. Sound pickup may be 'boxy' and lighting tends to be steep.

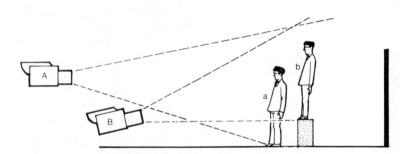

Fig. 8.15 part 3   Proportions of settings—staging height
The height necessary for staging depends on the longest shot to be taken (A), may need to be the lowest camera height (B), and the height of the subject (e.g. action on a staircase). A 3 m (10 ft) high flat is adequate for general use, a 3·6 m (12 ft) one being needed for larger settings. High backgrounds should be avoided to prevent steep lighting.

Space economies

In the small studio space is at a premium, and you may find that there is insufficient room for all the sets you need for a production. You can overcome such problems by discontinuous recording (altering or striking sets during recording breaks), by using prerecorded inserts (on film or videotape), use chroma-key insertion processes, or introduce versatile staging techniques.

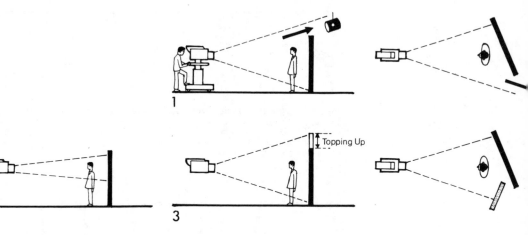

Fig. 8.16   Shooting off
1, Cameras can easily shoot past the edges of a set, and see the studio beyon
2, Solutions include modifying the shot; 3, extending the set; 4, introducir
appropriate masking pieces to prevent shoot-off (e.g. screen, wall, drapes).

Multiplane techniques

Our impressions of *scenic depth* are comparative. They come fro
our relating the appearance of progressively distant planes, ove
lapping, scale, and a series of similar visual clues. The more plan
visible between our viewpoint and the horizon, the stronger is th
illusion of depth. When there are few spatial clues in a shot (e.
an open stage), the picture seldom conveys a convincing thre
dimensional effect.

These impressions of depth derive primarily from the for
ground and middle distance. So you can arrange your stagin
techniques accordingly. Foreground pieces particularly enhanc
the feeling of depth and dimension. These can be such divers
objects as tree branches, furniture, tracery screens and columns
all used to provide progressive scenic planes.

When carefully organized, foreground planes come into sho
quite naturally as the camera moves around. But the effect ca
appear contrived, particularly when the viewpoint is unfamilia
Overdone, this technique could leave cameras continually weavin
amongst foreground obstacles, with the viewer failing to get
clear view of the action.

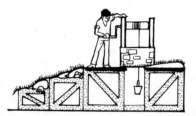

Fig. 8.17   Height and depth in floors
Covered platforms (parallels, rostra)
can be used to build up floor levels
where height or depth are required.

The illusion of four walls

It is not necessary to have *four* visible walls to an 'interior' settin
to convey a convincing impression of completeness. Even a singl
background plane may suffice; the rest being imagined by ou
audience. You can even build a room in *separate sections*, and b
careful intercutting imply that they are contiguous.

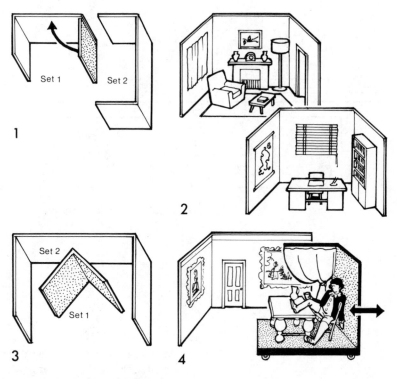

Fig. 8.18   Space economies
1, Where congestion is acute, flats may be moved aside for camera access (hinged folds, wheeled, man-handled, or hung). 2, Settings' appearance may be changed by re-vamping (retain structure, alter dressing), rearranging scenery, and changing lighting. 3, Nesting sets, where the inner unit is used and struck to give access to the second set. 4, Part of a setting can be built on a low wheeled platform (stage wagon, truck, float) and moved aside for access. Combined with chroma-key backgrounds (Chapter 19), considerable flexibility is possible.

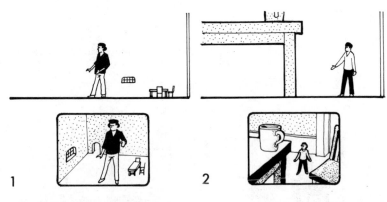

Fig. 8.19   Scale changes
1, Where the set is built with everything proportionately scaled down, people assume gigantic proportions. 2, If everything is scaled up, people are dwarfed.

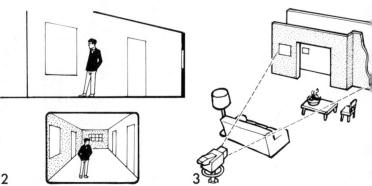

Fig. 8.20   False perspective
The background can deliberately convey a false impression of space. 1, A flat background (photographic, painting, chroma-key) may contain an illusion of depth. But give-away shadows, wrong scale, or wrong foreground-background perspective can destroy the effect. 2, The whole setting may be built in false perspective (exaggerated size-reduction with distance), but then action must be kept downstage. 3, By deliberately arranging to have large objects in the foreground and progressively smaller items further from the camera, perspective can be cheated.

An actual four-walled room can be very restrictive to shoot, for the camera has difficulty in getting far enough away from subject to provide wider (longer) shots. This occurs even when using wide-angle lenses or shooting in through doors or windows. Also where several cameras are being used, they are liable to get in each other's shots if space is very restricted. On location, you just have to make the best of circumstances.

Where four walls are essential in a studio setting, you may use intermittent videotaping to move in, or strike walls to shoot each viewpoint. Or you can actually build a complete four-walled set, arranging to raise, hinge, or slide *wild walls* out when needed to give access to cameras and booms. In these situations lighting may become a compromise.

Fig. 8.21   Multiplane techniques
Foreground planes enhance the impression of depth. 1, They may arise naturally from the camera viewpoint. 2, Items can be deliberately positioned in the foreground. 3, Even a localized foreground piece can be effective.

Fig. 8.22   The illusion of four walls
1, The fourth wall is usually assumed to be at the camera's position. Where the foreground wall must be seen, a mobile scenic piece (door, window) is often used. Designed as a *break-through piece* if camera dollies 'through' it. 2, Cameras can shoot over a low four-walled set for group discussions. (Beware of reverse cuts!)

It may be simpler to provide peepholes for cameras—through doors, windows, fireplaces; *camera traps* as hinged wall flaps, e.g. behind pictures or camera-concealing drapes. When shooting a four-walled set, you must particularly guard against *frame jumps* or disorientation by inadvertently cutting over the *imaginary (action) line*.

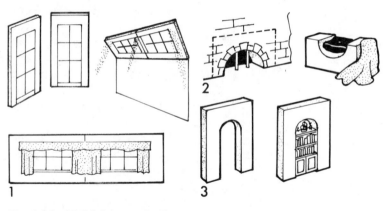

Fig. 8.24   Multiple use of units
Individual scenic units can be transformed. 1, Two glazed doors become a long window, or a ceiling light. 2, A low window can be turned into a throne. 3, An arch becomes book-shelves.

Fig. 8.23   Partial settings
An environment can be implied by building a complete but abbreviated section.

Partial settings

Why build more than the camera needs to see? Although some production situations require a total overall viewpoint or extensive construction, many shots need only a sectional view—a *partial setting*. Two deliberately restricted staging approaches are widely used. The first implies the whole by showing a *complete but localized part*—the doorway shot suggesting that the rest of the building is there—if we could see it. Audio effects help to create the illusion. The second method creates an impression by *judiciously placed foreground pieces* suggesting that the staging is more extensive than it really is, e.g. shooting through a foreground bookshelf, you imagine an entire library.

Admittedly such restrictions may not allow the director much latitude to change his mind, but the economies achieved in cost, space and materials are considerable. The resultant effect can be totally convincing for short, relatively static scenes, and provide greater elaboration and scenic variety than would otherwide be practicable. Although, when unskilfully used it can lead to a cramped, restricted quality in the staging.

Realism of settings

The artificiality of theatrical scenery is generally unacceptable in television wherever realism is intended. Yet paradoxically, where artificiality creates a genuine-looking effect, deception can go to any lengths. You may successfully use a fishline-activated curtain to suggest billowing winds, or have a leafy branch shadow indicate a nearby tree. Yet a genuine tapestry may look less realistic than a dye-painted canvas replica.

Realism can derive from quite subtle touches. A living room set must look lived-in. Too tidy, too pristine, and it loses conviction. Surreptitious *dirtying-up* (*antiquing*, *blowing down*) with a fine film of dark water paint can suggest wear, grubbiness and discoloration around switches and door handles. Hand-sprays are useful for rapid localized treatment; although for large-scale effects (a slummy tenement) broad spray gun work becomes necessary.

Deception must be discreet. You must not expect photobackgrounds with foaming breakers that never move, to fool anyone. Occasionally still backgrounds containing motionless trees, even static people will be acceptable, but sooner or later they look suspiciously inert; particularly when accompanied by sounds of associated movement.

1   2   3

Fig. 8.25 Visual detail
1, Excessive scenic detail is distracting, confusing, fussy. 2, If all but essentials are excluded, the result can be meager, cheap, empty; although suitably applied it becomes open, expensive. 3, Coordinated shape and line produce a stimulating arresting effect.

186

Sometimes the most convincing results come from using the real thing (e.g. real turf, tall grasses) while at others, grass matting (suitably discolored), plastic flowers, and foliage, would be preferable to natural growth wilting under hot lights. Real water pools may be no more realistic on camera, than clear plastic sheeting.

Scrim (scenic gauze)

Scrim or *gauzes* consist of thin cotton or synthetic net with a mesh of around 1·5–3 mm (1/16–1/8 in) diameter, in white, light gray or black. Although it can be used in draped form, the material is generally stretched taut without visible seams or wrinkles. It needs to be of sufficiently open weave to prevent excessive light absorption or obscuring subjects beyond, yet robust enough to stand handling and stretching.

Stretched before a light-toned cyclorama, a scrim diffuses and obscures irregularities, and enhances the impression of spaciousness. Over a scenic background, a scrim reduces contrast and artificiality. The material may be used as a glass substitute in scenery; it is safer, lighter, and gives no reflection problems. Lettering or decorative shapes can be attached to 'window panes' of scrim.

By adjusting its lighting, you can alter the appearance of the scrim in several ways. When frontally lit, the scrim will look like a flat 'solid' surface. Lit from the rear, it becomes translucent with surface details silhouetted on it. Unlit, the scrim appears transparent, softening the distant scene.

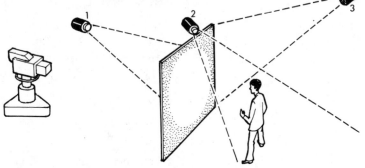

Fig. 8.26   Scrims/gauzes
Lift from the front alone (A), the scrim appears a solid plane. Surface painting or decoration shows up brightly against a plain white or black background. Unlit subjects behind the scrim are invisible. By reducing brightness of front illumination (A) and lighting the subject (B) (behind scrim), the subject and setting are revealed with outlines and contrast softened, surface painting having almost disappeared. A third lamp (C) added to rear-light the scrim, increases the mistiness over the scene, while silhouetting details on the surface of the scrim.

Mobile scenic units

■ *Mobile vehicles*   To create a convincing illusion that a studio vehicle is moving, you need to introduce a variety of 'clues': moving passing backgrounds; parts of the vehicle shaking; people being swayed or jogged; lights or shadows passing over them;

perhaps wind, or dust clouds; sounds of progress (engine nois
hoof beats, wheel noises, etc.); and the sounds of other passir
subjects.

You have to interrelate these various features carefully, an
above all ensure that there is compatible action between th
subject and its background scene. If the vehicle stops, the bacl
ground must stop too! This requires close coordination. Whe
moving film or videotape backgrounds are used (rear projectic
or chroma key) their timing and duration should be anticipated
prevent run-out or incompatibility.

A *sprung platform* when suitable handled, can provide realist
movements for a variety of applications—from rowboats, buggie
hansom cabs, to automobiles.

For a large structure such as a ship, you may need to imita
movement by gently tilting the camera, introducing moving ligl
(e.g. a key light rising and falling), and even having people swa
in unison.

■ *Movable scenic sections* When you want to move around
subject and space is restricted, or to reposition staging to form ne
composition arrangements, you can often do this by usin
movable scenic sections. These may be flats or solid units o
wheels, low wheeled trucks, (floats, stage wagons), turntables, an
similarly adaptable staging.

Set dressing

Dressing is the process of furnishing and decorating the bui
setting . . . the arrangement of furniture, ornaments, drapes. Th
is the treatment that personalizes an environment.

These properties (props) are of several categories:

1. *Dressing props* used solely as decorative features.
2. *Action props* used in the course of plot action; e.g. a newspape
   or telephone.

3. *Personal props* used or worn by specific performers; eyeglasse
   wallet.

Where the item is *non-practical* it is not functional, for instanc
a revolver that cannot fire or is simply not loaded. A *fully practic*
article is completely working—the revolver is loaded, and used i
the course of the action.

A setting soon acquires a filled-up look, cluttered with bric-a
crac that never registers on camera. So appropriate selection rathe
than profusion is the aim.

Where foreground space has been left clear for performers an
cameras, you may need to introduce (*set-in*) extra props o
furniture there for longer shots, removing (*striking*) them fo

camera moves. Although camera cranes can extend over fore-ground furniture to some degree, even slight obstacles impede pedestal dollies.

Furniture should not only be environmentally suitable, but physically practical. Deep well-sprung armchairs, for instance, can encourage people to slouch, to look leggy when sitting, and to have difficulty in rising. Cushions may help where unsuitable design in unavoidable. So too, seats that allow guests to swivel round, or that perch them uncomfortably, are best avoided.

If there can be said to be any specific tool for set-dressing, it must surely be the *staple-hammer* or the *staple-gun*, which can shoot wire staples through prepared board and soft wood up to 3 mm ($\frac{1}{8}$ in) thick. These versatile tools can attach pictures, drapes, posters, foliage, wall coverings, plastic moldings to flats; felting to flooring . . . A close runner-up is *double-sided adhesive tape*, which prevents articles from slipping or being moved out of position, attaches decorative motifs, holds arranged drapes in place, and fixes graphics. Wide gray *gaffer tape* too, is a tough many-purpose material, sticking firmly to many surfaces, yet removed without marking them. Its main purpose is to attach lightweight lamp fittings or cables to walls on location.

Electronic and lighting considerations

Production treatment and the camera's idiosyncracies place a number of restrictions on set design. Occasionally you can disregard them—but only occasionally.

■ *The final effect* It is important to remember that the audience only sees what the camera *shows*. Many a delightful skeletal setting has resolved itself into just a series of 'pole-through-the-head' shots as the director has concentrated on close viewpoints. Many a setting has looked fine in an establishing long shot, but been quite ineffectual in closer shots, or from other camera positions.

Color and tonal proportions in the picture can vary considerably with the shots selected. There may be only one small area of bright

Fig. 8.27 Movable units
1. The movements of a vehicle can be imitated by mounting it on a sprung platform or truck. 2, Using movable scenic sections, the large staircase splits to leave the solist isolated.

Springs
1

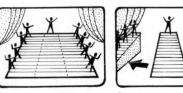

2   3

red in the set; but this could totally dominate and distract in close-up.

Various hues are quite distinct in color, yet may merge indecipherably in monochrome; and for graphics (titles, graph maps) this can be a real embarrassment.

■ *Tonal values*   Marked tonal contrasts give a long-shot a more definite emphasis than subtle pastel shades. But when you pan in closer shots over sharply contrasting background tones, or intercut between them, pictorial quality and visual continuity can be badly upset.

Conversely, if subject tones are too similar to their background the overall effect is unsatisfyingly flat or confused. So the tonal contributions of the setting, lighting, costume and make-up need to be coordinated for optimum results.

To ensure good tonal reproduction, it is best to work within a fairly restricted scenic contrast range for most purposes—around 10:1 to 15:1 (40–60% reflectance) is sometimes quoted. Unless lighting is diffused, it will usually extend this effective contrast to the system's limits.

■ *Surface detail*   Plain surfaces, unbroken by detail or modeling have limited pictorial appeal, but they can be considerably enhanced by lighting (shading, dappling, color variations).

Bold details such as a strong pattern or elaborate motif, are attractive as long as they do not draw attention from the action. They are most likely to do this if isolated within a plain background, or appearing unexpectedly in isolated shots.

Very small detail easily becomes lost in longer shots, or defocused where depth of field is restricted. So an attractive pattern that is quite clear in closer shots, may dilute to an unexciting overall gray background at a distance. Consequently, where action moves to and from such a background, we find pattern prominence altering as the focused plane changes.

Very close vertical or horizontal stripes are best avoided at all times—whether they derive from distant coarse patterns or close fine ones—for they can produce violent line beating (strobing).

■ *Surface brightness*   Bright surfaces leave the TV camera at a disadvantage where a full tonal range is to be reproduced. Even if no technical defects arise, the effect is generally distracting. So it is a good working principle to avoid overlight or shiny surfaces unless for a specific effect.

Whereas in a *monochrome* system excessive highlights simply reproduce as white areas within grays, in *color* TV they block-off to become white blotches on colored surfaces—and are therefore more prominent.

Highly reflective materials such as decorative aluminium foil or shiny plastic are notorious for causing troublesome hot spots, flares or trailing. Providing any defects are not obtrusive, their attractive sheen can introduce vitality to a shot. But the borderline between pictorially attractive effects and disturbingly eye-catching blobs can be very slight.

■ *'First-aid treatment'* Exposure is normally adjusted to achieve the most realistic *face* tones. If there are any disproportionally brighter surfaces in the shot, they will block-off to a detailless white. The result may be quite unobtrusive. But if it is unacceptable, the only solutions are to modify the lighting, remove the offending subject, or apply 'first-aid treatment'.

Although there are times when excessive lighting has aggravated the situation (as when strong back light burns out the shoulders of light-toned garments), such problems are more often due to high-reflectance or smooth surface textures.

The obvious way to improve any over bright surface is to reduce or cut off the light falling onto it. But this may be impractical, especially where it robs adjacent areas of light. Flares or strong reflections (speculars) from a shiny surface can usually be cured by reangling or tilting it, or changing the camera position slightly, or even by masking it with another subject.

Various aids can be applied to reduce such reflections (on automobiles, glossy paint, plastics), including dulling spray/anti-flare (wax-spray), water paste, latex spray, putty and model-

Fig. 8.28 Costume problems
1, Avoid costume tones merging with the background. 2, Beware detailed or fussy patterns in both costume and background. 3, Close stripes and checks in costume flicker (strobe) at certain distances. 4, Low necklines appear topless in close shots.

ing clay. Spurious reflections in glass or clear plastic can be similarly treated, although we may well obscure the surface by doing so.

Sometimes the only way to reduce overbright surface tones, is to spray or spatter the surface lightly with black water-soluble paint or dye. Overdone though, this can produce a dirty lifeless result, and destroy surface pattern.

Overdark surfaces may be improved similarly by light-toned treatment, picking out (edge-painting) molding or detail, or perhaps by increased localized lighting.

Care is needed too, in avoiding white or near-white set dressings; clothing, drapes, table coverings, papers, bed sheets, etc. You may substitute cream or light gray materials, or lightly dye them, or have them strategically sprayed or powdered down. Carefully done, the results on camera are well-modeled light-toned surfaces, reproducing as white. Overdone, though, you may finish up with people in dirty shirts reading gray newspapers!

Staging for color

Staging for color TV tends to fall into two broad categories:

1. Situations in which the excitement, vigor, and persuasive potentials of color can be fully exploited (singers, music groups, dance).
2. Situations in which color, unless carefully controlled, can inadvertently create false associations, appear tawdry, over-glamorize, or provide an inappropriately vivid background to action (talks, newscasts, demonstrations).

In excess, color defeats its purpose and becomes visually tiring. Too subdued, and the picture lacks vigor.

For various psychological and physiological reasons, the TV screen can often exaggerate—even caricature—reproduced color. Certainly the general practice is to underplay color, rather than have color emphasis.

As a good working principle, it is best to choose the surrounding décor to match the unalterable objects in the scene. Strong color is easily achieved; but subtle, sensitive color staging without simply resorting to gray-scale neutrals, requires interpretive skill. Somber, slummy, drab surroundings can be particularly difficult to stage realistically in color.

Large areas of even, unrelieved color are best avoided; preferably being broken up with lighting, scenic elements, or set dressing.

Emphasis is generally achieved through the use of color in costumes, key props, and furnishings, rather than in the set walls or drapes. Strong background hues easily dominate.

Costume (wardrobe)

Costume effectiveness is strongly influenced by its background. In larger TV organizations, performers' clothing (costume, wardrobe) is the responsibility of a specialist. But for many productions peaople wear their own clothing, and diplomatic guidance may be necessary to ensure that unsuitable attire is avoided.

You need to be sensitive to talents' feelings and taste when suggesting changes; particularly when you want them to wear an item from stock (an off-white shirt, or a different necktie) to replace their own. Experienced talent may bring along alternative garments for selection on camera.

A costume that looks attractive full-length, may be less successful when seen as a head-and-shoulders behind a desk. Color matching that looks good to the eye, can reproduce quite differently; for example 'reds' that differ due to their having dissimilar brown or blue proportions. Colors that have a strong bold appeal in long shots, can seem harshly crude in closer views.

Table 8.2   Costume problems

White shirts, blouses, etc.	Details and modeling lost where surfaces block-off to white.
Glossy materials—satins, etc.	High sheen, especially from shoulders, blocks-off to white or reflects color of incident light.
Light tones	These emphasise size, but if loosely cut, light garments can appear formless.
Dark tones	These minimise size, but modeling is easily lost in reproduction; particularly with dark velvets.
Strong, vibrant colors	Usually appear oversaturated and reflect onto neck and chin.
Fine stripes, checks, or herringbone patterns on clothing	Pattern strobes causing localized flicker, or produce blue-colored fringes (cross color), or creates color break-up. Color detail is liable to be unsharp and lost in longer shots.
Shiny, sequinned or metallic finishes	Blocked-off highlights. Reflects onto nearby surfaces.
'Noisy' jewelry or ornamentation— e.g., multi-string bead	With a personal mike these cause extraneous clinks, rattles, or rustles.
Rhinestones and other highly reflective jewelry	Reflects bright spots of light onto chin, neck, face; and flashes obtrusively, especially when using star filters.
Very low necklines	In close shots, can create a topless look.
Color fidelity	Certain colors can become emphasized or distorted (blues, magentas, deep reds) due to deficiencies in electronic or film processing.

9 Make-up

Types of make-up

Television make-up treatment follows three general forms: *straight*; *corrective*; and character make-up.

Straight make-up

Straight make-up is a basic compensatory treatment, affecting th. performer's appearance to a minimum extent.

■ *Skin-tone adjustment* This provides a good tonal balance i: the picture: darkening pale faces and lightening swarthy com: plexions.

■ *Routine improvements* These subdue blotchy skin tones shiny foreheads; strengthen lips and eyebrows; remove beard lin. (blue chin); darken overlight ears; lighten deep-set eyes; an: lighten bags under eyes.

For many TV productions, performers require little or n: make-up, with minimum correction and brief last-minut: improvements often being sufficient. Regular performers may d: their own.

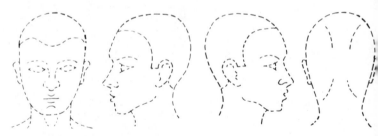

Fig. 9.1 Make-up chart
Outline chart is used to record detailed notes of make-up treatment, materials hairwork, etc.

Corrective make-up

Corrective make-up seeks to reduce less pleasing facial character istics, while enhancing more attractive points. Actual treatmen. can range from slight modifications of lips, eyes and nose; t: strapping sagging skin or outstanding ears, or concealing baldness

The general aim is to treat the person without their appearing 'made-up'. Skin blemishes and unattractive natural color mus: normally be covered; preferably by using several thin application of increased pigmentation, rather than trying to obscure wit: heavy mask-like coatings. Arms, hands, necks, ears may need blending to an even tone (body make-up).

A person's skin quality will modify the make-up materials used

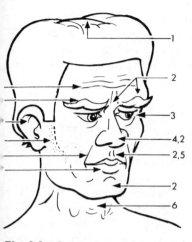

Fig. 9.2 Corrective make-up
Make-up can improve or disguise various effects that may appear emphasized on camera. 1, Shiny bald head; untidy hair; scalp showing through thin hair; hair too light, dark, dense, to show well. 2, Perspiration shine. 3, Deep eye sockets; eyes appear too prominent or small; eyes lack definition. 4, Shiny nose; nose coloration prominent. 5, Beard-line prominent (blue chin). 6, Neck scrawny. 7, Lips need definition or shaping; normal lipstick too dark or light on camera. 8, Age-lines; wrinkles over-prominent. 9, Ears too prominent; different color from adjacent skin. 10, Eyebrows barely discernible, over-prominent, untidy.

While coarser skin textures provide more definite modeling, finer complexions tend to reveal veining or blotches that the camera may accentuate. In color, ears can appear reddish and translucent. Complexions become flushed with exposure to heat (or hospitality). Make-up can become disturbed in the course of action (e.g. lip colors 'eaten off'). Also, bleached hair can exhibit alarmingly greenish shading, while blue-tinted white hair can look startlingly overcolored, only improved by corrective rinses.

Character make-up

Here, emphasis is upon the specific character or type that the actor is playing. By facial reshaping, remodeling, changes in hair, etc., the subject may even be totally transformed; for example Frankenstein's monster. But most character make-up involves less spectacular, subtle changes. Theatrical make-up treatments appear rather too broad and crude under the camera's scrutiny, except for such stylized characters as clowns, ballet, pierrot.

Conditions of television make-up

The principles and practices of television make-up are almost identical with those of motion pictures, except that in television the tempo and continuous performance usually prevent the elaboration or shot-by-shot changes that are possible in film.

A long shot ideally requires more defined, prominent treatment than a close-up. A similar situation, in fact, to that found in lighting techniques. But such refinements are seldom possible under typical conditions. You may not even be able to do anything about such distractions as perspiration, or disheveled hair, when the actor is on-camera for long periods—except to correct them for any retakes, where time permits.

For the very exacting demands of television drama, careful planning and presentation are essential. At a preliminary meeting with the program's director, the supervisory make-up artist will discuss such details as character interpretation, hair styling, special treatments, and any transformations during the program (ageing, etc.). Actors who need fitted wigs, or trial make-up, are then contacted.

■ *Camera rehearsal* For camera rehearsal, either of two approaches is common. In the first, the performers are made-up beforehand as experience suggests. When seen on camera, this treatment enables the make-up artist to judge more exactly the eventual detail work and tones needed. It also allows the lighting director and video control operator to assess tonal balance, contrast, and exposure.

While watching camera rehearsals on a picture monitor, the supervisory make-up artist notes any changes that seem desirable, to guide the individual make-up artists handling the performers. The other make-up approach (especially for straight or corrective forms) is to see performers on the screen first, treating them as time and facilities allow.

■ *Make-up treatment*   Generally speaking, a straight make-up for men may take around 3 to 10 minutes; women require 6 to 20 minutes on average. Elaborate needs can double or even treble these times.

After a few hours, cosmetics tend to become partly absorbed or dispersed through body heat and perspiration. Surface finish, texture and tones will have lost their original definition, and fresh make-up or refurbishing becomes necessary. Apart from on-the-spot retouching and freshening (mopping off and applying astringents), performers are normally treated in make-up rooms near the studio. Miracles of makeshift quick-changes have been achieved on the studio floor amidst the turmoil of production. But unless time limits or body position necessitates this (as with wounds), a more leisurely procedure is preferable.

There will always be problem occasions. Some performers cannot have make-up, owing to allergy or temperament. There are times when make-up has to be done immediately before air time, without any opportunity to see the performer on camera beforehand—a situation that the wise director avoids!

■ *Varying conditions*   Apart from artistic considerations, many technical factors affect make-up treatment:

1. Lighting—intensity, balance, direction, etc.
2. Scenery—relative face/background contrasts (simultaneous contrast effects).
3. Video adjustment (picture control)—exposure, gamma, black level, color balance.
4. Costume—relative to face/costume tonal contrast.

Such variations help to explain why the same performer's treatment may need to alter from one show to the next. Sometimes an astringent and light powdering may suffice, while at others more particular make-up becomes necessary.

Principles of make-up

The broad aims of facial make-up and lighting are complementary. Make-up can sometimes compensate for lighting problems, lightening eye sockets to anticipate shadowing cast by steep lamps. But, whereas the effect of lighting changes as the subject moves, that of make-up remains constant. This distinction is important when we consider corrective treatment.

1. Large area tonal changes—making the entire face lighter.
2. Small area tonal changes—darkening part of the forehead.
3. Broad shading—blending one tonal area into another.
4. Localized shading—pronounced shading to simulate a jawline.
5. Drawing—accurately lineated lines and outlines.
6. Contour changes—built-up surfaces.
7. Hair work—moustaches, wigs, etc.

Localized highlighting by slight color accents will increase the apparent size and prominence of an area, while darkening reduces its effective size and causes it to recede. By selective highlighting and shading, therefore, you can vary the impression of surface relief and proportions considerably. But you must take particular care to prevent shading looking like grime!

You can reduce or emphasize existing modeling, or suggest modeling where none exists; remembering though, that the deceit may not stand close scrutiny.

A base or foundation tone covers any blotchiness in the natural skin coloring, blemishes, beard shadows, etc. This can be extended, where necessary, to block-out the normal lips, eyebrows, or hairline, before drawing in another different formation.

Selected regions can be treated with media a few tones lighter or darker than the main foundation, and worked into adjacent areas with fingertips, brush or sponge. After this highlighting and shading, any detailed drawing is done using special wax lining pencils and lining brushes.

Make-up materials

The make-up media used in television and motion pictures include:

1. A dry matt cake of compressed powder (Pancake).
2. A non-greasy base of creamy consistency, in small stick-containers (Pan-Stik) or jars.
3. A greasepaint foundation, contained in tubes.
4. Powder and liquid base.

Before applying a foundation (base), the skin is first prepared by thoroughly working it over with a cleansing cream or lotion (or cold cream) to remove any traces of existing cosmetics. Wiping this off with paper tissues, an astringent (e.g. eau de cologne) is patted on, to close skin pores, reduce absorption and perspiration, and generally freshen. Following an after-transmission clean-off, the skin is cleansed with special removers and washed with soap and water.

Which of the various make-up media is used, depends on the effect required, the degree and nature of the treatment and personal preferences. Each material has its particular features.

■ *The cake foundation* This base is worked up with water-moist sponge, and applied thinly over the whole face. Its covering power (ability to overlay other tones, blemishes, etc.) is excellent. Its finish is predominantly matt. When too thickly applied, it can produce a mask-like appearance, flattening out facial modeling. Highlights and shading can be introduced with a sponge edge, using shades two or four tones different from the foundation color.

■ *Cream base*   This is dabbed on to small areas, and worked by fingertips evenly over the regions being treated. Localized highlighting and shading with different tones is pat-blended or merged with flat-topped sable brushes. Surface finish depends upon any subsequent powdering to set the foundation. Unpowdered, the skin has a soft sheen. Powdering reduces shine, leaving a smooth, satin finish. The cream has reasonable covering power and may be retouched easily. Leaving the skin's natural texture visible, it permits lighting to reveal more subtle half-tone modeling.

■ *Greasepaint*   This is similarly spread from small dabs. Although it has little covering power and may need refurbishing sooner than a heavier foundation, it is easily worked and retouched (by brush or fingertips). Powdering modifies its distinct shine to a silky gloss.

Such after-powdering, when applied to any completed make-up, softens contrasts and binds the materials. Any excess can be removed with a powder brush. A *beardstick* using a grease base can reduce a prominent beard line.

■ *Powder bases*   These are supplied in compacts and resemble reinforced face powder. Applied with a puff, their covering power is moderate. They have general use for broad shading and tonal improvement, emergency repairs, shiny foreheads, and the

Fig. 9.3 part 1   Basic make-up techniques—general treatment
1, Hard hairline, forehead lowered by shading, face width reduced, cheeks depressed. 2, Softened-off hairline, forehead width reduced by shading, deep-set eyes lightened, cheek and chin modeling increased. 3, Hairline reshaped and extended, forehead height reduced, chin made to recede, 'apple-cheeks' highlighted.

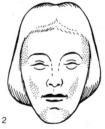
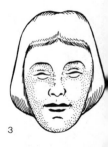

1   2   3

like. Most liquid-type media have poor covering power, and tend to become streaky and patchy.

Any powdering, whether to 'set' a foundation, reduce shine, or to absorb perspiration, must be applied judiciously. This is because the powder often tends to change color in use, or obscure base coloring.

Several accessory preparations are common to the beauty salon and the make-up room alike, e.g. mascara, eye shadow, and false eyelashes. A few, like rouge, have less value on-camera and are better stimulated by other means; while materials such as tooth enamel (to blacken or whiten teeth), artificial blood capsules, are more suitable to the television and film studios.

The make-up artist's tools range from brushes (for modeling, lining, applying lip color, etc.), sponges, wax pencils, to palette knives. But for many workers, their fingertips are their most-used aid, with which they blend foundation media into a homogeneously molded complexion.

Surface modeling (prosthetics)

Manipulation of the subject's own skin or by sticking on new formations can change the physical contours of the flesh.

■ *Manipulation*   To produce scars, ridges, etc., *non-flexible collodion* may be brushed on to the skin. This mixture of proxylin, alcohol and ether, contracts the flesh; the painless contraction increasing with successive layers. Although readily removed, collodion can irritate sensitive skins. In such cases, quick-drying, liquid-plastic sealers may be preferable.

Fish skin can be used to contort the flesh within limits. The slected area of flesh is drawn into its new position, and the fish skin spirit-gummed to hold it in place. Apart from imparting an oriental slant to eyes, this material is mostly used to emphasize or flatten out folds of superfluous flesh.

■ *Surface contours*   Surface contours such as warts, eye-bags, wounds, nose modeling can be built up by several methods. Wax nose putty, mortician's wax, and modeling clay (Plasticine), can all be molded into shape and stuck on. A latex rubber-base solution, *flexible* collodion (resin, castor oil, ether, alcohol), or various plastic equivalents, can be coated over a selected area until it is sufficiently extended. For larger protruberances, pads of absorbent cotton wool or sponge can be attached first by spirit gum. Surfacing is then carried out over this substructure. Because normal foundation media do not take to these surfaces, plastic sealers are finally brushed over to key them on and prevent any interaction.

For more drastic changes, partial or complete latex masks are

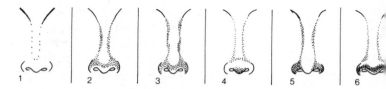

Fig. 9.3 part 2   Basic make-up techniques—nose treatment
Variations in nose shape achieved by shading (side shading, ridge highlighting)

molded and attached. They provide us with anything from double chins to grotesques. The advantage here is not only that more extensive alterations are possible, but that they can be prepared beforehand, and the fitting/removal process is quicker and easier. All prosthetics are best camera tested to ensure color and textural suitability.

Hair

Hair may be treated and arranged by the make-up artist, or by a separate specialist. Such hair work includes: alterations to the performer's own hair; the addition of supplementary hair pieces and complete wigs covering existing hair.

■ *Hair alterations*   In television, a certain amount of restyling, resetting or waving, may be carried out on the performer's own hair, but where extensive alterations such as cutting or shaving are needed, complete wigs are more popular. Hair color is readily changed by sprays, rinses or bleaches. Hair whitener suffices for both localized and overall graying. Overlight hair can be darkened to provide better modeling on-camera, while dark hair may need gold or silver dust, or brilliantine, to give it life.

■ *Supplementary hair pieces*   Where the performer's natural hair is unsuitable for treatment, hair pieces or full toupees can be attached to the scalp, and unified with existing hair. Where staining of the scalp with masque cannot hide baldness, these made-up hair pieces may be necessary. For women, postiche work (pinned-on hair) may provide flowing tresses, buns, ringlets, etc., augmenting a short coiffure.

Beards, mustaches, side-burns (side-boards), stubble and the like, may be 'home-grown', preformed by using prepared hair goods, or built up from cut hair lengths. Prepared hair goods are, undoubtedly, most popular. These are obtained pre-fabricated by wig specialists. Here, hair has been tied strand-by-strand to a fine nylon silk or nylon netting (lace), and can be dressed to suit beforehand. This method is less demanding of the time and skill of the make-up artist.

On the other hand, treatment built up from cut hair is a lengthy and skilled business, but more versatile. Human and yak hair are the most-used materials. (Crepe hair has limited application.) Such hair may be laid in spirit-gummed sections, until the required area has been covered. The result is trimmed, waved, and fixed (lacquer sprayed), as necessary.

■ *Wigs* They cover the performer's own hair entirely and are again formed from hair tied to a shaped foundation net. The front and sides of this net may be stuck to the forehead and temples and, where necessary, hidden by overlaying it with the base medium.

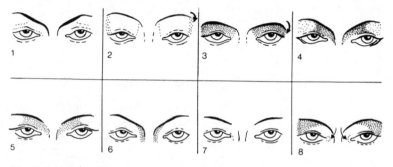

Fig. 9.3 part 3   Basic make-up techniques—eye make-up
1, Untouched. 2, Area above eye protrudes as a result of lightening, eyebrows raised and thinned, making eye seem smaller. 3, Eyebrows thickened and lowered, and area above eye shaded. This opens up the eye and make it appear larger. 4, Method of lengthening eyes. 5, Method of broadening eyes. 6, Distance between eyes reduced by shading. 7, Distance between eyes increased by shortening brows. 8, Detailed example of eye treatment: base of top lashes underlined thinly from center to just beyond outer corner, end being upturned. A short upturned line is occasionally drawn under lower lashes, but this tends to reduce the eye size. Eye shadow light towards the inner corners, slightly heavier at outer corners. A white penciled line along inside edge of lower lid.

10 Audio

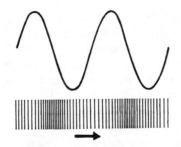

Fig. 10.1   Sound wave
When any material vibrates, its movement creates compressions and rarefactions in the surrounding air. In its simplest form, the air-pressure fluctuates at a sinusoidal rate. Sound in this pure form (*sine wave*, tone) is emitted by certain sources.

It is easy to take TV sound pick-up for granted. A desk stand, a tie clip mike, or the ubiquitous baton microphone seem to meet most needs. We adjust audio amplification so that the reproduction is loud enough. Where is the problem? We can hear what people are saying.

The trouble with casual sound pick-up is that we often cannot. The audio quality alters so much; it does not sound like the original. It becomes amazingly sibilant, muffled, distorted, volume varies, we hear random noise and distracting background sounds. The scale or proportions of the audio do not match the picture (close-up sound for a distant shot). Voices can become inaudible or confusingly jumbled together. The result is unrealistic and tiring to listen to. Little wonder that the professional audio man takes so much trouble to achieve the 'obvious'.

Sound quality

The simplest sound vibrates regularly in a sinusoidal fashion. Its oscillatory motion traces a *sine wave* of a certain frequency. Some sources (tuning forks, flutes, oscillators) can produce such pure tones. But most emit more complex sounds comprising a main

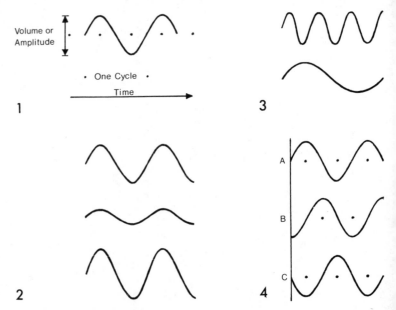

Fig. 10.2   Defining sound
1, *Audio frequency*: The number of complete vibrations (cycles) made per second in hertz. The distance the wave travels while completing a cycle, is its *wavelength*. Its amplitude (strength) is measured in decibels (dBs) or phons. 2, *Loudness*: Quiet sounds produce slight fluctuations, loud sounds strong fluctuations. 3, *Pitch*: As the number of vibrations increases, the sound's pitch rises—from a low to high frequency. 4, *Phase*: The relative time displacement between two sounds (in degrees, one cycle=360°)

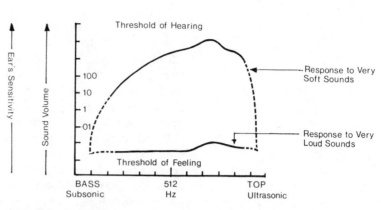

Fig. 10.3 Audibility
The ear's coverage of the audio range varies with sound volume. Highest and lowest extremes are less audible for soft sounds.

note (the *fundamental*) accompanied by a mixture of multiple tones (*harmonics, overtones*). Their proportions alter with the source; e.g. the type of instrument, its design, how it is played. Sometimes, as with the oboe, these harmonics may even be louder than the fundamental note. Transient sounds on the other hand (clicks, crashes, bangs) contain a broad random mixture of frequencies.

Fig. 10.4 Frequency range
Here the audible frequency range of various sources are compared.

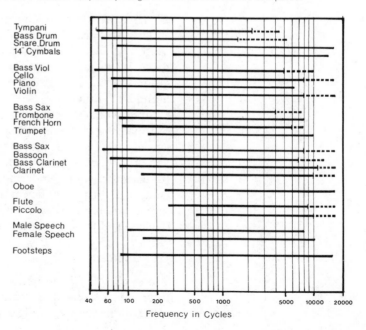

We come to recognize the characteristic ratios of fundamental and harmonics from various sources as their particular *quality*. Ideally we want to reproduce the original proportions exactly— but that is only possible within limits. Most audio systems modify quality due to their emphasizing or reducing parts of the audio spectrum, or adding spurious frequencies that are heard as various forms of *distortion*.

Paradoxically, while our brain is astonishingly tolerant and can interpret even wild travesties as substitutes for the original sound (as small radios continually demonstrate!), yet we remain quite critical of certain distortions—sibilance, overload distortion, severe loss of higher notes (top loss), strong resonances.

Reproduced sound

TV audio is *monaural* or '*mono*' (one-eared), and like a two-dimensional picture, lacks the ability to convey spatial information directly (although this is possible in stereophonic and quadraphonic systems). Mono sounds can only be segregated by volume, pitch and quality differences.

■ *Monaural problems* Our ears demand a wider frequency range for mono-sound reproduction, than is necessary to achieve similar fidelity in stereo. Also, we become more aware of (and confused by) reverberation in monaural reproduction.

The mono microphone can give equal prominence to any sounds in its pick-up range. So careful microphone positioning is often necessary to avoid one sound masking another, or sources merging in reproduction, or our overhearing extraneous noises.

■ *The audio system* The *dynamic range* (volume range) that any audio system can handle is limited. When too loud, sounds will cause *overload distortion*, producing spurious discordant over-tones. If too quiet, wanted sounds will become merged with background noise of comparable level (volume), such as: tape noise, hum and ventilation.

So to avoid exceeding these limits it is essential that you do not overload the microphone itself (too near a loud source), or over-amplify the signal (overmodulation, 'overmod'). Conversely, you must prevent the audio signal from becoming too weak (undermodulation, 'undermod') by placing the microphone close enough and using sufficient amplification. But at the same time, as you will see later, you must not destroy an impression of the dynamics of the original sound source.

Acoustics

Although acoustic design is the concern of the specialist, we are interested as users of the studio in how acoustic characteristics can affect audio quality, studio operations and audio-visual matching.

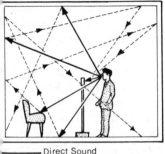

———— Direct Sound
- - - - - Reflected Sound

g. 10.5 Studio acoustics
s sound strikes surfaces a propor-
▪n is absorbed (frequency-select-
ely), or reinforced by structural
sonance. The reflected sound, now
odified, adds to the original (aug-
enting or partially masking it).
▸me coloration is artistically desir-
▸le.

■ *Reverberant studios* A *live*, highly reflective studio would emphasize all sounds—wanted (speech, music) or unwanted (camera moves, cable drag, footsteps, ventilation, scenery move-ment). Even using close directional microphones, extraneous sounds would intrude; particularly during quiet scenes.

No staging depicting exterior scenes or small rooms, can hope to be convincing if accompanied by echoing sound! It is always possible to *add* simulated reverberation to any sound, but impossible to remove it to deaden the effect.

Strong acoustic reflections can mask the original sound, reducing intelligibility. And, due to quality changes caused by frequency-selective absorption, the overall effect may become hollow, hard, boomy, 'woofy'.

■ *Dead studios* Given sufficient sound absorbing surfaces, most acoustic reflections can be suppressed at source, to produce *dead* (non-reverberant) surroundings. Only in special anechoic chambers or free space can we totally avoid reflected sounds. Such treatment reduces pick-up of extraneous noise, and improves acoustics for studio 'exterior' scenes. However, sound does not carry well in such surroundings, for it lacks the reinforcement of reflections. Consequently, such conditions are very trying for performers to work in and extra-close microphone positions become necessary.

■ *Practical acoustics* A degree of reverberation enriches and strengthens sounds, conveying an impression of vitality and spaciousness. Therefore, most studios have quite carefully chosen acoustics approaching neither of these extremes.

In practice, quality and sound absorption become modified by staging, equipment, audience, etc. Local acoustics within a setting can introduce considerable quality changes (sound becoming harsh, hollow, boxy), especially is there is a ceiling.

On location, the audio engineer has to contend with a wide variety of acoustic conditions and encroaching environmental noises, and must rely heavily on selective microphones and careful positioning to obtain optimum audio quality.

Microphone characteristics

The main thing you want to know before using a microphone concern:

1. Physical features.
2. Audio quality.
3. Sensitivity and directional properties.
4. Installation suitability.

Which aspects are most important to you, depends largely on the type of sound pick-up involved and operating conditions. For example, ruggedness may be at the expense of fidelity.

■ *Physical features*   While *size* may be unimportant for a slung microphone, it can matter where the microphone is to appear in shot or to be handled. *Appearance* then counts too. *Ruggedness* is a consideration where rough or inexperienced usage is likely. *Handling noise* too can be a distraction, for some sensitive microphones. *Stability and reliability* are features that only time and experience reveal, and most high-grade microphones can be relied on if given careful treatment.

■ *Audio quality*   *Ideally* a microphone should cover the entire audio spectrum evenly. Its *transient response* to brief sharp sounds (from musical triangle to jingling keys) should be impeccable. Audio should be accurately reproduced without coloration or distortion. Fortunately, such parameters are less critical in practice!

■ *Sensitivity and directionality*   A microphone's *sensitivity* determines how large an audio signal it produces for a given sound volume. Although audio amplifiers can compensate for even the least sensitive microphones, excessive amplification adds spurious hiss and hum to the audio signal.

Any microphone will normally have to work closer to quiet sounds than louder ones, but less sensitive mikes must be positioned that much closer. However, they are less liable to be overloaded or damaged by loud sounds, so that in certain applications (percussion) they may be preferable.

A sound-boom microphone needs to be pretty sensitive, or it would have to work too close to sources (casting shadows); and yet it must not suffer from *rumble* (when racking the boom arm in and out), or wind noise (as it moves through the air). So condenser or dynamic mikes are generally used.

The *directional properties* of the microphone (*polar diagrams*) are simply its sensitivity pattern in space. Sometimes you will need an *omnidirectional* response, that hears equally well in all directions. At others you require a very *directional* response, able to pick out a selected sound source and ignore or suppress others nearby. Certain microphones have adjustable directionality.

■ *Installation suitability*   When you connect together various pieces of audio equipment interfacing problems can arise. Inter-unit cable plugs are not standardized, and you must use the appropriate connectors or adapters. (Types include: Mini, phone, RCA, Din, Cannon.) As with all audio equipment the electrical *impedance* of your microphone must match the line or amplifier into which it connects. Although a low impedance unit may sound

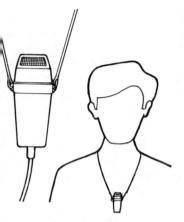

**ig. 10.6 part 1   Lavalier
anyard, neck)**
his microphone is slung on a neck
ord 15 cm (6 in) below chin.

fine in a higher impedance input, the reverse is less likely. Low impedance mikes (50–600 ohms) are most versatile for you can use longer cable lengths without top-loss; unlike high impedance mikes (2000–10000 ohms) which are also susceptible to hum/ electrical interference pickup.

In European designs the amplifier input should be at least five times the mike's rated impedance (typically 200 ohms) to avoid high frequency distortion.

■ *Choice of microphone*   All audio men have prejudices about the right mike for the job and exactly where to place it; for no two situations are *identical*. Listening to a piano performance, we find the instrument's tone varies considerably with its manufacturer, tuning, performer; and even changes with temperature, humidity and acoustics. While most specialists would agree to use a condenser or ribbon microphone, its positioning is influenced by many subtle factors.

While there is no 'universal' microphone type, one design may fulfil several different purposes. A certain condenser mike may well serve for a boom, floor stand, hung, or desk stand. But it would be unsuitable beside a boisterous drummer or as a hand mike at a remote, due to its susceptibility to overload.

Personal microphones

One widely used method of audio pick-up employs a personal microphone attached to the speaker. Audio quality is generally good, even under fairly noisy conditions. The main disadvantages are that: the mike can be muffled by clothing; loud noises can occur from contact with clothing (especially personal ornaments), or objects being demonstrated, or from handling; quality and volume vary as the wearer's head turns; and the mike may overhear other nearby sources.

Small windshields should be fitted for use outside the studio. And remember, leads and connections are particularly vulnerable and should be thoroughly checked.

**ig. 10.6 part 2   Miniature
apsule mike (electret or
lynamic)**
his is clipped to necktie, lapel, or
hirt, in single or dual form (with
tandby mike). Presence filters may
mprove quality. Plastic foam wind-
hields reduce wind-noise.

Although a lavaliere or a clip-on mike is itself unobtrusive enough, its cable (visible, or unseen beneath clothing) usually connects the wearer via an audio plug, to a long umbilical audio cable stretching away to a distant outlet. If a guest is seated, you will have to link him up once he is in position. If he moves around he has to drag the cable along behind. Some people quickly accept the idea, other feel tethered, others foul up their cable on obstacles . . . In any case, this cable must be unplugged after program contribution, before the guest can depart.

■ *Wireless (radio) microphone*   This is simply a miniature mike operating into a small pocket or belt transmitter. Its signal is picked up by the nearby antenna (aerial) of a sensitive FM (frequency-modulated) receiver, and fed into the audio mixer.

Although modern units are now reliable, transmission pick-up
not entirely predictable. Radio black-out, fading, distortion
interference can arise from multi-path reception or screening du
to metal structures or equipment. Diversity reception usin
multi-antennae arrangements can improve reliability. Th
system's range is often limited to around 400 m (440 yd) and th
transmitter's battery should be regularly checked (life abor
4 to 6 hours).

Each microphone needs its own transmitter/receiver system (
its own frequency), and there may be interaction. For lon
distance reception, a more sensitive antenna array may l
essential. A preliminary site survey is generally desirable.

■ *Hearing-aid earpiece* A technique that is being increasingl
used, is to fit experienced commentators or presenters with
hearing-aid earpiece (*deaf-aid*). This may be fed by a trailin
cable or a small pocket receiver. A combination of wireless mik
and earpiece can provide the wearer with considerable freedom o
movement, while enabling him to have continuous contact wit
production staff for instructions, up-to-date information, editori;
guidance, or prompting.

His earpiece can be fed with private line information, interco
(talkback), program audio, or remote sources (distant interview
ees). This arrangement is known variously as *interrupted feedbac
(IFB), program interrupt, switched talkback (STB)*.

Hand microphone (baton mike, stick mike)

The hand microphone, which is usually of the omnidirection;
type, must be rugged, permit free handling, and have an efficien
windshield (*wind gag, blast filter, pop filter*). This cloth or foam
plastic shield suppresses the rumbles and thumps from win
noise or close explosive sounds.

Quality is usually satisfactory, but unless users are reasonabl
competent results can be unpredictable, and there is always th
risk of misuse or careless handling. Although most hand mikes us
a trailing lead, the cordless *wireless microphone* incorporates a tin
radio transmitter—perhaps with a brief 'hanging tail' antenna.

Many hand microphones are fitted with a *spherical ball mesl
This permits very close positioning when used outdoors in nois
environments (to exclude extraneous sounds), or to avoid *howl
round* (*audio feedback*) where the mike feeds a PA system (publi
address, studio address loudspeaker) for a studio audience
It is a safeguard, too, where singers traditionally '*mug the mike
holding it close to the mouth to add presence to their performance

■ *Lip mike/noise canceling microphone* This specially designe
hand-held mike is invaluable for off-camera commentaries i
noisy surroundings. Its mouthguard is held against the upper lip
Equalized for close speech (attenuating its normal bass excess)

Fig. 10.7 Hand microphone
The baton microphone is widely
used for interviews. Directional response (cardioid) reduces extraneous
noise pick-up. *Techniques*: 1, Held
just below shoulder height. Pointing
at speaker improves quality but can
be daunting, and mike may then be
too close. Avoid omnidirectional
mike, which needs closer use (23 cm/
9 in). 2, Avoid low mike position (it
picks up background noises). If
visible microphone is unacceptable
then use personal, fishpole, or rifle
mike.

Fig. 10.8 Lip microphone
Used under high-noise conditions; the bar is held against upper lip to maintain constant mouth-distance.

the environmental noise becomes inaudible. Because it obscures the user's face, the lip mike is seldom used in shot.

Stand microphone

Made in various sizes and of adjustable height (0·3 to 1·5 m/1 to 5 ft) the stand microphone is used for soloists/speakers and for instrumental pick-up in orchestras, groups, etc. Various offset tubes (swan-necks) permit precision positioning and small slim fittings prevent the stand from being obtrusive. A quick-release top-clip enables the mike to be removed and hand-held where necessary.

The main disadvantage of the stand microphone are that the performer (talent) may kick it causing loud resonant noises, or displace it, and its cable can prove a hazard (floor tape it if necessary). Remember too, that it is essentially a *static* pick-up arrangement and people may easily move off-mike.

Desk microphone

Microphones can be clamped into small supports to enable them to stand on tables or desks. Formerly used to give an 'authoritative air' to a broadcaster, today the mike is small and unobtrusive, or sometimes hidden. A foam-plastic pad may prevent table-thump pick-up. Some audio engineers prefer to use stand mikes, hidden by a false table-front or through table-top holes, to circumvent this noise.

When audio pick-up is difficult, a desk mike is sometimes used in drama productions to augment other microphones—hidden at a vantage point within staging, furniture props.

■ *Desk discussions* Although personal mikes can be used for this situation, desk microphones offer better quality and are preferable. Microphones should be positioned in the direction of head movement. Directional mikes reduce unwanted pick-up of acoustic and environmental noise or audience loudspeakers, but placement may be more critical. Do not intermix different types of microphone because quality will be inconsistent. Cables should be concealed, tidied and taped (to desk and floor). Never run them alongside power cables, but if necessary, cross them at right angles.

Prerehearsal checking includes: connections, microphone routing to the audio control board (scratch mikes, never blow on them when identifying) and *phasing* (connections). If mikes are not all similarly connected, the quality will change when their outputs are mixed together (bass loss).

Fig. 10.9   Stand microphone
Floor stands are height-adjustable and may be fitted with angled extensions. Microphone clips into quick-release head. Heavy base prevents instability.

Slung microphone (hung, hanging mike)

Where space is restricted or orthodox mountings cannot be used,

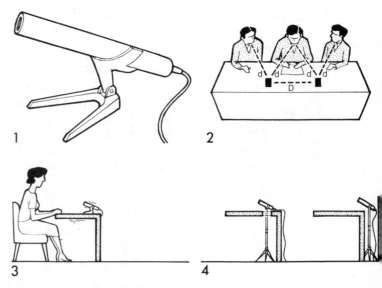

Fig. 10.10   Desk microphone
1, For desk discussions, mikes in small fittings are often used. 2, Careful positioning is essential for good sound pickup; typical distances D=0·9–1·2 m (3–4 ft)
and d=0·3–0·6 m (1–2 ft). 3, Desk-top mikes may be vulnerable (desk thumps,
displacement). 4, Alternatively the mike may be isolated on a floor-stand, either
using holes in desk-top or behind false desk front.

you can suspend a microphone over the performance area. Slung
microphones should not be used for well-balanced pick-up of
individuals in a seated group especially for close-ups. Although
they are useful for such broad group sounds as orchestral sections
or an audience. If the slung mike is to be kept out of long shots, its
positioning can become a compromise.

In a limited-access area (e.g. a narrow hallway), a slung microphone may be effective providing the performers play to the mike
and there are not excessive reflections from nearby walls. Sometimes the microphone can be disguised as a light fitting.

If a *directional* microphone is used, you reduce pick-up of
acoustic reflections and environmental noise, but run the risk of
people moving 'off-mike'. A non-directional (omni) pattern is less
restrictive but needs to be closer to avoid reflected sound.

A slung microphone must be rigged securely with safety lines
and placed to avoid distracting shadows or overheating by nearby
lamps.

Fishpole (fishing rod)

Here a microphone is attached via a rumble-insulating cradle,
to the end of an aluminium or bamboo pole (2–2·5 m, 6–9 ft long).
The mike angle is fixed. Suitable for short-duration use under
difficult conditions or to save a sound boom, the device is extremely flexible but tiring to operate. The fishpole, like the shotgun
(rifle) mike, is very widely used for fieldwork.

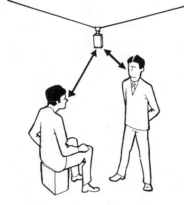

Fig. 10 11   Slung microphone
Suitable for area pickup or nearby
sounds. Problems arise when people
move, or are at different distances, or
are seated.

Fig. 10.12   Fishpole operation
The fishpole is fitted with the mike (preferably in shock-proof cradle) at one end, and a counter-balancing audio cable connector-box (intercom to operator, audio to console) at the other. The fishpole can be held well-balanced above head or tucked under arm. 'Pole-in-belt' methods are unbalanced, particularly tiring, and hazardous.

Sound booms

■ *Small sound boom (small giraffe, lazy arm)*   Extensively used in smaller studios, this lightweight supplementary sound boom is used for static pick-up or limited movement over a smooth floor. Due to its fixed (preadjusted) arm length, the mounting must be pushed around to follow action; an operation that can be cumbersome and preoccupying.

As its central column height is raised the assembly becomes increasingly top-heavy and less stable (particularly during moves), so you normally tilt the boom arm for higher mike positions, keeping the column low. The boom's closeness to the subject makes it correspondingly more difficult to keep out of shot.

Cable guards on wheels prevent accidental cable overrun. Pushing over floor cables can be hazardous. The microphone's cable is laid along the floor away from cameras and lamp cables. The boom operator often plugs his headset into a nearby camera's auxiliary outlets for program and intercom.

■ *Large sound boom (big boom)*   Operationally, the large boom provides the most flexible method of audio pick-up, for you can easily follow movement and favor selected sources. This ensures optimum sound quality while keeping just out of shot and avoiding shot shadows.

The boom's microphone is held in a swiveling cradle at the end of its long telescoping tubular arm. This counterweighted arm is pivoted on the central vertical column of the wheeled boom perambulator (pram).

With his left hand the standing operator controls mike tilt and direction (gripping and turning respectively) and swings the boom arm, while extending/retracting the boom with his right (hand crank). The platform's operational height is adjustable—useful for seeing action past cameras. The microphone cable may be suspended or run along the floor to a wall outlet.

Fig. 10.13   Small studio boom
The pre-adjusted telescopic arm (counterweighted) is pivoted on a column of adjustable height in a swivel-castered tripod stand. The boom-arm can be tilted and panned, the stand wheeled around, and the microphone can usually be turned and tilted. Boom stretch is typically 3–4·5 m (10–15 ft).

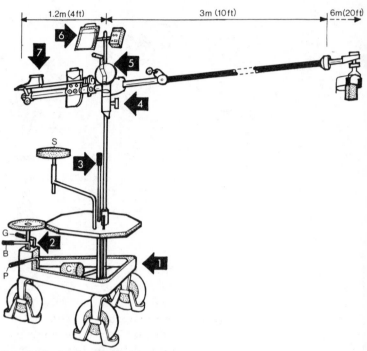

Fig. 10.14 part 1   Fisher boom
1, Three-wheeled (1·1×1·6 m/3 ft 9 in×5 ft 3 in). 2, Brake and gear change.
3, Control to hydraulically raise/lower platform (moves with column), max
column height 2·4 m (7 ft 10 in), pump handle to recharge cylinder, Operator's
seat for standby periods. 4, Pan and tilt lock screws (pans over 360°). 5, Right-
handed crank (max mike height 4·5 m/15 ft). 6, Script board. 7, Left hand-grip
(squeeze for mike tilt, turn for mike turn).

The large boom's main disadvantage for the smaller studio lies
in its size. Also, unless the operator descends and moves his boom
around, a second person is needed to reposition it. Boom tracking
(moving the pram while in use) is only normally necessary for
large-area working, to follow widespread action, to move aside
for cameras, or for fast relocation.

Unidirectional characteristics

The directional response (polar diagram—directional or directivity patterns) of a microphone is determined by its type and
design. Its behavior may not be constant throughout the audio
range. If it is more directional at higher frequencies, sources in
line with the microphone axis will reproduce well. But as they
move to one side, or the mike turns, the audio loses *definition*
(brilliance, highs) as the high frequency response deteriorates.
The unidirectional microphone is sensitive over a particular angle
and relatively deaf in other directions. This characteristic helps
you to concentrate on certain pick-up areas and suppress background sounds, or to favour quieter sounds (weak voices) against
louder ones. The exact response-angle needed, depends on its
intended application. A unidirectional mike is invariably used in a
sound boom. The more directional the mike, the further it can
work from the source without extraneous pick-up.

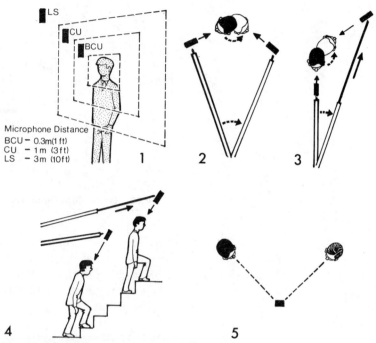

Fig. 10.14 part 2 Typical boom operations
1. *Microphone position* is adjusted to suit shot, remaining just outside frame limits. Boom arm is raised/retracted as shot widens. Operator must anticipate cuts to longer shots. 2. Head-turn: Either a swing and mike turn (slight arm-length change), or a static compromise position with mike turn. 3. Reverse turn: Extend and swing arm, with mike turn. Aim to avoid over-reaching subjects and turning mike back (any face shadows cannot be seen by operator). 4. Ascending: Boom arm tilts up and extends, mike tipping down during walk. 5. Split sound: Boom can accommodate separate sources up to about 2 m (7 ft) apart, by turning mike and/or swinging boom. Otherwise sources must move closer together, or use a second mike.

Omnidirectional characteristics

This microphone design provides pick-up of equal sensitivity and quality in all directions. So it cannot discriminate between wanted and unwanted sounds. It responds to nearby speech, acoustic reflections, and environmental noises equally well! It is useful for all-round pick-up of a group or where subject moves are unpredictable.

■ *Cardioid* This broad heart-shaped unidirectional pattern covers some 160°. Used for general purpose sound pick-up, it readily accepts spaced sources, unlike a half-eight pattern which is more restrictive at close distances. Rear pick-up is limited; although some designs are more *omni*directional at lower frequencies.

■ *Hypercardioid (supercardioid, 'cottage loaf')* A modified cardioid response with a narrower frontal *lobe* (pick-up pattern), and a weaker rear pattern. Generally considered as providing the best balance between incident and ambient sound.

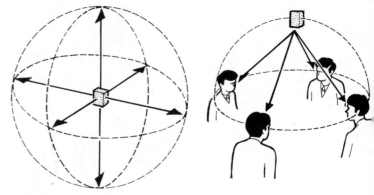

Fig. 10.15   Omni-directional response
Equally sensitive in all directions.

■ *Bidirectional*   Both faces of the microphone are live, providing a two-directional figure-of-eight pick-up patterns of about 100° each. To either side are two dead (deaf) areas. A typical traditional ribbon-microphone characteristic, this pattern is occasionally used for musical set-ups.

Table 10.1   Checking out the sound boom

Microphone	Speak to the audio engineer to identify mike and check audio quality. *Phase* the microphone if necessary, with others.
	Check mike connector is secure (waggle cable end).
	Check mike cradle—suspension OK? Free movement?
	Turn and tip mike. Operating strings taut and free moving?
Boom arm	Extend/retract boom arm (*racking* in/out). Any rumble? Controls and movement smooth? Is arm balanced throughout movement? (Adjust weights to suit mike weight.)
	Lock off the arm's pan and tilt.
Pram (perambulator)	Raise/lower platform (hand crank or hydraulic pump).
	Check wheel brake, lock, and steering.
	If steering mode is adjustable (steer/crab) check action.
	Tires OK? (Not overinflated, no stuck-on floor tape?)
	Check the seat.
Cable	Check plug into wall outlet.
	Is cable routing satisfactory? *Slung cable*—avoiding lamps, anticipating moves. *Floor cable*—not obstructing, sufficient available, free for boom moves.
Additional	Check intercom/talkback and private line (reverse talkback) to audio engineer.
	Check script board is secure and script available.
	Check picture monitor being used by boom operator (fixed to boom, or slung).
	If a loudspeaker is attached to boom (for *foldback* of effects, music, talkback instructions, etc.), check cables, connections, and reproducing levels.

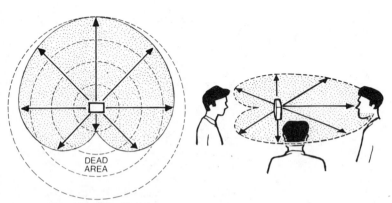

Fig. 10.16   Cardioid response
A broad heart-shaped pickup pattern, insensitive on its rear side.

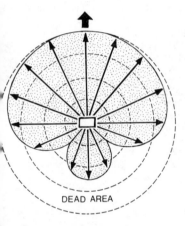

Fig. 10.17   Hpercardioid response
A modified cardioid response—more directional with a prominent rear sensitivity lobe.

■ *Highly directional microphones*   These are used wherever you want extremely selective pick-up, to avoid environmental noise or for distant sources.

Line microphones use a slotted interference tube(s) with an attached electrostatic mike. The smallest version (0·54 m, 22 in long), the *shotgun* or *rifle* mike (50° coverage), can be hand-held in the studio to isolate people within an audience; and in the field, hand-held or attached to the top of a camera. For exteriors, a foam-plastic windscreen/windshield is fitted.

The largest *machine-gun* versions (2·5 m, 8 ft long) use combined tubes of varying lengths and are tripod-mounted. Although rather cumbersome and producing inferior quality (particularly off-axis), the device is useful for isolating individuals in a distant group. Because directivity worsens below 2000 Hz, a bass cut is often introduced to reduce noise pick-up.

Parabolic reflectors provide a sensitive highly directional system (10–40°) in which a parabolic shaped metal reflector (0·6–0·9 m, 2–3 ft diameter) focuses sound into a central cardioid

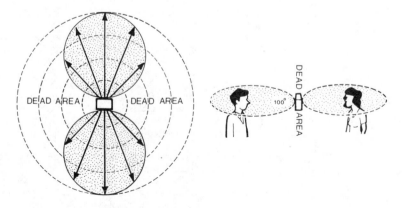

Fig. 10.18   Bidirectional response
Sensitive on either face, but deaf to sound pickup at the sides.

Table 10.2 Sound boom operation

Preliminaries

Check the boom is in the planned floor position relative to lighting treatment, and angle it to provide maximum use.

Check platform height to give good viewpoint over cameras, but avoid hanging lamps when operating.

Check boom arm at max/min reach relative to area coverage, obstructions, accommodating possible action spread.

Note key-light positions, look for potential shadow problems.

Operation

Mike position	Positioning depends on several factors: microphone sensitivity, and directivity, acoustics, ambient noise, source loudness and shot size. Closeness varies with the ambience required; a close mike minimizes studio acoustics, provides deader sound and has better presence. It also avoids extraneous sounds. Very close mike positions coarsen sound quality, reveal breath noises, etc. Close microphones come into shot and cast shadows. A more distant mike may provide inappropriate sound quality.
A steady boom	For a stationary situation, hold the boom arm still and the mike location steady towards the speaker. Avoid overhead positions—except when an omnidirectional mike covers random groups.
Swiveling the mike	A directional boom mike may be turned successively to favor each person speaking (gunning the mike), but someone can inadvertently be left *off-mike* unless the show is scripted. Benefits with closely positioned people are arguable.
Anticipate action	Lead rather than follow talent as they move around. Do not rely on script, memory, or instructions alone, but watch people for preliminary signs (e.g. before they rise or sit).
Locking off	Beware locking off the boom arm (tilt and pan) except for a static subject, or when leaving or moving the boom. People may unexpectedly move away, get up, or sit!
Fully extended	Despite counterweighting, the boom arm becomes less manageable when extended. Swinging (arc) and raising the arm is less controllable, particularly when done quickly. By retracting during wide swings, it is easier to move and avoids causing shadows.

Divide your attention between:

Performers	What performers are doing—or going to do.
Headset	The audio is 'split' on your earphones—one ear has *program sound*, the other has *intercom* from director and audio engineer. You can only judge audio quality and balance approximately, so be guided.
Shots	Be aware of picture monitors showing line (transmission) picture; check shot size. Camera tally lights show which is switched on-air. Monitors are usually too distant to detect whether your mike is appearing in shot. If in doubt, during rehearsal carefully dip it into shot; and when told of the shot limits remember these for subsequent mike positions.
Boom shadows	Watch for mike shadows on people or backgrounds. Move the mike and check if sound quality is still acceptable. If not, indicate that it cannot be cleared. (Changes in the volume, shot, or lighting may be necessary.) During transmission (or a 'take'), hold the shadow still, or creep it out stealthily, or slowly raise the boom, or try to hide it, or wait until off-shot before repositioning.

Boom moves to a new floor position may require a separate operator's aid (*tracker*) under instruction. Coordinate boom moves with cameras; watch for cable or boom arm catching in lamps. Take care when running over cables (damage, noise, instability).

Script or show format

Keep up to date on deleted or extra shots, or moves. Mark *take-over points* where action is to be covered by another boom.

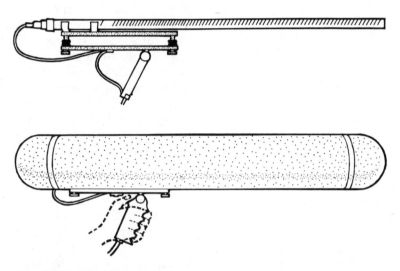

Fig. 10.19 Line microphones
A form of highly-directional microphone. The popular shotgun (rifle) form is shown here with and without its windscreen (windshield). Usually hand-held, it can be affixed to a camera, or used (without shield) in a sound boom.

microphone. Ineffective below about 300 Hz, the reflector is bulky and liable to wind rock unless firmly mounted. Whereas the *line microphone's* directionality is achieved by rejecting sounds outside its pick-up angle, the *parabolic reflector* actually concentrates sounds by acoustical reflection, and so has increased sensitivity.

Dynamic (moving coil) microphone

A very robust design, the dynamic mike is not sensitive to handling and has a wide range of applications: baton, fishpole, small boom, desk or stand, slung, and field (location) use. It is often used for less critical sound quality (close singer) or close loud sources (e.g. drums). Small versions are used as lavaliere and clip-on personal microphones.

The microphone tends to be unidirectional to higher frequencies, but non-directional to low frequencies. Consequently there is a top fall-off as subjects move off-axis (center line). This feature is useful with sources that are shrill, toppy or sibilant; when one deliberately positions them off-axis.

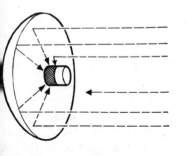

Fig. 10.20 Parabolic reflector
Extremely directional and sensitive, sound is focused by the reflector into the central microphone.

This type of microphone works by sound wave pressure varia‑
tions vibrating a diaphragm which has a small coil of wire at‑
tached. Coil movement within a magnetic field generates audio
currents. A typical frequency range is 40–16 000 Hz.

Condenser (electrostatic) microphone

The condenser microphone produces the highest audio quality
(typically 20–18 000 Hz), and has an excellent response to
transient sounds. It is used for high grade pick-up (moving or
static), especially when some distance from source (e.g. slung over
an orchestra). It can be fitted to a boom, stand, fishpole, or slung.

The condenser mike has the disadvantages that it is relatively
large, and being highly sensitive, it is liable to input overload
distortion by very loud nearby sources. In addition, the unit
requires a polarizing voltage supply and an adjacent preamplifier.

In the condenser microphone a light flexible metallic membrane
is stretched close to a flat metal plate and a polarizing potential
(e.g. 60 v DC) is applied between them. Sound wave pressure
fluctuations alter the intervening space and hence their inter‑
capacity. Varying current continually taken to recharge this
changing capacitance constitutes the audio signal.

■ *Electret capsule* The electret capsule is very extensively used
as a personal or a concealed mike, providing good quality pick-up.
This relatively cheap miniature microphone incorporates a
plastic film diaphragm with an inbuilt 'permanent' electrostatic
charge; thus eliminating the need for an applied polarizing
voltage. A tiny associated amplifier (battery powered) is enclosed
in the mike housing.

Performance deteriorates with ageing, resulting in a loss of
higher frequencies (top, highs), reduced sensitivity, and increased
background noise. Deterioration can be accelerated by high
humidity, moisture, heat and dust. Rapid unexpected failure is not
uncommon.

Ribbon (velocity, pressure-gradient) microphone

This microphone is suitable for static pick-up of speech and
music, and for stand and slung applications. It has an excellent
transient response and produces the highest audio quality. The
ribbon microphone has an even frequency response (e.g. 30–
18 000 Hz) over its pick-up field—although there is some top-
loss for pick-up oblique to its ribbon. Bass notes are emphasized
when the source is very close. Its normal figure-of-eight direc‑
tional response is often made unidirectional (asymmetrical) by
internal design.

On the debit side, the ribbon mike is not generally robust or
compact. It is not normally suitable for mobile boom use (or hand

218

use), being susceptible to wind rumble and wind damage. Also it is subject to overload from close loud sounds.

In the ribbon microphone, a thin corrugated foil strip held between magnet poles vibrates with air pressure differences on either face, so generating an electric audio current in the foil.

Crystal microphone

Although very cheap, highly sensitive and of small size, the crystal mike is fragile and its audio quality is restricted. Only adequate for less exacting use (field interviews), it is very suited to *contact* applications (electric guitar).

In operation, sound impinges on a diaphragm connected to a small bimorph (two-sliced) crystal, which through the piezo-electric effect, produces a corresponding audio voltage when twisted.

Audio control

Located at his *audio control console* (board, desk, mixer panel) the *audio control engineer* (*sound mixer*, *sound supervisor*) selects and blends the various program sound sources.

His attention is divided variously between:

1. Selecting and controlling audio sources.
2. Keeping the flickering needle of his volume indicator within system limits.
3. The director's intercom (talkback) information.
4. The audio quality from his high-grade monitoring loud-speakers.
5. Picture monitors showing line and preview shots. (Maintaining audio-visual scale. Detecting visible mikes or boom shadows.)
6. Guiding and cuing audio operators (booms, and audio equipment).
7. Possibly operating disc turntable, and audio tape recorders.

Fig. 10.21 part 1   Basic audio control
1, *Microphone* plugs into 2, studio wall outlet and 3, the audio is amplified. 4, A side-chain *audition/pre-hear circuit* provides check-point before further processing. 5, A pre-set *attenuator/pad* adjusts channel amplification (*gain*) and 6, a *channel fader* (slide, quadrant, or knob) controls it overall.

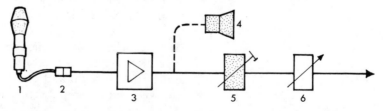

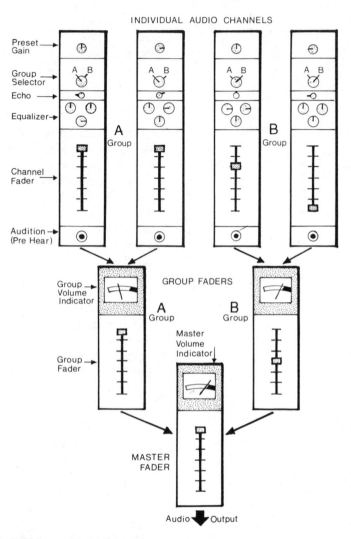

INDIVIDUAL AUDIO CHANNELS

Preset Gain

Group Selector

Echo

Equalizer

Channel Fader

Audition (Pre Hear)

A Group

B Group

Group Volume Indicator

GROUP FADERS

A Group

B Group

Master Volume Indicator

Group Fader

MASTER FADER

Audio Output

Fig. 10.21 part 2   Audio console
Audio control desks may use modular construction (plug-in sub-units), or integrated form. Each channel passes to a group fader for combined control of several mikes, then to a master fader. Meters indicate audio signals' volume. Additional facilities include: echo, equalizers (filters), audience address (playback) system, audio limiter/compressor.

Dynamic range control

As mentioned earlier, it is important to prevent the audio signal from exceeding the system's limits. Adjustments of the audio gain (amplification) to achieve this can be done manually or by automatic electronic circuits.

■ *Manual control*   Although our ears are tolerant of a wide volume range, tensing to very loud sounds and relaxing to receive quiet ones, straight audio systems do not have this property. If you keep audio amplification down to prevent peak overloads, quietest sounds will be inaudible. Increase amplification to hear these and even fairly loud sounds now reach upper system limits and distort.

You can overcome this dilemma by imperceptibly altering the audio gain manually, to compensate for these volume peaks and troughs. Some sound sources need little or no control, while others (symphony orchestras) require continual sensitive readjustments. The trick is to anticipate, and to make any changes gradually, so that the listener is unaware of the subterfuge. You must not 'snatch back' peaks, or boost quiet passages, destroying dynamics completely through overcontrol (riding the gain).

A VU meter/PPM shows the audio signal's strength relative to the system's available range (Fig. 10.22).

■ *Compressors/limiters* Automatic circuits can be used to adjust audio amplification; restricting the volume range, controlling unanticipated peaks or wildly varying sound levels. They can substitute for manual judgement—but electronics make no artistic evaluation and can create unnaturally strangled crescendos, or inappropriately boost quiet sounds. Most audio engineers regard such equipment as a useful safety facility for the unexpected moment, rather than an artistically viable control method.

While both variable gain systems prevent peak overloads above a selected volume, the compressor also increases amplification of quieter passages, and can be used to compensate for unwanted volume variations, e.g. fading as a vocalist's hand-mike position changes. In radio, DJ's deliberately use (misuse?) such facilities to make quiet background music surge in whenever they stop speaking. Similarly, the surging beat of *peak-limiting* is used as an audio effect.

1

2

Sound balance

When you *balance* various sound sources, you are adjusting their relative volumes, clarity, quality, scale, perspective, and the ambience of direct to indirect sound. The audio engineer is concerned in practice with two forms of program balance, that we can consider as *continuity balance* and *internal balance*.

Fig. 10.22 Volume indicator
1, *Volume Meter (VU Meter)*: The upper scale, marked in decibels (dBs), is for calibration and line-up purposes (using steady tone). The lower scale shows percentage modulation, where 100% is system's max limit. 2, *Peak Program Meter (European)*: Designed specifically to indicate sound volume *peaks*. Has fast-rise/slow-return characteristics, so its needle fluctuates less than VU meter. Its easily read seven section logarithmic scale is akin to ear's loudness response (unlike VU meter).

■ *Continuity balance* Here he controls a succession of sources; ensures that one person is not disproportionately louder or quieter than another; that the background music or effects are proportional to the dialogue; that other audio sources (film) match with studio sound. (Preliminary *level tests/voice levels* assess typical audio control settings needed.)

■ *Internal balance* Internal balance involves combining and blending a group of sound sources; e.g. arranging a microphone set-up for a musical group, an orchestra or a choir. Actual tech-

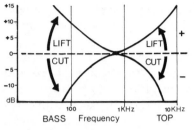

Fig. 10.23 part 1   Equalizers, audio filters, response selection amplifiers (RSA)
Filters for compensatory or dramatic effect. Progressive accentuation or reduction of top or bass.

niques vary with personal preferences, the sound sources, acoustics, and the final effect required.

A *single microphone balance* places a single highly sensitive (often non-directional) mike at an optimum point for communal pick-up. At best this method avoids the phase distortion created when several microphones pick up and blend a particular instrument, and blur its attack. It avoids the variations in perspective and proportion we get when giving varying emphasis to selected instruments, while backgrounding others. But its effectiveness depends upon acoustics, layout and suitable positioning.

A *multimicrophone balance* uses a series of pick-up points. It is seldom possible in imperfect acoustic conditions to find one good mike position to suit a variety of instruments having different quality and attack. Distant pick-up of soloists when using one communal microphone can sound rather thin. The blending of a multimicrophone set-up offers better balance flexibility when rehearsal time is short.

Close balance techniques now widely used for musical group recordings are virtually a 'sound processing' operation. Each instrument's (or sub-group's) output is recorded on a separate track of a multitrack audio tape (24 to 32 parallel tracks on 2 inch wide tape). Instrumentalists or singers (using earphones) may perform with others, or combine with prerecorded tracks. Individual tracks can be *lifted-off*, replaced, treated selectively (adjusting volume, quality, reverberation, pitch), synchronized and blended with the others to achieve exactly the required combined sound. This complex process can be computer assisted in the *dub-down* to a final track!

Sound perspective

At first sight it seems self-evident that the sound's proportions should match its associated picture. Close-up sound for a close-up shot; more distant sound for a long shot. This is a good working maxim. But there are exceptions where it is preferable to cheat this relationship and deliberately introduce nearer or more constant perspective than the actual shot suggests.

For example: the scene intercuts between shots of a bedroom where an old lady is dying, and shots of a street parade showing her long-lost soldier son's band returning from overseas. Audio cuts from the quiet death bed, to the blaring military band. Should we avoid the rapid volume jumps that would actually take place—and either reduce the differences, or spread the same sound over both scenes? The optimum treatment is a matter of taste, the effect you want to achieve, and the need to avoid ambiguity.

If a singer wanders around a spacious set, moving from close to long shots, do we really want the volume to vary? Can the mike maintain a constant relationship to the singer? It may be necessary to prerecord the song and have the performance mimed on camera to its playback.

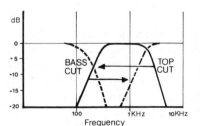

Fig. 10.23 part 2   High and low pass filters
Suppress rumble, hum, or high-pitched noises respectively.

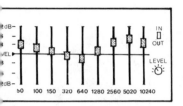

**ig. 10.23 part 3   Audio baton,
ctave filter, shaping filter**
series of slider controls boost or
duce chosen frequency segments
the audio spectrum. Slider posi-
ns show effective response.

We normally adjust sound perspective within a scene by altering the mike distance to suit the shot size by boom movement or intermicrophone mixing. As closer sounds tend to seem louder, with stronger bass and top than more distant pick-up, *presence filters* (*dialogue equalizers*) may be introduced to simulate or emphasize differences between close and remote sound quality.

Sound quality

Although high fidelity is our normal aim, there are good technical and artistic arguments for using audio filters to modify the system's frequency response when necessary:

1. Cutting low bass to reduce boom rumble, hum, improve speech clarity, improve the hollow or boomy quality in boxy scenery.
2. Cutting upper parts of the audio spectrum to reduce sibilant speech or S-blasting, quieten studio noise, reduce disc or tape noise.
3. Cut bass and top to enhance the illusion of exterior settings.
4. Slightly accentuate bass to increase the impression of size and grandeur of vast interiors.
5. Effects filters to simulate telephone (300–3000 Hz) intercom, public announcements. Phone filters (*distort*) are often controlled by the video switcher equipment, to provide automatic *audio-follow-video* switching as the picture intercuts between the two ends of a phone conversation.
6. By matching different types of mikes' frequency responses, we can reduce audible quality differences.
7. Quality matching between different sound sources (e.g. studio and film) can be improved.

Filtering can provide better audio continuity, and helps you to remedy such defects. Where the director is using *electronically inserted backgrounds* (e.g. by chroma key), or studio settings have inappropriate acoustics (e.g. a 'cave' constructed of timber and plastic), the audio quality can be made much more convincing by audio filtering and reverberation treatment.

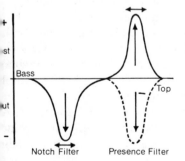

**ig. 10.23 part 4   Notch filter,
resence circuits**
otch filter puts very steep localized
p into audio spectrum to diminish
terference (e.g. hum) without sub-
antially affecting overall quality.
resence circuits (*mid-lift*) provide
oost or cut (up to 6 dB) around
electable frequencies (2 to 6 kHz) to
nprove clarity and audio separation,
nprove sibilants lost with distance,
r bring sounds 'forward' (presence)
om others.

■ *Equalization*   When disc or audio tape recordings are made, the frequency spectrum is automatically modified (top is boosted, bass is reduced). This anticipates or overcomes various technical difficulties such as background noise. There are several 'standards' in use. Strictly, therefore, a disc or tape should always be reproduced with compensatory equalization matching the standard used for recording (*Audio tape*: NAB, NARTB, CCIR, IEC, DIN. *Disc*: RIAA, Standard Spectra, ANSI, BSI, DIN, IEC, ISO, NAB, etc.). Some reproducers have selectable filters to suit a variety of standards.

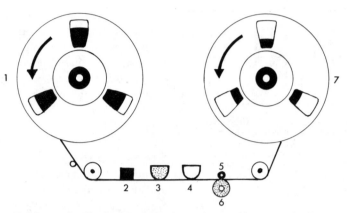

Fig. 10.24 part 1   Typical layout of a reel-to-reel tape recorder
1, Supply spool. 2, Erase head. 3, Record head. 4, Replay head. 5, Capstan roller. 6, Pressure idler roller. 7, Take-up spool.

Audio tape—recording and replay

Operationally, audio magnetic tape recording is simplicity itself. You have only to avoid over- or undermodulation, to achieve results virtually indistinguishable from the original.

■ *Audio tape*   Audio recording tape consists fundamentally of a magnetic coating (e.g. ferric oxide, chromium dioxide, metal tape) on a plastic base (acetate, polyester, PVC). Thinner base tapes enable more tape to be contained on a given spool size (therefore longer playing time), but at the risk of print-through between adjacent layers and greater vulnerability to damage.

■ *Recording*   Before recording, the tape must be magnetically neutralized (wiped, erased). This can be done initially by a bulk electromagnetic eraser (bulked), but is usually a regular part of the recording process. In the recording mode, the tape first passes an *erase head* where strong ultrasonic currents eliminate any previous recording. The result when played would be a barely audible rustling—*tape noise*. This *clean* tape passes against the *record head*, which is fed with the audio signal you wish to record. As each particle of the tape contacts the head's gap, it becomes magnetized in proportion to the audio at that instant. The modu-

Fig. 10.24 part 2   Splicing audio tape
After marking edit point at replay head gap, the tape is cut at 45° on splice groove. The tape is then butted together and adhesive jointing tape is attached to the *back* of the splice.

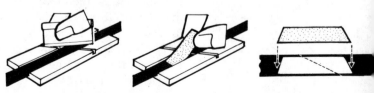

lated magnetic pattern is invisible. To assist this recording process the record head is also fed with an ultrasonic *bias current*, which substantially reduces distortion and background noise.

■ *Replay* The tape can be replayed (monitored) during recording and subsequently reproduced (and/or recorded on) several hundred times. To reproduce the tape, it is passed over a *replay head* (playback, reproducer); the tape's magnetic pattern inducing weak electric currents in the head, corresponding with the original audio. An interlock mechanism usually prevents your activating the recording circuit accidentally and wiping the tape during replay.

■ *Tape speed* The tape must be reproduced at the recording speed, or you will get a marked pitch change; fine for chipmunk effects, but otherwise to be avoided. Higher tape speeds provide a better high-frequency response, lower background noise, less obvious *drop-out* (momentarily silent gaps due to tape surface irregularities or damage), easier audio editing (the audio pattern is more spread out), but it uses more tape per minute.

■ *Editing* You can edit magnetic tape by physically cutting (45° diagonally, with an unmagnetized razor blade) and joining sections with an adhesive patch on the back of the tape. This quick flexible process produces a silent joint but disrupts the original recording, particularly if it contains several different tracks. Instead, you can rerecord (*dub*, *dub-off*) sections onto a new tape; but this can result in cumulative distortion, hum, background noise, etc. Colored non-magnetic tape is often interspliced as lace-up *leader*, or as mute *blanking* between sequences for identification.

■ *Production use of audio tape*
1. Providing a premixed track of music, effects, background, to simplify handling during the show.
2. Bringing back sounds from location shooting, to provide audio background to studio action.
3. To cover speech from unseen speakers (other end of phone, outside a door).
4. Providing taped commentary while a lecturer moves to another exhibit, or is out of microphone range.
5. Providing total prerecorded commentary to film or videotape voice over, voices off, VO; sound over vision, SOV).
6. Preselected effects or music dubbed from disc for cuing or treatment.
7. Recording atmospheric sounds (*room tone*), or other non-synchronized sound (*wild track*) for use when laying audio tracks during subsequent videotape editing. To provide overlays (bridging sound) between scenes.
8. Checking what was said during impromptu discussion (to guide editing).

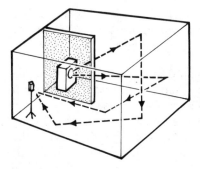

Fig. 10.25 part 1   Echo room
Sound to be treated is reproduced over loudspeaker in echoing room (hard walls, baffle)—reverberant sound being picked up by microphone and mixed with direct original sound.

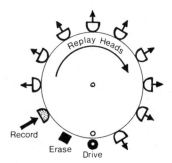

Fig. 10.25 part 2   Magnetic tape delay
Time lag as audio tape travels from record to replay head, can be used to produce a delayed reproduction of original—an echo when mixed with incoming sound. If this delayed reproduction is fed back into recorder input, *flutter echo* develops. The multi-replay head system shown here produces a succession of delays, which can be mixed with original sound or fed to series of studio loudspeakers to modify effective acoustics (ambiophony). The tape loop or magnetic disc records continuously, erasing each revolution.

Playback (foldback)

By reproducing sound effects or music over studio floor loudspeakers (*playback*, *foldback*) performers can relate or synchronize their cues or action. However, this reproduced sound becomes colored by studio acoustics, and if picked up by the action microphones it often has a falsely hollow quality.

■ *Miming and postsynchronizing*   The idea of having someone mouthing words silently to a recorded song is a familiar enough device. Where a singer is to be shot under unfavorable sound pick-up conditions (particularly on location with noise, wind, traffic, etc.), or where strenuous action impairs performance, you can either prerecord the song and have them mime to playback (*prescoring*), or shoot the performance and have them provide the song later (*postsyncing*).

Although complete operas have been performed in these ways—for good voice production and attractive character close-ups are not always compatible—exact *lip-sync* is not easily maintained over long periods. The accurate synchronism achieved in *dubbing* for motion pictures (*dialogue looping* in foreign language translations,.and regular postsyncing of exterior shooting) comes from repetition of brief sections. Any loss of lip-sync is distracting and may appear ridiculous.

Music and effects (M & E track)

As television has spread throughout the world, it has become increasingly profitable when taping productions to anticipate foreign-language markets. Arrangements are now often made to supply videotape or film programs without speech, but with a *music and effects track* (M & E) which includes all atmospheric sounds, effects and music. The purchasing organization then adds its required language commentary or dialogue, based on the original scripting.

Audio effects

You can modify audio quality in various ways to create or enhance a particular impact; for dramatic, environmental, or novelty effect.

■ *Filtering*   As we saw earlier, audio filters (equalizers) enable you to emphasize or reduce parts of the audio spectrum to any degree—either correctively or to modify quality. So you can make sounds seem shrill, edgy, thin, tubby, hollow, harsh, brittle, etc.

■ *Distortion*   To simulate the audio distortion from shortwave radio pick-up, public address systems, or intercom, you can hold a microphone near an earphone or earpiece, or small loudspeaker.

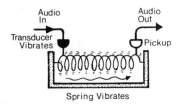

Fig. 10.25 part 3   Magnetic spring delays
The original audio signal causes transducer head to vibrate the spring(s). The mechanical resonances detected by pickup transducer head have a reverberant coloration and are converted into corresponding audio signal.

You can alter tonal quality further by placing the assembly in a box, jug or pipe.

■ *Acoustic feedback*   When a microphone's amplified output is fed to a nearby loudspeaker, and then overheard by the same microphone reamplified, a loud oscillatory howlround (acoustic feedback) can build up. This situation is a hazard when using public address/sound reinforcement systems. The cure is to reduce the loudspeaker output, move the mike or loudspeaker, use a more directional mike or loudspeakers; or even use digital processing, which moves successive repetitions away from the original frequency. As an *effect*, you may deliberately introduce controlled acoustic feedback to produce a hollow, echoing sound quality.

■ *Reverberation*   Artificial reverberation can be produced in various ways, and mixed with the direct untreated sound as required (Fig. 10.25). Certain digital processors have considerable audio-flexibility, including reverberation, delay, echo, spatial effects, and source duplication (chorus) facilities.

Audio effects using tape
Audio tape provides a valuable range of sound treatment opportunities.

■ *Feedback*   During recording, the tape output is controllably fed back to mix the replay sound with the input. The result can be varied from a drawn-out hangover effect, to distinct echo, or stutter effects (rapid multiechoes—flutter echo).

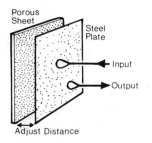

Fig. 10.25 part 4   Reverberation plate
A similar principle to part 3, using a suspended steel sheet as the vibrating medium. Adjusting the closeness of a parallel sheet of porous material modifies the reverberation time (1½–5 sec). *Digital processing systems* create reverberation effects by selectively switching electronic delay elements.

■ *Speed change*   Replaying at *slower* speeds than the original recording lowers pitch and tempo, increasing reverberation. Most sounds take on a deliberate, sinister character. Xylophones, bells, thunder, gongs, become strikingly powerful. Others are totally transformed (e.g. slow-speed wind or water) and can be used for 'synthetic' or atmospheric effects.

Reproduction at a faster than recording speed raises pitch and tempo. About one-and-a-half to two times, sounds become 'miniaturized,' appearing perky and comic—the audioland of singing mice, insects, household utensils, etc., of cartoon. Reproducers with special rotating multihead turrets can provide pitch changes at normal tape speeds, or correct the pitch for tapes reproduced at the wrong speed.

■ *Reversed sound*   Tape reproduced backward (a normal recording fed in tail first) can modify the character of sounds. Sustained sounds tend not to change, but percussive or transient ones have a characteristic 'indrawn-breath' quality.

Table 10.3   Disc and tape reproduction

	Disc	Tape
Advantages	Lightweight, easily stored (vertically) and occupies little space. Any section is rapidly selected and Disc is readily excerpted. A series of brief effects is easily extracted from several discs. Extensive library sources available. Two pick-ups can play one disc (for continuous repetition, echo effects, simultaneous play of different sections). Speed adjustment permits pitch change (for effects). Two turntables provide continuous playing, mixing, crossfades. Only two basic disc formats (plus obsolete 78 rpm format).	With moderate care, not easily damaged (but rapid stop-starts can cause tape snarl-up!) High audio quality, consistent throughout tape; permitting many replays. High reliability and durability. Tape length can be chosen to suit duration required. Recording can be revised, edited. Audio tape can be erased and reused. Background noise remains generally low. Tape recording is a simple reliable process. Process can be automated.
Disadvantages	Surface noise (static attracts dust, fingermarks) worsens with use. Playing life limited (lightweight pick-ups are not robust, vulnerable), distortion and defects increase with use. Quality deteriorates towards disc center (waveform compression). Disc is easily damaged (scratches, blemishes, warping). Pick-up stylus is easily damaged; can ruin recording. Equipment may have speed instability (wow, flutter); and rumble. Unwanted sections cannot be removed. Disc recording involves precision processing (results vary); or *direct recording* onto lacquer/acetate discs (short life).	Larger reels are bulky and pose storage problems. Recording is damaged by magnetic fields, heat, humidity. Tape is susceptible to damage. Tape *drop-out* can disrupt the recording. Tape is not rapidly excerpted. *Unwanted* passages necessitate play-thro, fast run-thro, editing out, or dubbing on to another recorder. Tape has to be rewound before reproduction. Various incompatible tape formats. Recorder heads must be demagnetized with small electromagnet to obviate background noise build-up.
Cuing	Requires skill (pick-up output is checked at a prefader/prehear monitoring point with channel faded out). *Methods*: 1. Play silent up to cue point, then fade in channel. 2. Play to cue point. Switch off turntable with stylus in groove. Hand-reverse disc to cue point, then back a further $\frac{1}{2}$ revolution (to provide run-up time at switch-on). When disc is required, switch on and fade up. 3. Use felt *slip mat* under disc. Find cue point as in 2 (but do not reverse $\frac{1}{2}$ rev.). Pinch-holding the mat, switch on turntable—disc remains still, turntable revolves. When required, release slip mat and fade up.	Precision cuing is simple. (Check on monitoring system.) Equipment provides immediate start at any required points. *Methods*: 1. Play to cue point. Stop. Wind back by hand to exact cue point. (Rear mark tape at replay point for future cuing.) 2. Wind tape to footage indicator and check precise cue point. Cue point seeking can be automated: metallic cue tag; photostop (wipe off magnetic coating with solvent—clear backing-gap stops tape); microprocessor memory. Identifying color leader can be spliced in.

Table 10.4 Audio tape and disc formats

Reel-to-reel

Uses tape 6·35 mm ($\frac{1}{4}$ in) wide, of any length up to maximum spool capacity.

Spool diameters 127 mm (5 in); 177 mm (7 in); 267 mm (10$\frac{1}{2}$ in). Capacity 760–1100 m (2500–3600 ft).

Tape speeds (per sec) 2·37 cm ($\frac{15}{16}$ in); 4·75 cm (1$\frac{7}{8}$ in); 9·5 cm (3$\frac{3}{4}$ in); 19 cm (7$\frac{1}{2}$ in); 38 cm (15 in).

Manual threading. Readily edited, and fitted with inter-section leaders, self-stop (foil or photostop).

Tracks 1 (full track); 2 (half-track); 4 (quarter-track) formats.

Running time 1 minute=11·6 m (38 ft)
30 minutes=342 m (1125 ft)   at 19 cm/sec (7$\frac{1}{2}$ ips)

Other lengths and speeds in this ratio.

Tape thickness Standard play—0·05 mm (0·02 in); long play—0·04 mm (0·0015 in); double play—0·025 mm (0·001 in); triple play—0·013 mm (0·0005 in).

Cassette

A small *enclosed* two or four track reel-to-reel system. Tape width 3·5 mm; speed 4·75 cm/sec. Self-threading. Rapid loading and instant start (no wow or slur on run-up). Runs to maximum duration of *twice* 30 min, 45 min, 60 min. (Turn over cassette at end of first play through.)

End of track stop may *not* include auto switch-off.

Not readily edited.

Cartridge (cart)

Its container holds an *endless* tape loop. (Tape width 6·3 mm; speed 9·5 cm/sec.)

Tape enclosed in a container feeds from inside a single spool, winding onto its outside.

Self-threading. Automatic rewind by *running on at normal speed* to the start.

Rapid loading and instant start.

Eight tracks can be recorded (or four stereo).

Duration of larger spools: 40, 60, 90 min.

Short-duration versions: 10-30 sec minimum; 4 min maximum, used for identification announcements, brief inserts of music, speech or effects.

Widely used for automatic replay of spot commercials in a *carousel* playing e.g. 24 carts in succession. Other equipment can play several carts simultaneously.

Disc formats

78 rpm Obsolete. *Coarse-groove* reproducing stylus. Mono only.

45 rpm Modern fine-groove (microgroove), reproducing stylus (tip radius 0·0005–0·0007 in), in mono or stereo forms. For wide-hole 45 rpm discs only, a special center adapter is necessary.

33$\frac{1}{3}$ rpm Similar to 45 rpm format. Available in mono, stereo and quad forms (several quad standards). Stereo discs can be played on fine-groove mono pick-up. Quad replays in stereo.

■ *Repeated sound* A tape loop enables you to repeat a soun
over and over again, for example to heighten tension, impl
hysteria or mental derangement.

■ *Fading* Slightly misguiding a tape during replay ca
simulate fading and top loss. Alternatively, two identical disc
played slightly out of step produces fading and distortion.

Synthetic sound

There are many fascinating program opportunities for synthesize
sounds including abstract background noises, voices for th
inarticulate or mute, newly created sounds for monsters an
science fiction.

■ *Treated sounds* You can use one audio signal to contr
another, modulating its volume or pitch by using voltage
controlled amplifiers, ring modulators, hybrid coils, etc. So yo
can articulate sound effects; a locomotive whistle cries 'A
aboard,' or an organ sings its own lyrics.

Taking fragments from existing recordings, you can modif
their quality by juxtaposing them in any order (reversing
modulating or repeating) to produce sounds quite dissociate
from their origins. The resultant *musique concrète* may have
musical character (for jingles, title music) or provide imitative o
humorous noises.

■ *Dissociated sounds* Many sounds lose their identity whe
divorced from their original context. Some are unfamiliar anywa
(straining pack-ice, roaring gas flames, insect noises). You ca
often transpose sounds for dramatic effect—jet engine noise t
accompany typhoon scenes. Some take on an entirely nev
significance—a crushed matchbox simulating a collapsing barn.

Musical instruments can be used to produce a fantastic rang
of sound effects—disembodied waverings from a musical saw
squeaks, grunts, whines and creaks from a violin. These noises ca
in turn be treated and modified. Such devices are more than audi
stunts. They enable you to stimulate the imagination, ofte
improve on available sounds, and to materialize the unknown o
the impossible.

■ *Synthesized sounds* Entirely synthetic sounds can be create
electronically by an *audio synthesizer*, which enables a keyboar
or interconnected control circuits to manipulate specially gener
ated and processed sounds. Oscillations of various shapes
frequency, repetition rate, can be selected and treated; volum
can be varied (*vibrato*, *crescendo*, *diminuendo*); build-up and die
away adjustments can be used (attack and decay times); qualit
can be modified (selective filtering) and so on. The resultan

synthesized sounds may deliberately simulate recognizable sources, or provide a new audio impression.

Dolby noise reduction

Very quiet sounds can merge all too easily with inherent background noise (hiss, hum, rumble). A sophisticated companding process developed by Dolby is now extensively used in audio recording, to reduce spurious background noise further than other methods (audio control, equalizers).

In the more complex *Dolby A* system used for professional recording, the audio spectrum is filtered into four frequency bands. Each is individually treated, boosting very quiet audio signals while leaving louder ones virtually unaffected. When decoded on replay the relative audio levels are restored, the background noise being proportionately reduced.

In the simpler *Dolby B* system widely used for audio cassette recordings, quiet high frequencies are boosted on recording (the frequency at which boosting begins, depending on the volume at that moment). On replay levels are restored to normal, suppressing tape hiss correspondingly.

The value of the fixed bias current used in audio magnetic recording is normally a compromise between introducing distortion at the higher or lower ends of the audio spectrum. (Lower bias improves top, but worsens bass distortion, and modifies the system's overall response.) The Dolby *Headroom Extension System* (Dolby HX) adjusts the tape bias automatically, according to the high frequency energy content of the program, to produce lower distortion levels overall.

11 Film

Although it can be difficult to discern on the screen whether program originates from film or television cameras, there are very practical differences in the production mechanics of these media. Paradoxically, as the TV system becomes more flexible the edges blur between these differing techniques.

Basic filming process

Film making involves a series of processes, each one hopefully taking into account later stages—the cameraman shooting in anticipation of the film editor's continuity problems, days or weeks later.

It is not possible to examine results while filming and you have to wait for proof prints (*dailies, rushes*) to show whether action camerawork, lighting, etc., was successful. Occasionally, a small TV camera is fitted to the film camera (electronic viewfinder) providing an *approximate* monochrome check on a nearby TV monitor. Otherwise only the camera operator can know, for example, whether the mike dipped into shot—and he might have overlooked it while preoccupied with the action. As a result of retakes to correct faults or errors and selective editing, about 1/10 to 1/20 of the overall film exposed is finally used (a 10:1-20: *shooting ratio*). To save processing costs faulty takes (NG) are not printed.

Synchronized audio may be recorded separately on a portable magnetic recorder; or in the sound film camera itself, on a magnetic stripe alongside the picture. Entire scenes are often shot with a *silent camera*, the eventual audio track being devised later in the dubbing suite.

Fig. 11.1   Basic sound filming process
Picture: Film is shot in a series of separate camera set-ups (re-takes, multi versions); clapper-board being shown at head or inverted at the tail of each take (for identification and sound/picture sync mark). Potentially usable takes are processed and a *Master negative* is obtained. An uncorrected positive check print is made (*rushes, dailies*). This is used for review and, later to devise editing *cutting copy*. *Audio*: Location sound is recorded on $\frac{1}{4}$ inch magnetic tape. It is later dubbed selectively onto full-coat sprocketed magnetic film, to correspond with cutting-copy picture selections. *Editing development*: 1, All information on material is logged (edge numbers, take numbers, descriptions, lengths, footages). 2, Picture and sound-track is then synchronized. 3, Sections are selectively assembled (rough cut, work print). 4, Revision, exact cutting prints and transitions are chosen (cutting copy, fine cut). 5, Various additional sound tracks are prepared or assembled (commentary, music, effects). 6, All sounds are then dubbed (using *dubbing mixer*) onto one mixed master soundtrack to match the edited picture. 7, Master negatives are prepared by the laboratory which correspond to editor's cutting copy (with his wax-pencil code marking the transitions required). Master negatives edge-numbers are used for section, frame identification. 8, Specimen test-strip print-sections supplied by laboratories (timing, grading, cinex), for approval of processing color balance, exposure variations, corrections, etc. 9, Final print: This is either a *combined print*, which has the picture united with optical or magnetic sound track; or is *double-headed*, having separate picture and magnetic track, which are run synchronously.

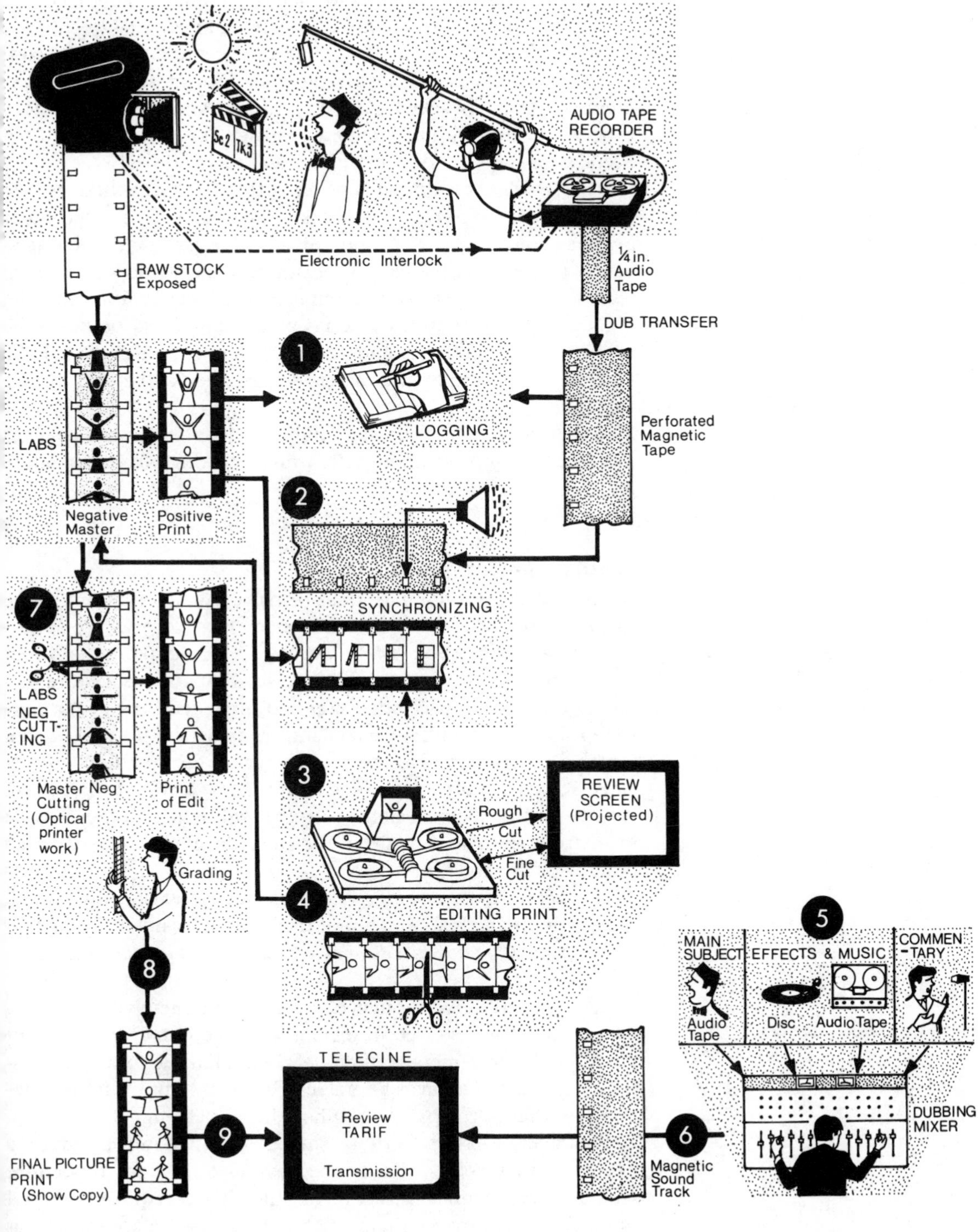

AUDIO TAPE RECORDER

Electronic Interlock

RAW STOCK Exposed

¼ in. Audio Tape

DUB TRANSFER

1 LOGGING

LABS

Negative Master

Positive Print

Perforated Magnetic Tape

2 SYNCHRONIZING

7

LABS NEG CUTT-ING

Master Neg Cutting (Optical printer work)

Print of Edit

3

ROUGH Cut

Fine Cut

REVIEW SCREEN (Projected)

Grading

4 EDITING PRINT

5

MAIN SUBJECT

Audio Tape

EFFECTS & MUSIC

Disc

Audio Tape

COMMEN-TARY

8

TELECINE

Review TARIF

Transmission

9

FINAL PICTURE PRINT (Show Copy)

6 Magnetic Sound Track

DUBBING MIXER

Scripting

There are several approaches to shooting a film, ranging from the hopefully empirical to the meticulously planned.

In *unscripted* off-the-cuff shooting, the camera tends to record available events; perhaps working from a rough outline (*shooting plan*). Material is selected during editing and compiled into a program format with added commentary. This technique is often used when shooting documentaries and news stories.

Fully scripted productions on the other hand, are first broken down into their component shots or scenes, and the mechanics involved for each shot or group of shots are assessed. The staged action is then methodically shot out of story sequence (*shooting schedule/plan*) for maximum economy and efficiency. For example, if the story line requires a series of intercut shots between the heroine at the cliff-top and the hero below, you shoot all the sequences involved at one camera set-up, then move to the other viewpoint.

This 'out of sequence' shooting prevents unnecessary changes in camera set-up, location, relighting, resetting, and uses actors more economically.

Single-camera shooting

Single-camera shooting is the normal method of filming—the main exception to this being the once-only event where simultaneous multiviewpoints are required, e.g. derailing a train, explosions, collisions, etc. This discontinuous procedure involves out-of-sequence shooting, and often requires performers to repeat their action for each change in camera viewpoint or shot length (and remember, their performance is probably out-of-story-sequence anyway). That introduces *continuity problems.* Action, gestures, expressions, costume, lighting, etc., must match in intercut shots and relate to the story development—although filmed at different times.

■ *Flexibility of viewpoint*   After a sequence is skilfully edited, the continuity appears so natural that one can easily overlook that it was originally shot as *a disjointed series of individual set-ups.*

In a typical filmed sequence, we may find ourselves close to a person as he walks—yet an instant later see him from a distant viewpoint. The camera is often 'left behind' as he moves away—yet is at his destination as he arrives. Shots intercut from his viewpoint, to that of the person to whom he is speaking.

The overall effect here is smooth-flowing and provides a variety of viewpoints. To do the obvious and shoot this entire sequence as *continuous* action would have required several cameras, spread over an appreciable distance—an approach that might not be practical anyway. Usually it would not provide the same effect.

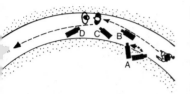

Fig. 11.2 Single camera shooting
To achieve apparently smooth-flowing action *with one camera*, a series of camera set-ups are often necessary. *Example*: A, Girl out walking (comes up to camera . . . which pans as she moves past). B, Cut—she sees friend in distance (over shoulder shot). C, Cut—CU of friend smiling. D, Cut—friend's point of view. Girl arrives, speaking (LS to CU). D, Intercut close-shots during conversation. C, Girl says good-bye (MS). E, Girl walks on. The action taken for each camera set-up will be enacted/repeated out of sequence to suit each camera set-up in turn.

Filmed shots tend to be of briefer duration than those in continuous TV production. Watching a well-made film, we see that most shots (the period between picture transitions) last only a few seconds (typically 3–15 seconds). A good serial film unit may achieve as many as 120 shots a day (about 18 minutes running time), although this is untypical. A live TV production might last 25 minutes, and contain 100–200 shots, edited 'on the hoof'.

Production decisions

■ *Film making* In *film production*, the director works in very close proximity to his single-camera crew. He normally discusses shots, treatment, action, with the cameraman, who then interprets these instructions. The director will only see the results later, and must rely on his judgement and his crew. The brief action segment is rehearsed, shot; and then retaken if necessary. Additional cover shots (protection shots) may be taken, 'just in case'.

Positive prints are made of the good takes (checked as the *dailies/rushes*), which the film editor separates, examines, selects, relates and interconnects. He uses a script or brief, cue-sheet and *continuity-report* information (forms giving action detail, slate numbers, etc., identifying each *take*). So he forms the individual clips of film into a cohesive and appropriate continuity (the *cutting copy*). The editor evaluates processing and grading (print density and color quality). Having decided on shot relationships, cutting points and suitable transitions, the editor submits his *rough-cut* for the director's approval. Subsequently he may modify the precise cutting points, tightening scenes, making final revisions to produce the final *fine-cut*.

Later the processing labs match the stored master negatives to his cut version, using the edge-numbers printed along the original film edge to guide them. An optical print is then made, using the transitions he has wax pencil codemarked on his cutting copy.

The sound track is compiled (*laying tracks*) and the *dubbing mixer* skilfully blends dialogue, commentary, effects and music. A *rock-and-roll* facility enables picture and sound tracks to be run synchronously forward/backward for rapid correction during dubbing.

There are opportunities to examine and reconsider at each step (time permitting!), and to remedy or improve results. Each stage in this exacting process paradoxically leads to a product with seemingly absolute freedom in time and dimension.

■ *Television techniques*   For the *TV camera* to achieve this degree of flexibility, it has to be used segmentally, in exactly the same way. It is not surprising, therefore, to find that many TV directors are aiming to do just that, now that more mobile facilities are available —despite the complexities of continuity and postproduction editing that consequently arise.

In practise, TV production approaches now range from the discontinuous 'near filmic' shooting alternative material for later postproduction decisions), to continuously performed multi camera presentations that provide a completed product at the end of the taping session.

For the most part, tight studio scheduling means that the majority of production decisions have to be made by the time you are due to videotape the program. Most transitions (cuts, mixes) are introduced while taping. Retakes are essentially corrective rather than experimental. So scenes are seldom reshot with several variations of action or treatment, for subsequent selection in editing. Background sounds (music and effects) are mainly included during the take. The greater time pressure of TV studio production usually precludes the opportunity for polishing precision and experiment, that film making offers. There is little use of postsyncing. The elaboration of postproduction editing dubbing in television depends largely on available facilities.

Uses of film in TV

Film has always had a regular place in TV studio production. And although videotape is progressively used for similar purposes, it remains a convenient medium for many applications:

■ *As a total presentation*   Studio talent introduces or provides commentary for a predominantly filmed presentation. Audio comes from studio and from the film's soundtrack.

■ *As illustrative material*   Film is often used to illustrate lectures, demonstrations and talks. It may be specially shot for the program, or *library (stock) shots* may be used. The latter are obtained from organizational archives, or rented from specialist libraries. These short film clips of news events, locations, processes, personalities, etc., provide both historical records and sequences where it would be impractical or uneconomical to shoot new material.

The picture quality of library shots can be variable in color blemishes and definition—they may not cut into studio shots unobtrusively. Also, available library shots can become too familiar, particularly when they are the only record of an event.

Where only monochrome shots are available and the program content is appropriate, these clips may be tinted (sepia, blue, red-orange). This is achieved by using a color synthesizer, by adjusting color balance, or using 'tarif' controls. Snow scenes and night shots may still prove convincing in black and white, although transmitted in a color system.

Library shots are often run mute (silent), and new effects or music substituted for any original soundtrack.

■ *To extend action* By using a recorded insert, we can have a person leave an apartment (studio set) and walk into the street (on film). The technique is used regularly in TV drama productions. Continuity can be important both visually and for audio treatment (background sounds). Location wild-track may be used in the studio, or 'dropped into' the final videotape during dubbing/editing. However, the wise director keeps matching needs to a minimum.

■ *To extend settings* Shooting a *partial setting* in the studio (e.g. a shop doorway); we cut to the complete street location in the film insert.

Fig. 11.3 Location filming
The location film team needs to be self-sufficient, anticipatory, well organized and adaptable.

■ *Reducing the number of studio settings* Where studio space i
too limited for the required number of settings, additional scene
can be insert-filmed.

■ *Once-only action* Prefilming or videotaping is essentia
where action might prove unsuccessful in the studio (for exampl
action involving animals); or take too much time during th
videotaping period (an elaborate make-up change); or is to
dangerous (fire); or is non-repeatable (an explosion); or is ver
critical (an accurately thrown knife).

■ *Visual effects* To produce time-lapse effects, reversed action
transformations, etc.

■ *Animation sequences* These may be cartoon or animate
still-life.

■ *To include performers not otherwise available* Filming person
who could not attend the taping session (e.g. overseas guests).

■ *Situations where effective studio staging is impractical* Filming

Table 11.1   Typical equipment for small-unit filming

16 mm camera lens	Arriflex, Eclair, Auricon Cine-Voice, Bolex.
	Zoom e.g. focal length 12 mm–120 mm (hand-crank or rod-operated).
	'Normal' angle at 25 mm.
	Or turret lenses—12, 16, 25, 50, 75, 100, 120 mm focal length.
Camera support	Body brace, shoulder brace, shoulder pad.
	Light tripod, with triangular spreader (spider).
	Sometimes a small lightweight dolly, for dollying (tracking) shots.
	Lens angles below 12°, i.e. focal length greater than 50 mm require a firmly supported camera.
Shooting duration	Maximum 11 minutes with 400 ft (122 m) magazine.
	Battery-powered camera runs typically $1\frac{1}{2}$ hours (3000 ft of film).
Lighting equipment	Small portable lamps on camera or hand-held (e.g. Sun-gun, Frezzi). Lightweight lamps, Red-heads, Lowell-lites, Colortran quartz lights; quartz broads, Mini-brutes; Photofloods in small scoops; reflector boards.
	Lamp supports—hand-held, gaffer-tape attached, screw-clamps, gaffer-clamps (spring-jaw clamps), clip-lights, lightweight tripod stands, telescopic spring-loaded support poles (polecat, barricuda) fitted ceiling-to-floor or wall-to-wall.
Microphones	Lanyard, electret or baton personal mike (cable connected, or wireless mike)
	Sound men with fishpole or shotgun (rifle) mike.
Sound recorder	Lightweight portable synchronized magnetic audio recorder (e.g. Nagra), earphone monitored.
	Synchronous sound using a camera-to-recorder sync connection (cable or wireless), crystal lock or time code.

location exteriors, spectacle, etc., where studio replicas would be impractical or less effective.

Sound on film

■ *Double-system* Using this method the picture and sound are recorded separately; so providing maximum editing flexibility. However, apart from general background sounds, close synchronization is normally essential between the film camera and its associated audio tape recorder; for speech we have to maintain exact *lip sync*.

To achieve precise synchronization, camera and recorder may be connected by a special sync-pulse cable, but this is physically restrictive. Instead, matched crystal controls can be fitted to independently maintain accurate motor speeds; or alternatively, wireless control interlocks or an edge time code may be used.

At the start of each *take*, an identifying *head slate* is filmed, giving all reference details to aid eventual editing. By matching up the sound and picture at its small clapper-board closing (*sync marks*), we ensure that film and soundtrack are in step for that take. Automatic slating facilities are sometimes fitted—these 'flash' a frame and bleep the soundtrack simultaneously. Alternatively, a *tail slate/end board* (upside down) is shot at the end of a scene.

■ *Single system* This method, frequently used for TV news filming, records the audio beside the picture on a magnetic *track*. However, the audio's 28 frame separation (in advance) from its corresponding picture (about one second difference), means that editing the picture will cut the audio too—but not at the same point.

An audio track may be recorded *optically* on film, using variable area or variable density light modulation systems. This optical track is printed photographically along one edge of the released film, in a *combined (married) print*.

Many TV organizations keep picture and magnetically recorded sound separate throughout, and reproduce them in synchronism (*double-headed*) in the telecine equipment.

Fig. 11.4   Film sound
Optical track—combined (married) print. 1, Variable area track. 2, Variable density track. *Magnetic track*—3, Magnetic stripe. Balancing stripe facilitates flat smooth roll wind. 4, Mute picture, separate 16 mm perforated magnetic film.

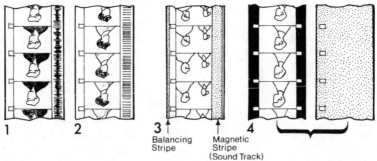

1   2   3   4

Balancing Magnetic
Stripe   Stripe
(Sound Track)

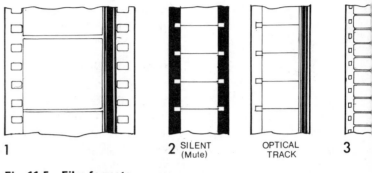

Fig. 11.5   Film formats
1, 35 mm gauge. 2, 16 mm. 3, Super-8.

Most TV organizations use 16 mm film almost exclusively; a gauge that provides an optimum balance between running cost and good picture quality. Equipment is reasonably lightweight especially when compared with costly, bulky 35 mm cameras. Editing, which becomes progressively difficult with smaller gauges, is still reasonably easy; although reliably unobtrusive splices require checkerboard (A–B roll) cutting.

In a search for further economy, a Super-8 format with magnetic stripe soundtrack is increasingly used; despite its inferior picture quality. To simplify editing, the film may be transferred to videotape; the stripe soundtrack being dubbed onto a separate magnetic (sep-mag) film.

Transmitting negative film

After the *negative stock* used for most filming is processed, a *positive print* is made. The original negative itself is generally considered too precious to risk damage through continual handling and projection, for any blemishes or dust (reproducing a white 'sparkle') can irremediably deface the picture.

However, to save the cost and time involved in printing, the negative film (electronically reversed) may be transmitted, e.g. for urgent news stories. The important compensatory color and exposure corrections (*grading*) normally made during printing have instead to be introduced operationally when the film is reproduced (on sight, or using programed electronic correction).

Negative film is sometimes used when dark or dense scenes are to be transmitted from a vidicon telecine system; this method overcomes lag (trailing) problems otherwise arising from low light conditions.

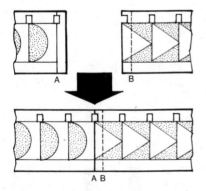

Fig. 11.6   Joining film
The two shots to be joined, are cut in the splicer and connected by: a *temporary splice* using translucent adhesive splicing tape (butt join), or a *permanent overlap weld* using film cement, (scraping off film emulsion for overlap join), or a *heat butt-welded splice* along frame-line (used for negative editing).

Reversal film

Film emulsions designed for *reversal*, produce a normal positive image with natural colors and tones on the original camera film

240

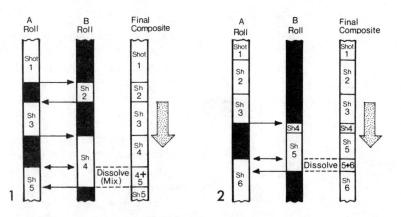

Fig. 11.7 Checkerboard and A/B roll assemblies
When printing narrow-gauge film, material is usually distributed between two rolls A and B, interspaced by blanking (black leader). By inter-switching/ dissolving between rolls, optical combinations (dissolve, wipe, etc.) and invisible splices (not intruding into picture area) become possible. The system avoids intermediate duplication stages for combined effects. 1, For *checkerboard assembly*, the output changes between A and B rolls at each cut; except where an overlap is required for a combined effect (dissolve, wipe, etc.). 2, *For A/B roll assembly*, shots are spliced consecutively on one roll, until an overlap is required for a combined effect. The output then remains with the other roll for successive shots until the next combination is required. These techniques are also used for rapid excerpting from complete programs (film and VT).

There is no negative involved. No print is necessary, so processing is quicker.

The advantages of reversal film include higher available sensitivity (*speed*), finer grain, improved sharpness, and better color reproduction. Against this, the exposure range is very limited, which may produce crushed highlights and shadows. Also, there is less opportunity to correct exposure and color errors. Most important, the original film is itself projected and is vulnerable. So a print may be needed, anyway, to provide a cutting copy for editing in order to preserve the master film.

Mechanics of film reproduction
We use film in television production in several forms:

1. As a *reel* of 1 minute to greater than 2 hours duration.
2. As a series of brief-duration sections (*clips*) joined by opaque *blanking*, or *cue leaders*.
3. As closed *film loops*, joined tail-to-head to provide continuous repetition (mist, snow, rain), or repeated cyclic motion (e.g. to study machine movement) using a centreless spool or take-up pulleys.
4. As short *film strips* containing a series of different still frames.
5. As individual *film slides/transparencies*—televised by a transparency projector (picked up by a TV camera), or a special slide scanner.

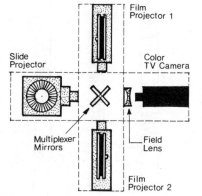

Fig. 11.8   Multiplexed telecine
A mirror or prism system enables one TV camera to selectively shoot a slide projector or either of two film projectors. The assembly is mounted on a film island.

Slide Projector

Film Projector 1

Color TV Camera

Multiplexer Mirrors

Field Lens

Film Projector 2

Telecine

The European television picture rate is 25 pps (50 Hz power supplies). If standard films (24 fps) are projected at this slightly higher rate (4%) there is no discernible speed-up. However, film shot specifically for European TV is normally exposed at 25 fps instead of the motion picture standard.

■ *Camera-type telecines*   This arrangement is extensively used in North America. On a *film island* installation, a standard TV camera faces a mirror or prism multiplexer system. Through this it can see selectively, the output of two film projectors or a 2 × 2 inches slide (transparency) projector.

A storage-type camera tube (vidicon, plumbicon) is used to produce the video signals, coupled with a film projector exposing each film frame for $2\frac{1}{2}$ TV fields. This compensates for the discrepancy between the standard film projection rate of 24 frames per second, and the standard TV picture rate of 30 pps (related to 60 Hz power supply frequency) in USA and Canada.

The intermittent projector moves the film in rhythmic 'pull-and-hold' jerks, to freeze each frame momentarily as it passes the picture gate; so giving the TV camera time to discern successive pictures clearly. The resultant video signals are then adjusted and processed.

■ *Flying-spot scanners* are widely used, and are capable of the highest quality film reproduction. The film travels at a constant speed (*continuous motion*) through the telecine equipment. On a special picture tube a constant-brightness raster is scanned. The focused image of this scanning light pattern explores the moving film. The resultant fluctuations, as the varying film densities interrupt the scanning light beam, are picked up by photocells. Their modulated electrical signal forms the telecine's video output.

■ *Digital telecine* systems use three full-line sensors (RGB) to scan the passing film. These charge-coupled devices (CCD) analyse its light variations. The resulting charges are held in a digital frame store to be read out as the video signal. This system provides flicker-free slow motion, jog and still frame pictures.

Video control for telecine

The video signals from any type of film scanner, as for regular TV camera channels (Chapter 21), are monitored and controlled. This is achieved by using manual adjustment, autocontrol circuits, or automated programs on control tapes (punched paper or magnetic) for picture correction.

Film is capable of superb color reproduction, but it is not an entirely consistent material. We shall find some variations in color balance between different batches of the same film stock; and between different types of color film. Moreover when filming,

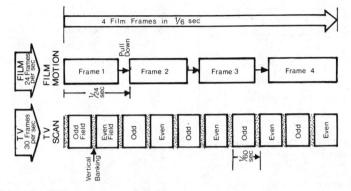

Fig. 11.9 Picture-rate conversion
To relate the standard film projection rate (24 frames per sec) to TV standards of 30 pictures (USA) per sec, several sampling methods are used (e.g. project one film frame 3 times, the next twice; fast film pulldown with light-source pulsing 5 times per film frame/4 times per TV picture).

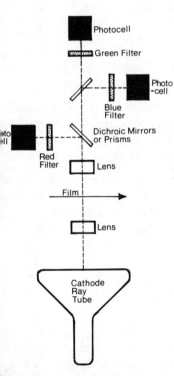

exposure or color temperature may not have been accurate or consistent. Where film has aged, or is a copy of the original print (dupe), further changes arise. So in printing, film labs adjust the intensity and color balance of their printer lights to compensate empirically for discrepancies (grading).

However, when televising a film, such errors are more apparent than in theater viewing, where eye adaptation and subjective effects tend to compensate. Consequently, a film that may appear satisfactory when theater-reviewed beforehand, often requires further compensations when televised. Where the film has not been finely graded, these may be considerable.

By adjusting video gain, black level (sometimes gamma), color balance, and saturation, considerable improvements become possible. Color correction may also be introduced by *electronic masking circuits*, compensating for film dye deficiencies, and certain *analysis errors* arising when televising color film. Video gain and gamma adjustments to the RGB color channels ('TARIF' operation) provide good correction for various errors. Similar correction can be applied to slide (transparency) scanners.

Non standard production

Depending on design, telecine equipment may be able to provide several kinds of non-standard projection:

1. Hold a selected single frame (*still frame, hold frame, freeze frame*).
2. Dissolve slowly from one frame to the next (*inching*).
3. Run up from still frame to full speed (mute) without frame-roll or flicker.
4. Run at non-standard constant speeds (e.g. 16–60 fps)—forward or backward.

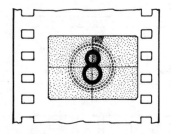

Fig. 11.11 TV film sync leader
Several types of film leader are in use. In the American National Standard leader shown (35 and 16 mm motion-picture theatre, and TV studios), a *start identification frame* ensures exact picture sync with the sound-head. A series of numbers from 8 to 2 (seconds) appear (a black background then a dark wedge moving clockwise in series of 15° steps) provide a visual count-down.

Fig. 11.12 Changeover synchronizing
At the appearance of a white (or black) circular dot in the top right corner of the picture area, the second projector is started (motor cue). After 7 sec, a similar dot cues the change-over switch-point. This mark is printed or punched in the print (during processing, or editing).

5. Fast wind-on (run-on) and rewind.
6. *Remote manual cuing* (from the production switcher), or *aut matic cuing* using attached metalized cue tags or tape strip, by microprocessor selection.

Cuing film

Telecine equipment cannot normally provide an instantaneou start for picture and audio. A brief run-up period is necessary fc audio to stabilize. By attaching a time-scaled *sync leader* (e SMPTE universal leader) to the head of a film, you can ensure tha the first picture frame with fully stabilized audio will appear predictable time after switch-on. The required picture syr marker is positioned in the projector's picture gate, the directc giving an anticipatory run cue 4 (or 8) seconds before cutting film channel output. This avoids inadvertently switching to stationary or unstable channel. To ensure that the film reprodut tion is neither clipped at its beginning, nor runs out immediatel after action finishes, a few *buffer (safety) frames* of the scene a included at the head and tail of each section.

Where several film lengths are conjoined, you can:

1. Use timed lengths of blanking between them; leaving th machine running while cutting away to studio (*roll-through*).
2. Introduce a series of sync leaders; recuing each section.
3. Cue the film at any time, switching to it when it appears.

Having previously ascertained the exact 'timing' (duration), th *in-cue* (first words) and *out-cue* (last words), precision insertic into the program becomes practicable. Occasionally, the exat entry/exit points for a videotaped film—insert are determine during subsequent VT editing.

Recording color TV on film

Before the advent of magnetic videotape recording, the monc chrome TV image could only be preserved by directly filming th face of a picture tube (Kinescope). Despite endless technic problems, remarkably satisfactory results became possible usin such *Kinescope recording (telerecording)* methods. Color recordin on film presented an even greater challenge; particularly relativ to the impeccable quality achievable with videotape systems.

However, film recordings have the particular advantages i world distribution, of overcoming the various TV standard differences that frustrate videotape sales. Films also perm direct large-screen projection and have wider education applications.

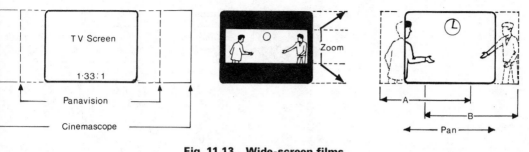

Fig. 11.13   Wide-screen films
Wide screen motion pictures do not use the TV screen's aspect ratio (4×3 shape; i.e. 1·33:1). If the picture is fitted to frame side-edges, so seeing all of film, overall picture size is small and areas are wasted at top/bottom of TV screen. If the *top/bottom* of the frame is fitted to the TV screen, then the sides of the film picture are lost. Solutions to this problem include zooming to full height, panning left/right to follow main action (top-fit) as shown (A to B). These operations involve scanning circuit adjustments (on-air or pre-coded), or compensated film prints.

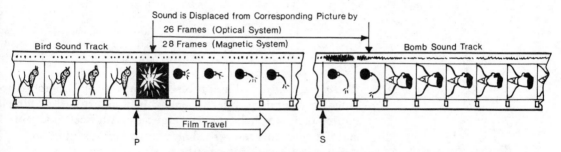

Fig. 11.14 part 1   Cutting combined prints—problem
In a combined print (optical or magnetic stripe) the sound is displaced ahead of its corresponding picture (to suit the picture-gate to sound-head separation distance). Physically cutting this combined print creates a dilemma: If the sound track is cut at a specific point (S), then over a second (26-28 picture frames) of unwanted picture is included in the selected film. When editing continuous speech, this can result in *lip-flap*, i.e. mouth movements left from the previous material. Cutting the sound-track (S) after the explosion will lose the picture of the event; while cutting on picture (P) will spuriously introduce sound from the next sequence.

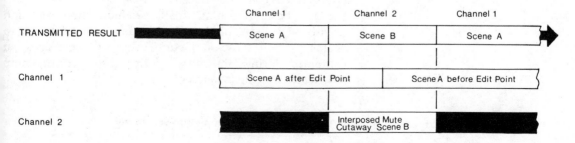

Fig. 11.14 part 2   Cutting combined prints—solution
A standard method uses a *double chain* or *A/B system*. A brief silent cutaway shot (reaction shot) is interposed to hide any lip-flap, jump-cut, etc., by strategically switching away.

Table 11.2 Film running times

| | | 35 mm (16 fr per ft) | | | 16 mm (40 fr per ft) | | | Super 8 mm (72 fr per ft) | | |
		24 fps	25 fps		24 fps	25 fps		24 fps	25 fps	
		ft	fr	ft	fr	ft	fr	ft	fr	ft	fr
Seconds	1	1	8	1	9	0	24	0	25	0	24	0	25
	3	4	8	4	11	1	32	1	35	1	0	1	3
	5	7	8	7	13	3	0	3	5	1	48	1	53
	7	10	8	10	15	4	8	4	15	2	24	2	31
	10	15	0	15	10	6	0	6	10	3	24	3	24
	30	45	0	46	14	18	0	18	30	10	0	10	30
Minutes	1	90	0	93	12	36	0	37	20	20	0	20	60
	5	450	0	468	12	180	0	187	20	100	0	104	12
	10	900	0	937	8	360	0	375	0	200	0	208	24
	15	1350	0	1406	4	540	0	562	20	300	0	312	36

		Seconds					
Feet	1	0·7	0·6	1·8	1·6	3·0	2·9
	3	2·0	1·9	5·0	4·8	9·0	8·6
	5	3·3	3·2	8·3	8·0	15·0	14·4
	7	4·7	4·5	11·6	11·2	21·0	20·2
	10	6·6	6·4	16·6	16·0	30·0	28·8
	15	10·0	9·6	25·0	24·0	45·0	43·2
	20	13·2	12·8	33·3	32·0	1 min 0·0	57·6
	30	20·0	19·2	50·0	48·0	1 min 30·0	1 min 24·0
	50	33·2	32·0	1 min 23·0	1 min 20·0	2 min 30·0	2 min 24·0

■ *Image transfer* The process of transferring programs from video to color film, offers even further opportunities; for no motion pictures can be shot economically with electronic cameras and the videotape rerecorded film for theater distribution.

Several image transfer methods are current:

1. A special film camera (using color film) shoots a color TV picture tube.
2. A similar set-up, filming a monochrome picture tube which displays in three successive runs, the red, blue, and green component images from a videotape. The three separate monochrome recording of the color information are combined to full color during laboratory printing.
3. *Laser beam* recording, in which three separate video-modulated lasers (RGB) combine optically on *color* film.
4. Electron beam recording (EBR) produces three separation negatives (RGB) on black-and-white film, scanned by electron beams akin to those in picture tubes. These records are combined during laboratory printing, using corresponding red, green, blue light sources.

Table 11.3 Displacement of picture and sound

	35 mm	16 mm	Super 8 mm
Magnetic track	28 frames behind pic.	28 ahead of pic.	18 ahead of pic.
Optical track	21 frames ahead.	26 ahead.	22 ahead.

12 Videotape

Videotape recording (VTR) is technically an extremely sophisticated process. But, happily for the user, equipment has become increasingly reliable, more compact, and operationally simpler.

Production flexibility

There are several approaches to recording program material:

1. *Continuously*—An uninterrupted record of events—'live on tape'. Minor corrections or revisions may be made by post-production editing.
2. In sections—A scene or act at a time.
3. In segments—Short takes, shot-by-shot; even one frame at a time, for animation build-up.
4. Changing time scale—Providing fast, slow, reverse, stop action, or time-lapse recording.

Many organizations record in duplicate, to provide a *back-up copy* (*backing, insurance copy*), in case technical faults develop in the master recording.

Problem of standards

Although audio signals can be recorded directly, *along the length* of a magnetic tape moving at a reasonably economic speed, this is not feasible with the much higher frequency spectrum of a video signal. Consequently, various ingenious systems have been developed, in which the video (transformed into an FM signal) is recorded *across* the tape as it moves. This produces an effectively high 'writing rate', at quite low economic tape speeds. The earliest effective videotape recording process was the *quad* (*quadruplex*) method, which was taken up by broadcasters to become the standard television recording system.

However, a simpler, cheaper and more transportable type of equipment was needed. So an alternative design approach— *helical scan* (*slant track*)—was developed, in several forms. These portable recorders proved suitable for many purposes, but picture quality was not as good as the more complex static quad installations. Moreover, although technologic developments have now made stable pictures of a high standard possible, designs still remain incompatible and helical scan recordings do not follow a universal format.

The major advantages of VTR

These advantages can be listed as:

1. At best, videotape quality is indistinguishable from the original video.
2. Immediate playback after recording. Sometimes monitoring (confidence playback) is possible during it.
3. The tape can be reproduced many times without deterioration.

4. The recording can be erased (*wiped*) totally or specifically, and the tape reused for another program.
5. Videotape can be edited; faulty or unwanted material being replaced by new. Performance impact can be improved by appropriate deletions (*cuts*) and retakes. Long pauses, ineffectual material, or lagging pace can be corrected— usually by cutaway shots disguising editing-out. *Continuity acceptance* (censor) cuts can be made unobtrusively.
6. Videotape facilitates scheduling. Program material can be recorded at convenient times and reproduced when required.
7. Taping can compensate for limited camera facilities; providing apparent mobility of viewpoint as separate recordings are edited together.
8. Problems of staging, time, and space can be overcome.
9. Any visual effects, including electronic insertion, can be videotaped and checked immediately. Recordings can be replayed for further electronic treatment.
10. Motion pictures are regularly transferred to videotape: to add identifying sub-titles; for language translation; for optimum presentation of wide-screen prints; when adding music and effects to silent (mute) films; to preserve a valuable or damaged film print; or to edit Super-8 mm film.

■ *Videotape* Videotape is designed to meet the standard with which it is being used; grain-orientated during manufacture to suit quad or helical format. The forms are not interchangeable. Also, the magnetic material is of several types, including gamma-ferric oxide, chromium dioxide, and cobalt-doped gamma iron oxide coatings.

Quadruplex (quad, transverse) scanning
In this recording system, the 2-inch wide magnetic tape (moving at 38 cm per second) first passes over an *erase head* that demagnetizes (*wipes*) its full width before reaching the transverse head wheel, housing the four recording/replay heads. These heads, mounted 90° apart on this head wheel, press against the tape and record across it in turn in a series of parallel tracks (effective writing speed 3800 cm/s or 1500 ips). Each track covers a 16-line segment of the USA 525-line television picture, or 20 lines for European 625-line systems. Interhead switching ensures that a continuous video signal is recorded. The video tracks cannot normally be checked during recording, but are locally spot-checked afterwards.

Further tracks are recorded along the edges of the videotape as it moves towards the take-up reel:

■ *The audio track(s)* This records the program sound. Although very satisfactory, the audio quality is unavoidably limited as the track is longitudinal, on a tape that is grain orientated for transverse tracks.

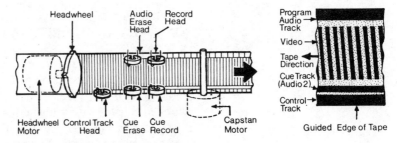

Fig. 12.1 Quadruplex system
The layout shows that the video heads (headwheel) are upstream of audio heads (by 23·5 cm/9·25 in). The second illustration shows the Quad tape track format. Each video track records 16 or 20 picture lines. In Quad II system, a modified format is used (audio 1, cue/code track, audio 2, video, control).

■ *A control track* This records the regular sync pulses that stabilize the scanning rate, and accurately synchronize the head-wheel and capstan (drive spindle) motors.

■ *A cue track* This often contains a special *address code/time code* recorded to facilitate electronic editing. Otherwise it could carry a second audio track (e.g. production information) at lower quality.

Each of these three tracks can be erased selectively, and replaced where necessary.

Helical (slant-track) scanning

In a helical scanning system, the videotape travels from the supply reel, past a fixed head (which wipes its entire width during recording), and moves in a helical path round a cylindrical *head drum/scanning drum*. From the surface of this drum protrudes a video record/replay head(s). As the upper section of the drum (or an internal disc) rotates, the head(s) on it turns, rhythmically pressing against the passing tape (tip penetration) and scanning across it. Due to the tape's slanting path, the head records an oblique video track.

Recorded along the upper and/or lower tape edges are the *audio track(s)*, a *control track*, and a *cue track* (or Audio 3); which may be selectively wiped.

Helical scan recorders may follow three typical formats: a 360° wrap (alpha or omega) using one head, recording one field per pass; a 180° omega half-wrap using two heads, recording one field per pass; and a similar arrangement recording a segmented (split) field each pass.

Two standards have been introduced for professional 1 inch (25·4 mm) helical scan recording:

■ *SMPTE type C standard*: continuous (non-segmented) field format. In this system, the video head sweeps across the tape to record one complete television field on each video track. (Two fields interlaced form one TV picture.) System used by AMPEX, SONY, HITACHI, NEC, MARCONI, RCA.

■ *SMPTE type B standard*: segmented field recording. Here the system uses a succession of 5-6 tracks to record each television field. As with the quad system, the video heads are sequentially switched to provide a continuous video output. (*BCN* models of BOSCH/FERNSEH, IVC, PHILIPS companies.) This has several technical advantages including enhanced video quality (despite head-switching problems), and a smaller faster head drum which aids speed stability, and improves portability.

The *continuous* (*non-segmented*) format has one particular advantage. Because it lays down one complete video field per track, it is possible to stop the tape and rotate the head wheel to continually rescan the same track, thus producing a *still frame* of one field for a short time (tape wear develops). Moreover, by adjusting the tape replay speed, you can reproduce a succession of complete fields at any chosen rate, and so obtain slow motion, fast motion, time lapse (ultra-fast motion), and inching/frame jogging (frame-by-frame). Such facilities are invaluable when editing the recorded tape. Furthermore, this format permits continuous replay during recording to monitor and check results (confidence playback).

The *segmented format* on the other hand, can usually only be played at the normal recording speed, if the heads are to reproduce the succession of tracks that comprise each field. In order to obtain freeze frame or slow motion facilities, it is necessary to include a *digital frame store*. This can retain a complete frame's worth of video information, which may be read out at any required rate.

Most helical scan systems have certain inherent weaknesses compared with quad systems, such as the information loss at the end of each scan (*noise bar*), picture disturbances and visual errors.

Fig. 12.2 part 1 Helical system (slant track) SMPTE one-inch Type 'A' standard (continuous field format)
Format introduced by Ampex in 1974. Recorded at 24·4 cm/sec (9·6 in/sec).

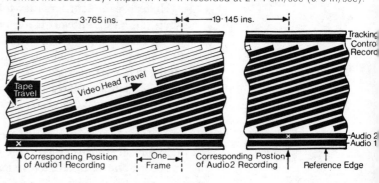

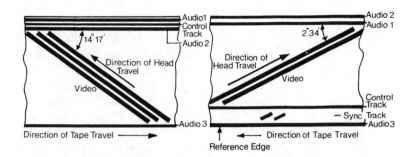

Fig. 12.2 part 2   SMPTE one-inch Type 'B' and Type 'C' standards
Left: Type 'B' standard (segmented field format). Each head-drum revolution
records 5 video tracks per field (525 line/60 field system), or 6 tracks (625/50
system). Tape wraps 180° round drum (50 mm diameter, 150 revs/sec). Tape
speed is 24·3 cm/sec (9·57 in/sec). Head drum contains 2 video heads and
control head. Three audio tracks along tape: 2 program, third for time-code or
cue-track. Video is recorded as FM carrier (blanking level 7·9 MHz, peak-white
10 MHz, for 525/60 system; 7·4 to 8·9 MHz for 625/50 system). *Right: Type 'C'
standard (continuous/non-segmented field format).* Each head-drum revolution
video track records one field (2 fields of $262\frac{1}{2}/312\frac{1}{2}$ lines comprise complete
picture); from line 16 of one field to line 5 of following field. This video track
contains all picture information (active lines), plus insertion signals in vertical-
interval (VITS, VIRS). The *sync track* records remaining 10 lines of vertical
interval, with a small overlap. Tape wraps 360° round head-drum (135 mm
diameter, 60 revs/sec). Tape speed is 24·4 cm/sec (9·61 in/sec). Head-drum has
6 *head-tip locations*: separate record and replay heads for video and for sync
tracks; both have simultaneous playback check heads; optional automatic
tracking control head. Three audio tracks along tape: 2 program; third for time-
code, cue-track, or extra audio. All longitudinal recording heads are aligned and
perpendicular to tape edge. Separate control track identifies odd and even
fields, and alternate frames. Video is recorded using standard high-band FM
technique.

Tape interchange between machines can pose tracking problems
(i.e. ensuring that the replay head exactly follows the recorded
track) unless automatically compensated (AST).

These shortcomings, together with lower detail resolution,
prevent many helical scan recorders from achieving maximum
broadcast quality requirements. However, technology continually
develops, for the format's advantages are considerable.

**Fig. 12.2 part 3   Typical helical
scanning arrangement**
The tape is wrapped in a slanting
path round a drum. Within the drum
a disc head-wheel supporting recording (and erase) heads rotates, pressing heads against the tape at an
acute angle as it slides past. (Designs
include rotating upper drum, or slot
scanner with narrow rotating center
section.)

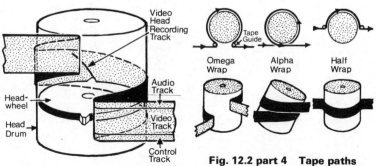

Fig. 12.2 part 4   Tape paths
Several tape-path arrangements are
used: omega, alpha, half-wrap forms.

Magnetic video disc

The magnetic video disc recorder has become a familiar facility for instant replay during televised sports events. In operation it continuously records a 30/36 second segment of the program (video only), progressively erasing and rerecording over the earliest material. If necessary, you may record only odd fields (i.e. half speed), but this results in jitter on half-speed replay.

Recording can be started and stopped instantly. The recorded program segment can be replayed at normal speed, in slow motion, fast or variable speed, in reverse, field-by-field, or any chosen picture can be held (freeze-frame, stop motion). So the video disc offers considerable production opportunities.

It is however, important not to overlook one limitation of the system. If you *film* action at an increased rate (e.g. 64 frames per second), the time for which each frame is exposed is noticeably reduced from the normal $1/48$ sec ($1/2 \times 1/24$) to $1/128$ sec ($1/2 \times 1/64$). Consequently, movement detail is captured more clearly.

Television on the other hand, is inherently a *single-speed* system ($1/30$ or $1/25$ sec). So a still frame of TV action will show noticeable movement blur. A very brief event (bursting balloon) may be too rapid for the TV camera to have recorded at all. So, although the video disc and some video recorders can provide slow-speed

Table 12.1   Typical videotape formats

Quadruplex (in Quad 1, 1A and 2 formats	Tape width: 5 cm (2 in).
	Tape speed: 38 cm/sec, 19 cm/sec (15, $7\frac{1}{2}$ ips).
	Tape duration: 60, 90, 96, 192 min on 35 cm (14 in) spool.
Helical scan	Various incompatible formats. Tape speeds 1·7–28 cm/sec (0·66–11 ips) 9·53 cm/sec—Type C. 24·5 cm/sec—Type B.
	Tape width
	5 cm (2 in)—highest quality video, 17·6 cm/sec (6·91 ips) tape speed.
	25·4 mm (1 in)—*excellent quality video—SMPTE 'B' and 'C' format standards.
	19·05 mm ($\frac{3}{4}$ in)—*very satisfactory video quality for many applications. U-Matic format. Tape speed.
	12·5 mm ($\frac{1}{2}$ in)—EIAJ/CCIR standard Type 1 Class 2. Typically 365 m (2400 ft) spool. Duration 35 min, 60 min at 19 cm/sec ($7\frac{1}{2}$ ips).
	6.25 mm ($\frac{1}{4}$ in). *Used for ENG
Cartridges (carts) and cassettes	Tape width
	5 cm (2 in)—duration 10 sec to 1 hour.
	25·4 mm (1 in)—duration 10 sec to 1 hour.
	19 mm ($\frac{3}{4}$ in)—duration 10, 15, 30 sec to 1 hour.
	12·7 mm ($\frac{1}{2}$ in)—duration $\frac{1}{2}$ hour to 3 hours.

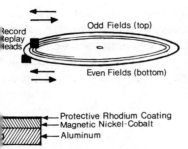

Odd Fields (top)

Record
Replay
Heads

Even Fields (bottom)

Protective Rhodium Coating
Magnetic Nickel-Cobalt
Aluminum

Fig. 12.3   Video disc
One or two 16 in diameter (0·46 m)
discs on a common shaft rotate at
50 revs/sec, (locked to vertical sync
reference). The record/replay heads
on either face, alternately move
inwards (later retracing outwards)
in steps, to form 450 concentric
video tracks total. Each track contains one video field. The replay rate
and order of reproduced tracks is
adjustable to produce fast, slow, or
reverse motion; and freeze frame.

reproduction of fast action, even freezing it, you should not expect to see sharp detail in such shots and motion may appear jerky. The only method of obtaining high definition of fast movement, remains the high-speed film camera. Above 64–100 frames per second, special film cameras become necessary (10 000 to 8000 000 being possible!).

■ *Further uses*   As well as its widespread use for instant replay, the video disc has other applications:

1. It can record a series of individual stills (graphics, titles), which are rapidly selectable for replay on cue, and can be augmented or replaced as necessary.
2. Time-lapse effects are achievable (e.g. recording one frame an hour).
3. Time-controlled freeze frame (holding a still for a specific period).
4. Compiling an animated segment (frame-by-frame build-up).

Computer disc packs with digitally recorded video have an extremely high still frame storage capability, with instant access.

■ *Program discs*   Non-broadcast video discs containing both video and audio program, use several incompatible techniques, including mechanical, capacitative, and optical processes. They are essentially packaged material for production, not a production facility.

Their potentialities for broadcast use, particularly for library stills, have yet to be explored, but as in one system over 50 000 different pictures can be stored on a single disc (of 30 minutes playing-time overall), there could be interesting possibilities ahead.

Tape formats

Video tape is found in open reel (reel-to-reel), cartridge (cart), and cassette forms. Both the latter enclosed forms are self-threading and not easily damaged. Quad and helical scan systems have used each of these tape transport methods.

Although quadruplex tape formats have been standardized (Quad 1 and Quad 2), helical scan systems have several incompatible variations. This necessarily limits tape interchangeability. Generally speaking, a wider tape and a faster tape speed are advantageous, but are less economic and require more storage space.

Both cartridges and cassettes are particularly convenient for brief playbacks, such as program promotions, commercials and announcements. Some editing systems use them to temporarily store material from longer tapes for subsequent selection and insertion. Film commercials too, are often dubbed onto 2 inch cassettes for more convenient handling. *Carousel* equipment

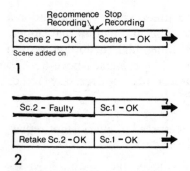

Fig. 12.4 part 1 Electronic editing—assemble mode
Program is built up either 1, A scene at a time or 2, shot continuously, retaking any faulty scene immediately from its opening point. Assembly involves switching from replay (of Scene 1) to *record mode* at a chosen cue point (manually or automatically). As assemble mode automatically lays down a fresh control-track for each take, sync pulses change at joins, thus disturbing replay.

designs, can be loaded with some twenty-four individual selections for *random access*, and automatic or semi-automatic presentation. Push-in cartridge systems can provide: immediate auto start on insertion, auto wind-on to the beginning, end-reject, fast forward wind to selected section, rewind at the end of pre-set time, and remote control.

Reasons for videotape editing

If your aim is to record an entire program 'live on tape', making all picture transitions on the production switcher and then reproduce it in that form, no videotape editing is involved. But most productions can be improved by retakes or changes that correct, remove, or replace unsatisfactory original material. So unless you are prepared to retape the entire program, some form of postproduction videotape editing becomes necessary.

In larger organizations, videotape editing is carried out by a specialist editor. For the most part, he is handling recordings that incoporate their own *integral editing*, made in real time using the production switcher 'on the fly', when the program was taped.

His main work usually involves *continuity editing*: removing faulty sections and inserting retakes, rearranging recording order to program running order, shortening, introducing inserts (e.g. cutaways), censor cuts, etc. This is quite different from working *ab initio* as the film editor does, with many isolated brief takes, differing versions of action, and continual continuity problems. Although, given sufficiently sophisticated equipment, the VT editor can work to precision limits, akin to those followed in film.

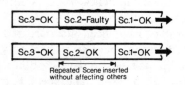

Fig. 12.4 part 2 Electronic editing—insert mode
If new material is to be inserted precisely between existing wanted shots, the 'in' and 'out' points must coincide with original, unwanted scene to avoid wiping wanted program. The original control track is not erased during the insert process, so no sync disturbances occur at the electronic splice.

Physical editing

It is possible but undesirable to physically cut and splice videotape. Although an exacting, time-consuming business, physical editing on quad systems can be quite unobtrusive on replay. But on helical scan systems a transverse cut interrupts several tracks, and so at best produces a vertical wipe transition. A physical splice always disrupts the tape (a dropout hazard), reduces tape life, and may even damage the recorder's heads.

The situation is further complicated by the physical displacement of audio from vision, preventing their tracks being cut simultaneously. To provide clean audio cuts, the sound track may need to be 'lifted off' and rerecorded on the edited videotape, over the join. Physical editing modifies the original master recording (first generation), and is best regarded as an emergency technique.

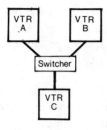

Fig. 12.4 part 3   A/B dubbing
Here duplicate copies of the program are used on VTRs A and B to permit quicker search for sections, than by shuttling a single tape to and fro for selection Video switcher permits various transitions on transferring to VTR C.

Electronic editing

Electronic editing is a rerecording process. It can be carried out *during* the actual production, or in a postproduction session.

■ *Editing during production* Here any changes or retakes are as you go, recording over and replacing faulty or unsatisfactory material. The result is a *first generation*, original master tape, ready for transmission.

■ *Postproduction editing* This is a cumulative method in which you record material in any convenient order, finishing up with videotapes containing both wanted and rejected sequences. Later, in an editing session, the required passages are selectively dubbed-off (reproduced, and rerecorded by another machine) with appropriate intersection transitions.

How complex this process becomes, depends considerably on the brevity of the sections and the facilities available. At best it provides production flexibility and polish. At worst, it becomes a salvage job! Extensive postproduction editing can be time consuming, demanding equipment and skills; so where possible, the majority of transitions are usually made during production taping using the production switcher.

A disadvantage of editing by dubbing is that the resultant composite is lower in quality than the master tape; because

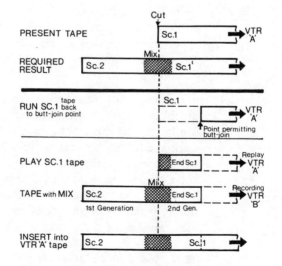

Fig. 12.4 part 4   Roll back and mix
Problem: Scene 1 has been recorded. You now wish to *mix* from Scene 1 to Scene 2 live in studio. Two VTRs available. *Solution*: Roll back Scene 1 on VTR 'A' to convenient cutting point before mix. Recording on VTR 'B', replay VTR 'A' tape, mixing to studio when required. VTR 'B' tape now incorporates mix. Result can be inserted within original VTR 'A' tape at the earlier butt-join.

deterioration takes place at each copying stage. Most organizations try to avoid dubbing copies beyond the third generation.
(Experimental *digital video recording* techniques promise well for
the future though, for they have produced 128th generation
copies, identical to the original!)

Dubbing down from larger tape formats (e.g. 2-inch quad) to
smaller formats ($\frac{1}{2}$ inch helical) invariably loses quality; although
the reverse (dubbing-up) is not necessarily the case.

A similar deterioration arises in film copying (*duping*), where
copies of copies show color errors, increased graininess, and
definition loss; all of which cannot be remedied. This occurs
when copying film library shots for program insertion.

Manual edits using electronic splicing

Quad VTRs, and helical VTRs *with editing facilities*, provide
electronic splicing opportunities. In this process, you play the
tape to be treated on the editing VTR, and at the required edit
point switch from replay to record mode by pressing the *edit
(mode change) button*; this records new material on the same tape.
The recorder must be up to working speed and stable (*locked up*).
If you try to do this with an ordinary helical scan VTR, the
resultant tape will normally show severe picture break-up at the
edit point.

Two forms of electronic splicing are widely used:

1. *Assemble (assembly) mode* New material is added consecutively to the *end* of existing shots, by recording new video,
   audio *and* reference control track pulses (that govern tape
   position and speed).
2. *Insert mode* A period of new material is slotted *within* an
   existing recorded sequence, replacing that section only. Video
   and audio are replaced, leaving the *original* reference control
   track pulses.

Because the erase (wipe) and record heads are physically
displaced, there is a time lag as any part of the tape moves between
the heads. In quad systems, this is electronically compensated
during editing, by erasing just before the mode switches. In
helical systems, the shallow slant track poses editing problems
and a *flying erase head* is used to make the erase coincide with the
seen edit point. Otherwise visible patterning arises (*edit moiré*)
due to brief overrecording of the old material. Even so, the edit
may cause some control track or audio interference.

■ *Laying a control track (helical)* When a helical scan videotape
is recorded, internal circuits generate special, regular motor
stabilizing pulses, which are continuously recorded on the *control
track*; therefore ensuring a stable constant speed replay. If these
pulses are interrupted, picture stability will be affected at that
point.

However, if you *re*record on this master tape (as when assembly dubbing) the editing VTR would normally create its own fresh pulses for the new material. And because the act of editing usually involves stopping/starting the editing recorder, this will interrupt the exact positioning of these new pulses in the edited composite at each edit point. So on reproducing the new composite tape, you are liable to encounter frame-roll or line tearing at the edits.

To overcome this problem, it is common practice *before* a helical scan editing session (whether you actually want to fit in, or add on the new material), to record an entire continuous control track alone on the clean editing tape. Then the actual dubbings are made using the *insert mode* of the VTR; for this mode leaves the prepared control track pulses undisturbed. So you have an entirely uninterrupted control track and optimum sync stability.

Editing controllers

Having located your edit point on the *replay VTR* (at slow speed) you need to set up the replay and recording machines so that they run up, stabilize and synchronize to make the edit at precisely that point on their respective tapes. The editing *controller* does all this 'housekeeping' for you automatically! It counts off the tape's recorded control-track pulses (and/or coded time signal) relative to your chosen edit point, and times operations accordingly. You can *simulate* the edit to check the effect, and the replay machine will automatically back to its pre-roll position, stop, restart and 'cut' at the chosen place. If you are satisfied, the controller will repeat operations and make a real edit. (The point is adjustable.) This simply operated system is reasonably accurate (±2 to 5 frames) but has the disadvantage that the edit point can become misplaced or lost during operations unlike the more expensive precision time-code methods.

Editing using a time code

The most sophisticated electronic editing systems make use of a *time address code/digital frame address system*. Operationally, this is simply a continuous record of a real-time 24-hour digital clock, showing precisely when each fragment of the videotape was recorded. The information is stored on the tape's *cue track* in binary digital code; as four sets of figures: hour, minutes, seconds, frame (0-29 SMPTE, 0-24 EBU) plus 32 user bits of information.

If necessary, the time code could be recorded on the tape simply as reference points, before or after program recording. But if it is laid down during production recording, it also offers you a real-time check of: when an event happened, its duration, available time left, and elapsed time. You can log clock time in anticipation of later editing; viz: 'We'll use the retake recorded at 15.25.30, not the one at 15.20.10.'

Hours Mins Secs Frames

Fig. 12.5   Time address code
Monitor displays shows the precise time at which the recording was made—hours, mins, seconds, frames —to facilitate identification, cuing and editing. (For special applications such as time-lapse recording—day, month, year may be indicated.)

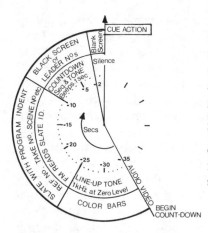

Fig. 12.6 part 1   Video tape leaders SMPTE standard
A video equivalent of the film leader, providing program indentification details, technical alignment, and synchronizing information. The diagram shows the time sequence.

■ *Random access editing* Time code pulses form a uniqu address, even identifying individual frames within the tota recording (although recordings made on different *days* could hav slmilar coding). Selected code locations can be programed int a memory system to instruct microprocessors controlling VTR: giving rapid *random access* to any point in the recording: t preview, replay, cue, select edit point, etc. So the address code ca be used as an identification and search aid, or form the basis c entirely automated dubbing according to prescribed editin instructions, producing a continuous, correctly assembled pre sentation.

Off-line editing techniques

One way of approaching production editing is to arrive at th videotape channel armed with notes made during the recordin, session, review the program, dubbing it progressively (with certain amount of experimentation), until a new composit videotape is obtained—the 'edited master tape'. But this *on-lin editing* technique uses costly facilities and personnel for consider able periods.

One economical solution is to make an identical copy of th original recording (recorded simultaneously or transferred after wards) together with its address code, on a portable helical scar video recorder. The director can scrutinize this copy at leisur (*off-line editing*) and, by scheduling the code references displayec

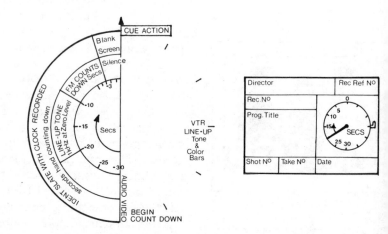

Fig. 12.6 part 2   Video tape leaders—typical British practice
Color bars and tone are recorded in prerecording line-up period. One-minute clock with seconds sweep-hand provides visual countdown. Retakes are identified by taped FM announcement (e.g. Retaking Shots 20-34, Take 1) Electronically generated numbered 'leader' counting down in seconds is used fo some.VT operations.

Table 12.2 Videotape—before recording

Precautions	Regular equipment maintenance checks anticipate trouble before it happens.
	Always ensure that the tape heads, tape path (transport), and head-drum are clean (using suitable solvents), free from tape-dust clog, and regularly demagnetized (degaussed). After use, clean and cover equipment.
Tape care	Avoid touching the tape surface (skin oils). Also avoid smoking, dust, crumbs, etc., near the VTR as these all cause drop-out. Do not hand-stop tape because it causes stretching.
	Cut off damaged leader on open-reel tapes, to avoid head damage.
	Check tape for uneven or loose winding (cinching, edge-damage).
	Store tape upright in boxes and away from extreme heat, cold, dryness, humidity, magnetic fields.
	Allow tape and equipment to normalize after transit (e.g. from warm interior to cold exterior) to avoid condensation friction problems.
Pre-recording	Check audio and video levels are within limits. Check tracking and tension adjustments.
	Record and reproduce a test sequence at the start of tape (black level, audio tone, tonal step-wedge or sawtooth, and color bars) for alignment on replay.

on his screen, decide on the excerpting he requires, the exact edit points, transitions, etc. Later, at an editing session, the chosen address-coded edit points are used to guide the VT editor, or fed into VTR equipment processors for automated editing.

■ *Freeze-frame problems* Where the time code has been copied onto a helical scan recorder's cue track for off-line editing decisions, problems can arise when *freezing frame* to identify action. In that condition the replay heads are moving over a stationary tape, continually reading out *picture information*; but the longitudinal edge cue track containing the code is not moving, so the address code disappears! Therefore, other frame identification methods have evolved, such as those counting the control track pulses (deriving 'user bits')—a simpler process than time code methods.

Videotape processors

In sophisticated video-editing processing equipment the editor can dub off sections of the program tape (audio and video) provisionally onto a series of cassettes, carts, or video discs (disc pack). These intermediate reference dub-offs, which are used solely for editing decisions and edit-simulation, can be played at any speed forward/backward, and freeze frame for repeated examination and selection. The entry and exit points for shots to be conjoined can be inspected simultaneously on dual monitors.

The consequent editing instructions are fed into a computer which operates broadcast VTRs, and edits the complete show, introducing cuts, dissolves (mixes), fades, insertions as required, to provide the *edited master tape*.

Audio dub

You can make minor alterations in the audio track, when dubbing off video to form the *edited master tape*. But where major audio changes are required—for example adding music or effects, post-syncing dialogue, etc.—this is best done as a separate operation. The revised audio track is then dubbed onto the picture-edited videotape afterwards. A multitrack audio tape recorder is often used for this purpose, carrying a copy of the original address code on one track for identification and synchronizing. A helical scan VTR may be used as an address code synchronized reference before the audio dubbing session.

Table 12.3   Videotape-fault tracing

	Check replay with a known good tape program.
No video	Check all units are switched on; correctly operated (facility switches and routing); connections OK. Check video input is OK. Is correct tape-type or system being used (compatibility)?
General noise	Adjust *tracking control* for maximum signal and minimum noise. Is head-tip penetration insufficient? Is head clogged? Is the record current/video level too low? Any head damage (chipped)?
	Adjust playback (replay) equalization for sharp undistorted pictures with minimum noise.
Noise band over top section of picture	If not corrected by tracking control, this may be due to an incompatible recording (misaligned tape path).
Horizontal noise bars across screen	Tracking control or control-track phasing incorrectly set. Capstan servo off lock.
Small black flashes on picture	Drop-out due to tape condition. Insufficient head-tip penetration. Ineffective drop-out compensator.
Close vertical bars (high frequency, RF pattern)	Incorrect adjustment of demodulator balance or limiter (interdependent effects).
Coarse beat-pattern across picture	Clogged erase head. Faulty tracking.
Picture flicker on playback	On a two-headed machine, imbalance between head outputs.
Breakthrough of previous recording	Badly clogged or faulty erase head, or low erase oscillator output.
Rhythmical picture break-up	Faulty servo system (responsible for picture-timing, speed, and synchronism). Other servo faults include noise or disturbance over a few horizontal lines (switching line fault), localized color loss, and inaccurate or noisy edit-points. Audio speed variations (rapid flutter/slower wow).
Hooking (skew), verticals bending over at top of frame	Adjust tape tension (for minimum).
Edge scalloping	Regular horizontal displacement due to low-level video (undermodulation).
Audio playback low level (quiet, high background noise)	Tape recorded at low level. Clogged audio head. Bias too low.
Audio distortion	Damaged tape edge. Overrecorded (overmodulated). Poor head/tape contact. Bias level wrong.
Buzz on audio	Video tracks are audible (poor erasure). Carrier frequency drift.
Wow or flutter	Dirty capstan or roller. Servo off-lock. Spool warped.

13 Graphics

Graphics (captions) are widely used in television and film production, taking a variety of forms: titling, decorative and pictorial illustration, explanatory charts, maps, graphs, diagrams, etc.

The purpose of titles

Opening titles establish the style and atmosphere of the production. The type face, its size, color, and its background, create a foretaste of the program itself.

Above all, titling must be appropriate. Good titling attracts, informs and excites interest. It should be neither confusingly decorative, nor severely characterless. Titling should be compact without being crowded; not spread out, to be lost beyond the frame edge.

Readability

Lettering is there to convey information, and if it cannot be read quickly and easily, it has failed to communicate effectively. The legibility of any graphic detail will depend on its size, elaboration, and background contrast. Lettering sizes below $1/10$ to $1/25$ of picture height are undesirable. Lettering should generally be bold and firm rather than slight. Thin outlines, serifs, fine striping or hatching, are best avoided. Such detail will either be lost or will strobe distractingly. Conversely, type faces that are too condensed will lose form, particularly if keyed into pictures or given edge enhancement.

Aim to keep titling information to a minimum, particularly if it is combined with detailed background. A screen full of 'printed information' can be daunting to most viewers, and tiring to read. People are easily discouraged from reading rapid titles. Leave information on the screen long enough to allow it to be read aloud twice, so that the slowest reader can assimilate it. If, for any reason, an announcer is to read out text that is clearly displayed, ensure that the reading is *accurate*!

Fig. 13.1   Frame proportions
1, To fit the TV frame shape, graphics must have 4 by 3 proportions. The graph shows relative height-width proportions eg. if a card is 12 units wide, it should be 9 units high. 2, To avoid edge cut-off on the receiver screen, graphic information should be confined to the *safe title area*.

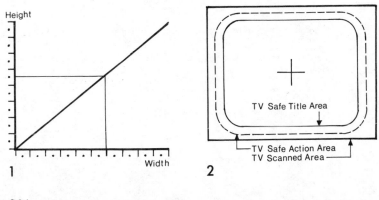

1

2

Backgrounds to titles

Whether plain backgrounds unobtrusively emphasize titling, or appear dull and uninteresting, really depends on the occasion. Ornamental backgrounds (texture, abstract design) may add visual appeal, or they can become fussy and distracting.

Lettering against a multihue or multitone background is invariably harder to read. If you insert titles over location shots (e.g. a street scene) the eye may have some difficulty in discerning information, and may also be tempted to wander around the background instead.

In most cases, by using larger type in light tones (white or yellow) with strong borders (black edged), legibility is considerably improved. But even plain titling can be unclear in certain color relationships (red on green, or red on gray); or in monochrome when gray-scale of lettering and background are similar.

As a general rule, avoid introducing lettering over backgrounds of similar tones or hues, or over printed matter (e.g. titles over a newspaper page). Obvious enough, but people do! Light lettering is usually more easily read than dark; and pastel or neutral backgrounds are preferable to saturated hues. Unless *superimposed* lettering has a black background (it then appears 'solid'), it is best keyed-in (electronically inserted) from white-on-black title cards or slides; or in film, by *burn-through* using masking or controlled overexposure. Black lettering cannot be superimposed on a white background.

Table 13.1 Television graphics using studio camera

	Advantages	Disadvantages
Graphics on stand or easel	Simple method. Camera can zoom, select any section, adjust framing. The original artwork can be used (avoids extra copying stages). Graphics can easily be modified or updated. Graphics can be animated.	Method ties up a studio camera. Graphics require an operator (opportunities for errors, wrong positioning, etc.). Graphics may shake during rapid operation or changes. If graphics sizes vary, camera may have to refocus and reframe. Positioning may vary, causing glare, specular reflections, or revealing surface unevenness.
Projected graphics		
Front-projected graphics	Camera can zoom, select, frame. Slides provide rapid, predictable changes.	Graphics must be prepared in a standard format (e.g. 35 mm slides). Spill light on screen may spoil image. Color quality variations are not easily compensated for. Apparatus usually requires an operator.
Rear-projected graphics	As above.	As above. Uneven illumination likely (central hotspot). Screen grain may be apparent in close shots.

If lettering on a title card proves to be too large, or needs to be offset, a larger black card placed behind it will enable the cameraman to take a longer shot or reframe without shooting off. Similarly, a chroma key title card (usually white lettering on a blue card can be extended with a large blue backing sheet.

Transparent plastic overlay sheets (cels) may be used for titling, which is then laid over background graphics. But there is always the danger of light reflections from the surface of the cel.

Although cameras can shoot titling on a foreground glass panel, depth of field is usually too restricted to obtain a sharply focused background scene at the same time. Instead, you would have to pull focus from the titling to 'dissolve' to the action.

Forms of lettering

Although lettering is usually taken for granted, nothing looks more amateurish than ill-proportioned, badly spaced, or poorly laid-out titles. Nowadays there are many short cuts to efficient titling:

1. *Dry-transfer (rub-down, instant transfer) sheets*—with a wide variety of rub-off lettering.
2. *Hot-press lettering*—which hot stamps regular metal type through pigment-faced plastic foil, to produce white, black, or color characters.

Table 13.2   Televising graphics using special apparatus

	Advantages	Disadvantages
Telop	Multiplexing apparatus enabling one TV camera to continuously shoot several stills or transparencies. Lighting changes provide cuts, mixes, supers, all on one camera.	Equipment requires specially prepared graphics. All graphics sizes require a fixed format (100×127 mm 4×5 in). Size changes or reframing not possible. Some images reversed.
Caption scanners	Designs vary considerably: slide transport by carousel or vertical drums; tray/box carriers, discs. Slide selection in sequential, random, or programmed order. Flexible color correction facilities included in equipment control. Can be remotely controlled.	Slides must be shot and mounted in proportions and framing required—position and size cannot normally be adjusted. Slides are not readily modified/corrected, and may acquire blemishes, dirt. Equipment reliability varies.
Rostrum camera	An adjustable overhead camera shooting horizontal artwork. Used to record a series of graphics (stills) on film or videotape; in order to simplify a rapid graphics sequence, to record a complex graphics presentation, or to animate a succession of shots. Sequence can be prerecorded, simplifying studio production.	Revision of individual segments requires editing or remake.

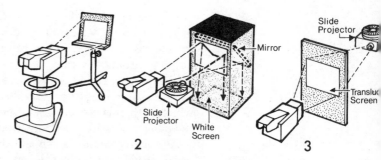

Fig. 13.2 Methods of televising graphics
1, Studio camera shoots graphic supported on easel or stand. 2, Caption projection unit: Slide is projected onto screen within box, via a mirror. Camera shooting via same mirror, sees and explores the picture. 3, Camera explores picture from rear-projected slides on translucent screen.
   Slides are usually transmitted using a multiplexed film-island projector (see fig. 11.8) or a flying-spot scanner (see fig. 11.10). Opaques may use similar systems.

3. *Plastic lettering*—stuck or clipped onto a background (slotted feltboard, magnetic board).
4. *Constructed lettering*—arranged from toy bricks; formed thumb tacks, toothpaste, rope; finger traced in sand or snow.
5. *Decorative lettering effects*—using shadows, silhouettes, reflections, stencils, etc.
6. *Handwritten lettering*—which is extremely difficult for the inexperienced (even using lettering aids), and best avoided. May be successful where unsophisticated effects are required (on walls, sidewalks, chalk boards).
7. *Video character-generators*—enabling electronically generated lettering to be 'typed into the picture', or presented as complete stored information.

Character generators (alpha-numeric generators)

Using programmed digital circuits, a range of lettering, numbers and symbols can be electronically generated in selectable fonts and sizes; including such variants as Roman, Cyrillic, Japanese, Arabic.

   Most titling and basic graphics are only required for a single TV program; So the rapidly produced low-cost display that the character generator provides, offers considerable advantages.

Fig. 13.3 Title positions
Titling is normally localized for maximum impact, avoiding important subject areas (eg. face).

ig. 13.4   Border Generators
, White solid lettering, 2, black-edge
order, 3, drop shadow effect, 4,
utline.

A standard typewriter/computer keyboard is used, and aided by a location cursor symbol indicating the next character position, the typed copy is prepared on a monitor screen display. Information could be inserted into the studio *line (master) picture* as it is being typed, but usually the prepared message is stored (magnetic floppy disc or tape, programmable microprocessors, or punched paper tape), recalled and inserted as required.

Lettering can be corrected or revised instantly, and moved around the frame for layout adjustment, or to create vertical roll or horizontal crawl titles. Lettering may be selectively flashed (blink). Refinements include autocentring and tabulation, varisize, and multipage storage.

Using a color synthesizer, the lettering and background (normally white-on-black) can be given push-button selected colors—sometimes colorizing individual words.

■ *Border generators*   To improve legibility or impact, lettering can be processed in various ways (whether from the character generator or other sources):

1. It can be given a black, white, or color edge *outline border ;* all round, or at one edge.
2. A *dropped shadow* can be produced to create pseudo solidity.
3. An *edge outline* can be developed from the original solid character.

■ *Electronic keying/matting (inlay)*   Electronic insertion, by keying or non-additive mixing, enables lettering to be 'punched solid' into an existing picture, so avoiding the interaction of hues and tones that arises when *superimposing* lettering. It is customary when adding titles to multitone color pictures, to use keyed-in black-edged characters for optimum clarity.

Table 13.3   Typical graphics sizes

Smallest size for convenient handling (smallest type size 24–36pt)	30·5×23 cm (12×9 in)
Commonly used sizes shooting a 30·5×23 cm (12×9 in) area	43×35·6 cm (17×14 in) 40×30 cm (16×12 in)
Flip cards—copy area 23·5×18 cm (9·5×7 in)	35·6×28 cm (14×11 in)
Largest size for elaborate artwork	61×46 cm (24×18 in)
Thickness of support art board	2 mm ($\frac{1}{16}$ in) minimum

Forms of graphics

Pictorial and diagrammatic graphics can make a valuable contribution to all types of TV program, for they show information succinctly and unambiguously in a few seconds. You can demonstrate developments, proportions, relationships; make statistics attractively meaningful, thus simplifying complex data.

Graphics can be presented by any of the methods discussed in Tables 13.1 and 13.2. You can also introduce them as part of the studio setting: as display screens, desk charts, wall maps, and free-standing panels. However, you must ensure that they do not become obtrusive or distract attention—you can do this by keeping them out of shot, or obscuring them (sliding panels, revolving units, etc.) until required.

Although it is possible to show graphics on an in-shot picture monitor, a better scale image is generally obtained from a wall display screen. This enables you to present a succession of graphics, while relating them to a demonstrator. Either *rear-projected* slides, or electronically inserted graphics can be used (Chapter 19). He will be able to see and point out detail in inserted graphics, by using a nearby picture monitor for reference.

There is a particular appeal in seeing someone draw upon or alter diagrammatic displays, particularly where information is to be built up progressively. Figure 13.10 shows several widely used methods of adding, removing or changing detail. Alternatively, graphics can be *animated* (usually out of shot) to introduce variations.

Three-dimensional graphics often make information more interesting, for example a giant ear showing all its component parts. Everyday objects such as coins or wooden blocks can be used to demonstrate principles. Elaboration may be ingenious, but it is often arguable whether a three-dimensional multilayer graphical model in styrofoam is any more informative than its humble flat counterpart. It is certainly more expensive!

Occasionally, you may introduce titling or pictorial graphics as 'natural features' of a scene: a book title on a shelf, wall poster, street sign, baggage label . . . even attached to people (a sweat-shirt).

Fig. 13.5 Graphics supports
1, *Caption stand* (tiltable shelf, adjustable height) suitable for title cards 2, *Card-pulls* attached to edges make removal easier. 3, *Title card box*—Top card pulled out to reveal next title. 4, *Strap easel* used for larger graphics Weighted webbing straps adjust to suit all sizes. 5, *Flat* displays various graphics of different sizes.

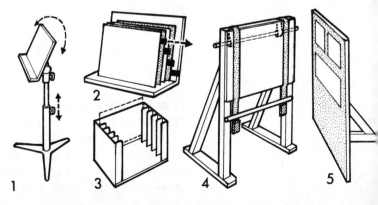

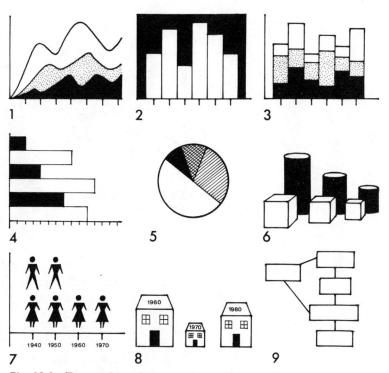

Fig. 13.6 Types of graph
Graphical presentation enables information to be assimilated and compared rapidly. Several forms are available, having greater visual interest than a routine line-graph: 1, Multi-surface or strata graph. 2, Column graph. 3, Combined column graph. 4, Bar graph. 5, Sector, pie, or circle chart. 6, Volume chart, height chart, area chart. 7, Pictogram pictorial chart. 8, Pictorial symbols. 9, Flow chart.

Digital graphics offer many interesting developments, for they can provide charts, maps, graphs, with instantaneous up-date opportunities. The equipment involved may incorporate a character generator, digital video effects, video processing, graphic color camera, and an electronic still store from which material can be recalled.

Computer graphics provide the ultimate in electronically generated imagery, developing entirely syntheisized pictures without the use of camera or videotape. Mostly used for animation effects, they have extensive applications in training simulators.

Animated graphics
Movement brings a graphic alive. But if you are not careful, it awakens interest only in the ingenuity of the animation!

The simplest methods of introducing movement into a graphic, are to pan over it from one detail to another, to zoom in/out, or to intercut between sections of it. Such techniques are used for illustrated children's stories—an inexpensive method of continually redistributing attention.

Graphics can be built up in vision, progressively adding details

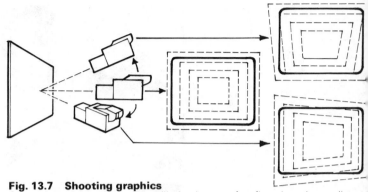

Fig. 13.7 Shooting graphics
Only when shot straight-on, will the image of a flat plane be undistorte
Shooting off-axis produces *keystone* distortion, e.g. when the viewpoint is t
high or shot from the side. In geometrical subjects (music, print, diagrams)
effect may be pronounced.

or sections (uncover wipes, lighting changes, superimpositior
etc.). Or detail can change or disappear. Graphics have been ma
more arresting by introducing such animation as falling sno
swirling mist (obscuring one title, and clearing to reveal the nex
falling rain, traveling light patterns; and even by setting fire
them! Puppets have 'operated' them or pointed out informatio
Ingenuity knows no bounds!

Graphics with movement (See pages 272–273)

■ *Card pull/pull-out/pull title* One title card is slid aside
reveal another underneath. A series of edge-attached pull ta
help to remove one at a time. The cards may be stacked on
caption stand (easel) or supported within a slotted box. Durir
operation, a rapid smooth pull is preferable to a quick jerk. Ta
care that cards do not become displaced or upset.

■ *Flip* Easily operated, but susceptible to jamming or falli
out of position. Ensure that card holes are of suitable size a
undamaged.
Flip-in/drop-in title cards are top-hinged on a ring binder (
sometimes tape hinges).

Fig. 13.8 3-D graphics
Three-dimensional graphics have a considerable visual appeal and provi
demonstrator-involvement. Examples shown are models (left) and a transpare
graph (right).

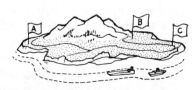

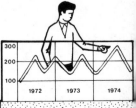

Fig. 13.9 Exploring graphics
A graphic can be 'animated' by exploring it with the camera, so creating continual visual change, e.g. from shot A, zoom out to include B, then to the full picture C, zooming in to select frame D, then E.

Flip-out/drop-out versions are bottom hinged.

Flip-over or flip-round types rotate on a central rod, but require a black backing behind them to prevent shoot-off.

Table 13.4 Precautions with graphics design

Always work within 4:3 proportions.

Provide an adequate border for camera framing.

Use matt materials. Avoid spurious reflections from shiny or glossy lettering, paint, plastic tape, overlay sheets, pencil marks, etc.

Avoid buckling, wrinkles, and similar surface unevenness. Attach thin copy to heavy illustration board, which can be reused.

A large graphic can look the same as a smaller one on the screen. But considerable size reduction will noticeably weaken thin lines; while magnification of small graphics coarsens detail.

Ensure that graphics are evenly lit, and shadows do not disrupt multiplane captions.

Coarse tonal variations, for example, five spaced gray scale steps between black and white, provide more dynamic reproduction than subtle half tones.

Always consider how colored graphics will appear in monochrome.

Color impact in titling and graphics is important. A vibrant hue (red, yellow) may be arresting, forceful, giving emphasis, or disturbing, brash, tasteless, or aesthetically inappropriate.

Simplification of detail, clear symbolism, and unambiguous approaches have maximum effect.

Keep information detail to a minimum for rapid assimilation.

Choose simple structural forms for diagrams, particularly for comparative graphics.

Maps should contain only relevant main features, with good tonal and color contrasts; water areas (sea, lakes, rivers) being shown in darker tones. Name labels may be provided by a plastic overlay, superimposition, keyed in, or manually attached.

Comparative information on a single graph is preferable to a succession of graphics.

Fig. 13.10 Attached detail
Regular methods of attaching detail, labels, symbols, sections to graphics:
Flannelboard—Both the rear-strip behind shape and the background have surface of felt, baize, flannelette, PVC (plastograph). *Teazlegraph* (Velcro)—Shape backed with small nylon multi-hook strip; background has multi-loop strip. *Adhesive strip*—Double-sided adhesive tape behind shape; smooth background. *Magnetic*—Magnetic plastic strip or small magnets, attach to steel-sheet background.

■ *Traveler/roller caption* This continuously moving *crawl title* enables a lot of printed information to be presented in an easily read form. Hot-press or photo-printed on a strip of stiff black paper, the *traveler* or *roller caption* is a convenient one-camera alternative to a series of brief title cards.

The machine may be hand operated, or motor driven at an even, adjustable rate. A motorized machine can be timed, and then cued to synchronize with a program play-out for end-of-show credits. Either upward or left-right drive directions can be used, but lateral word movement is less easily read. You can also use either method to display continuous graphics—for example the development of a manufacturing process, or a series of illustrative stills. This approach is invariably more convenient than long *pan cards*.

Table 13.5   Designing animated graphics

A number of ingenious techniques have been developed to create animation in graphics for cartoons, diagrams, etc. The cost is considerably lower than frame-by-frame film or videotape animation, and the action can be adjusted or timed to suit the production.

To add or remove information (names, details, parts, etc.)	Change part of the graphic using pop-ins, pop-outs, reveals, pull-outs, breakaways, or multiplane treatment.
	Use electronic insertion (chroma key, CSO).
	Manually attach or detach detail. Superimpose detail.
	Drop transparent cel detail over caption.
To show movement (moving wheels, heart-beating)	Have moving surfaces pivoted, hinged, or in runners; operated by tags, levers, strings, magnets, threads, or rubberband 'springs'.
	Intercut shots showing different stages of movement (e.g. hand up . . . hand down).
To show flow (liquid down a pipe)	Cut-out background areas requiring flow treatment. Simulate flow by turning patterns, pull-outs, roller-strip, interference patterns (e.g. a striped strip moving behind a striped transparent panel).

■ *Rotating titles   Rotating box, rotating strips, rotating turntable* —these devices turn to present new information on each face. *Drum titling* can be used to provide a crawl, or made to rotate round a central motif, or appear within a small display.

■ *Reveals*   These all *uncover* information in various ways:

1. *Breakaway*—The graphic slides apart to show new information.
2. *Shutters*—Sections slide under each other.
3. *Multiplane*—A sandwich of transparent planes with black cards between each. As the cards are moved, information is added progressively.

Fig. 13.11   Traveler/roller-caption
This continuous strip display may be rolled up, or rolled horizontally. Varying the frame position of titling produces a *crawl title*. It may also be used as a traveling path display.

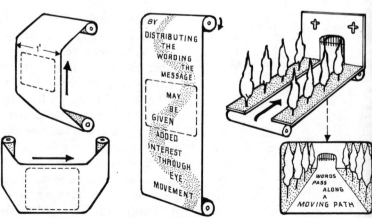

To show flow variation (pressure or density fluctuations)	Change density, tone or speed of flow pattern.
To show direction (aircraft track moving across a map)	Stenciled pull-out. Slide symbol on surface cel over background. Move lighting pattern. Use electronic insertion.
To direct attention to detail	Use indicator (ring, arrow, spotlight) superimposed or electronically inserted. Projected light. Rear-illuminated area.
To move detail about (a product along a conveyor belt)	Use rear magnets, surface cel, thread-operated animation, tags or wires in background slots. Superimpose or electronically insert detail, and pan/tilt its camera.
To show changes in size, volume, quantity	Build subject in a series of subdivided sections, animated simultaneously to change overall size. Intercut between captions showing progressive stages. Use superimposed or inserted detail, and zoom its camera in/out for size changes.
Removing surface to show subsurface detail or sectional views	Remove surfaces by pop-out/pop-in, or by detaching, unflapping, sliding apart.
To show enlarged portions of a subject	Use pop-in enlargement of detail. Electronically insert shot of detail.

■ *Stencils/cut-out cards*   These pierced graphics enable a background to be seen through a stenciled background. They can be used in several ways:

1. A rear adhesive strip is pulled off to reveal the pierced sections (a map route or lettering).
2. An animated surface is created by a rear moving-card pattern; or by sliding together sandwiches of fine wire-mesh photo-engraving screens, or plastic sheet interference patterns.

■ *Animation by lighting*   Lighting treatment can create apparent movement or change by:

1. Selectively lighting details.
2. Creating shadow effects that reveal, conceal or decorate information.
3. Producing reflections and appearance changes as the lighting angle varies.
4. Selectively rear lighting detail (words, points on a map).

■ *Letter movement*   Individual words or letters can be made to move, using pull-strips, magnets, pop-in or slide-in sections.

Unseen drawing
Seeing a drawing gradually form as we watch, has a persuasive fascination. It is a useful method of introducing complex information piecemeal, or making a little material go a long way. However, its extended use can become tedious.

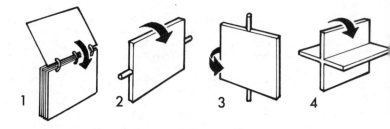

Fig. 13.12 part 1 Animated graphics—flips
1, Title-cards in ring binders, flip into shot (drop-in); or bottom hinged flip-out (drop-out). Margin finger-grip tabs aid operation. 2, Flip-over—double-faced title-card horizontally pivoted. 3, Flip-round. 4, Rotating flip.

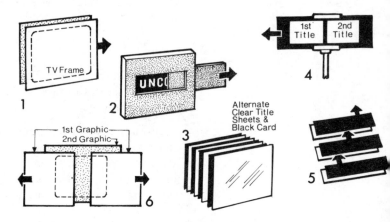

Fig. 13.12 part 2 Animated graphics—pull-outs (reveals)
1, Top card is slid aside to reveal the next. 2, Black slide-out section in black card, reveals titling underneath. 3, As each black card is pulled, titling on the clear sheet beneath is revealed. 4, Slide provides push-over wipe effect. 5, Shutters slide aside to reveal new information below. 6, *Breakaway*, where top graphic splits (slides apart or hinges) to reveal another.

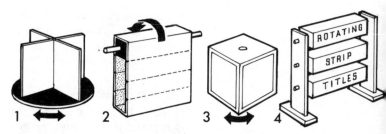

Fig. 13.12 part 3 Animated graphics—rotates
1, Turntable. 2, Flop-over or vertical slide. 3, Rotating box. 4, Rotating strips.

Fig. 13.12 part 4   Animated graphics—stencil and lighting methods
1, Stencil where pull-off tape (seen from rear) progressively uncovers rear-lit stenciled titling. 2, Animation by lighting—shadows grow, directing attention or creating atmospheric effect. Localized illumination can be used similarly.

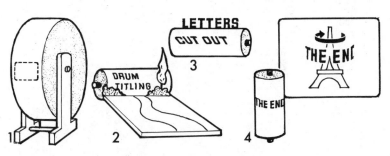

Fig. 13.12 part 5   Animated graphics—drum titles
1, Large vertical drum. 2, Surface-lettered drum as part of a model shot. 3, Stand-up lettering. 4, Drum lettering appears to move around an object (superimposed or keyed-in).

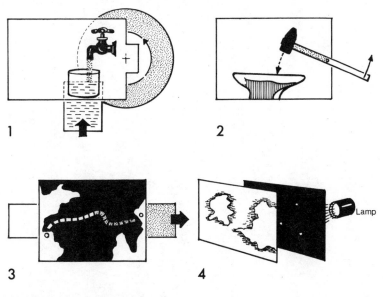

13.12 part 6   Animated detail
1, Rotating and sliding sections. 2, Hinged or pivoted sections. 3, Pull-out reveals. 4, Rear illumination.

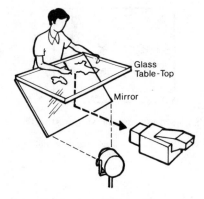

Glass
Table-Top

Mirror

Fig. 13.13   Animation table
The mirror reflects operations on the
glass-topped table. Sections can be
moved, removed, added, animated (in
vision or single-frame recorded). It
can be used for unseen writing.

Methods of obtaining unseen writing include:

1. Rear writing in white paint on a clear glass sheet backed by
   black drapes. The performer is totally dressed in black.
2. Rear writing with a felt-tip pen on a paper sheet; the camera
   shooting the bleed-through on the reverse.
3. Drawing on a surface of chroma key (CSO) hue (e.g. blue) for
   electronic insertion.
4. Writing on an overhead projector cel.
5. Special electronic 'remote-writing' equipment.

14 The background of production

The TV director's job varies considerably with the organization, and the size and type of production.

■ *Combined functions* For a smaller show he may be both producer and director. He is responsible for the entire business and artistic arrangements: origination, interpretation, casting, staging and treatment, and subsequently directing studio operations.

■ *Separate functions* The director may concentrate on the program's interpretation, staging and direction; while his producer as business head is responsible for organization, finance and policy. Sometimes a producer serves as artistic and business coordinator for several directors.

Fig. 14.1   Production team
Studio production requires the services and skills of a large number of people. Their exact job functions and titles vary between organizations.

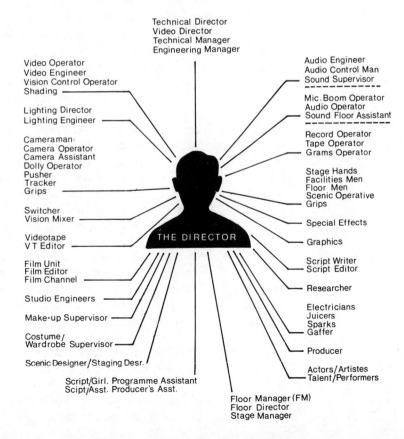

Technical Director
Video Director
Technical Manager
Engineering Manager

Video Operator
Video Engineer
Vision Control Operator
Shading

Lighting Director
Lighting Engineer

Cameraman
Camera Operator
Camera Assistant
Dolly Operator
Pusher
Tracker
Grips

Switcher
Vision Mixer

Videotape
V T Editor

Film Unit
Film Editor
Film Channel

Studio Engineers

Make-up Supervisor

Costume/
Wardrobe Supervisor

Scenic Designer/Staging Desr.

Script/Girl. Programme Assistant
Scipt/Asst. Producer's Asst.

Audio Engineer
Audio Control Man
Sound Supervisor

Mic. Boom Operator
Audio Operator
Sound Floor Assistant

Record Operator
Tape Operator
Grams Operator

Stage Hands
Facilities Men
Floor Men
Scenic Operative
Grips

Special Effects

Graphics

Script Writer
Script Editor

Researcher

Electricians
Juicers
Sparks
Gaffer

Producer

Actors/Artistes
Talent/Performers

THE DIRECTOR

Floor Manager (FM)
Floor Director
Stage Manager

The director's role

In all organizations, the director is the key figure who unifies and guides the studio team. A common practice with regular contributory productions, such as public affairs/current affairs, newscasts, magazine and talk programs, is for the director to have a *presentation role*. In this, he visualizes and coordinates a series of separate items (*stories*) that have been prepared by individuals in an editorial group.

In a *selective role*, the director heads a team of specialists (set design, lighting, audio, costume, make-up, graphics, etc.). After preliminary briefing/planning, they contribute their interpretations, which he subsequently appraises during rehearsal. Then, apart from indicating any alterations or preferences, he concentrates mainly on the dynamics of performance, camera treatment and switching (editing).

Occasionally the director, as *originator*, has written the script and initiated the staging. Then his specialist team aims to fulfil his ideas by organizing facilities, mechanics, labor and costing.

Production emphasis

The art of television direction is not *specific*. Emphasis changes considerably according to the type of production. While a drama production may benefit from painstaking shot-by-shot planning, a director on a live sports telecast relies on an almost clairvoyant ability to anticipate and switch shot opportunities from strategic camera positions.

In some shows (interviews), the staging and production treatment must provide an unobtrusive background to performance, while in others (spectaculars) these play a strong supporting role for action. Camera and audio treatment may aim at direct reportage or dramatically interpret mood.

While some productions have a relatively loose format, others require split-second timing with accurately cued inserts from live remotes, tape or film sources. Some types of production concentrate on action; others on *re*action. Dialogue may be all-important or incidental.

Away from the studio, the opportunities for coverage and treatment are invariably restricted. The considerable areas and distances often involved at remotes, the limitations of facilities, and the environmental problems, all influence production potentials. Lightweight cameras provide mobility, but various inherent problems (local acoustics, extraneous noise, weather, light variations, continuity, etc.) must affect treatment.

Production approaches

There is no 'standard correct way' of presenting subjects. Directors have tried most, over the years. But certainly, if we use an unsuitable approach our audience is quite likely to become confused, distracted, resentful . . . or simply lose interest.

Table 14.1   Production personnel

Producer	Organizational and administrative head of production group. May select program concepts, production personnel, and control budget.
Director	Responsible for staging the production, coordinating services, directing performers and crew; decides the final visual and audio treatment. May also operate video switcher.
Assistant producer/assistant director	Director's aide. May line up shots, prepare inserts, etc. Supervises pre-studio rehearsals. Advises director of up-coming cues.
Production assistant*/script girl	Checks performance against script. Times program.
Producer's assistant	As script girl, with certain cue preparation duties.
Floor manager (FM)/stage manager (SM)	Director's representiative and contact man on studio floor. Cues performers. Directs floormen. Responsible for general floor organization, safety, discipline. Assisted by AFM, who also ensures talent are present.
Technical director (TD)	Coordinates and supervises engineering facilities. Evaluates technical quality. May operate video switcher. May instruct operational crew during production.
Switcher/vision mixer	Specialist operating video switcher (and perhaps electronic effects).
Set designer/scenic designer	Conceives, designs, and organizes scenic treatment.
Make-up artist	Prepares and applies make-up treatment; aided by *make-up assistants* and *hair stylists*.
Costume designer	Designs/selects performers' costume (wardrobe). Assisted by *dressers/ wardrobe handlers*.
Graphics artist	Designs/prepares all graphics and visuals.
Lighting director	Designs, arranges and controls lighting treatment (technically and artistically).
Video operator/video control/ video engineer/shader	Controls picture quality by adjusting video equipment.
Audio engineer/audio control/ sound supervisor	Controls audio level, quality, balance; technical and artistic aspects of sound.
Camera crew	Cameramen, dolly operators, etc., responsible for camera operation.
Sound crew	Mike boom operators, audio personnel positioning microphones; operating audio tape and discs.
Floor men/stage hands/facilities men	Crew responsible for scenic changes, set dressing, properties, graphics operation, etc.
Electricians	Crew responsible for rigging and setting lamps, and electrical apparatus (including electrical props).
Special effects	Specialist designing and operating such mechanical illusions as fire, snow, explosions. Electronics effects specialist may devise/operate complex electronic insertion.
Announcer	Concerned with on/off announcements, commentary, continuity, etc.

*This term used in Britain for the FM on larger productions, involving some Assistant Director duties.

■ *Appropriateness* What is *appropriate*? Well, this is largely a matter of custom, fashion, convention:

1. *Informal situations* are often provided in naturalistic sur-roundings—the craftsmen in his workshop, the fireside chat, the country-walk interview.
2. *Formal situations* often follow a very stylized artificial format—carefully positioned chairs, desks, 'rules of procedure.'
3. *Display presentations* are generally unrealistic, with emphasis on decorative effects (singers and music groups).
4. *Simulated presentations* recreates a realistic environmental atmosphere, as in drama presentations; where any hint of the mechanics involved would destroy the effect.
5. *Actuality* has been emphasized (or imitated) by deliberately revealing mechanics. Somewhat unsteady hand-held cameras and microphones dipping into shot have supposedly added to the veracity of the occasion, as in 'vox pop' street interviews.

Routines

Some production techniques have become so familiar through convention or suitability, that it could seem strangely unorthodox if we presented them in any other way—e.g. a newscaster sitting in a lounge.

Certain approaches have become so stereotyped that they enter the realms of cliché; routine methods for routine situations. The sheer extent and repetition of TV productions has thrown up a number of such formats: newscasts, studio interviews, game shows, chat shows, etc. If we analyze such productions, we usually find that styles have evolved as the most effective, economic and reliable ways of handling their particular subjects.

If we regard such a packaging as 'a container for the goods,' then these routine treatments can free the audience to concentrate on the event. If however, we consider the presentation treatment as an opportunity to encourage interest and heighten enjoyment, then any 'routine' becomes unacceptable.

However, any dramatic or emphatic treatment would clearly be quite gratuitous for many types of TV production. Instead, it is best to aim for pictorial variety, coupled with clear unambiguous visual statements that direct and concentrate attention, rather than introduce any imposed 'style.'

Many non-dramatic subjects are *inherently* limited in their potential treatment. How many sensible meaningful shot variations can you take of people speaking to each other, or driving an automobile, or playing an instrument, or demonstrating an article—the range is small.

For certain subjects, the picture is virtually irrelevant. What a person has to *say* may be extremely important; while what he *looks* like, is immaterial to the message. It may even prove a distraction or create prejudicial bias. 'Talking heads' appear in most TV shows, but unless the speaker is particularly animated, the viewers' visual interest is seldom sustained. Changing view-point helps—but can appear fidgety.

Motion pictures inform us through continual illustration and commentary. Television's more economic but less enticing approach, is for people to tell us about matters instead. They sit in the studio with graphics or film clips, or stand at the scene of activities and talk directly to us, instead of using the camera itself to demonstrate. Here the difference in approach is not inherent in the medium, but the way it is being used.

Ambience

From the moment a show begins, we are influencing our audience's attitude to the production itself. Introductory music and titling style can immediately convey a serious or a jokey feeling towards what is to come.

We have only to recall how the hushed voice, quiet organ notes, slow visual pace, impart a reverential air to proceedings; or the difference between a regal and a 'show-biz' opening fanfare, to realize how our expectancy changes.

Surroundings can also directly affect how convincingly we convey information. Certain environments, for example, impart authority or scholarship: classroom, laboratory, museum, study. A plow shown at work on the farm will not only be more readily understood, but will carry a conviction that is lacking in the studio demonstration.

Selective tools

The camera and microphone do not behave like our eyes and ears, but substitute for them. Our eyes flick around with a knowledge of our surroundings, providing us with an impression of un-restricted stereocopic vision; in fact, we can only detect detail and color over a tiny angle (about $1\frac{1}{2}°$), and our peripheral vision is monochromatic and quite blurred. Also, our ears provide us with selective biaural pick-up, quite unlike the TV audio system.

In daily life, we build up an impression of our environment by personally controlled sampling; concentrating on certain details while ignoring others. The camera and microphone on the other hand, provide us with only restricted segments. And the informa-tion provided in these segments is modified in various ways, as we have seen, by the characteristics of the medium (distorting space, proportions, scale, etc.).

Selective techniques

If you simply set up your camera and microphone overlooking the action, the viewer soon becomes overaware of the small restricted screen and limited detail. He is prevented from seeing whatever is outside the shot. Go closer for detail and he progressively loses the overall view. Show more of the scene and particular detail becomes indiscernible. In the choice of suitable viewpoint and shot size for a particular purpose, lies the concept of guided selection; the beginnings of techniques.

Good production techniques provide *variety*: of scale and proportion; of composition pattern; of centers of attention and changing subject influence. You achieve these things by variation in shot size and camera viewpoint, by moving the subject and/or the camera, or by altering the subject seen.

Although you may sometimes encourage the viewer to browse around a shot, more often you will want him to look at a particular

Table 14.2   Typical reasons for production techniques

Artistically	To disguise the restrictions of the small flat TV screen and limited viewpoints.
	To guide and concentrate audience attention.
	To obtain visual variety and encourage continued interest.
	To emphasize, exaggerate, reduce or subdue information.
	To create a particular emotional impact (horror, tension).
	To beautify, glamorize, make more attractive (diffusion discs, soft focus, star filters).
	To expand/contract/distort space or form.
	To build up an illusion or effect purely by editing, viewpoint, camera-angle.
	To transform time and motion (time lapse, slow motion movement).
	To simulate an emotional or physical state (flying, drowning, unconsciousness).
	To achieve 'magical' effects (transformation, shrinkage, growth).
Technically	To disguise limited technical facilities (shooting on a single camera).
	To avoid visual discontinuity in time, place, position, angle, etc. (using cutaway shots to disguise time lapses, jump cuts, reverse cuts, etc.).
	To enable specific shots to be obtained (despite distance, terrain, confined space, etc.).
	To disguise or overcome spurious or intrusive factors.
	To permit editing cuts (shortening, correcting, covering lip-flap, etc.).
	To give time for moves (of performer, camera, boom).
	To provide space for moves.
	To overcome mechanical problems (prevent a camera appearing in shot).
	To overcome practical absurdities (shooting in the dark), or impracticalities (a succession of shots that would reveal the respective cameras if taken during continuous action).
	To create an illusion that would be impracticable or expensive if achieved directly.
	To provide picture (or sound) where none is strictly possible or available (see 'Visual padding').
	To analyze or synthesize (e.g. a study freeze-frame of action).

feature, and follow a certain thought process. The screen can all too easily become 'moving wallpaper', with our audience seeing—but not looking, hearing—but not listening!

The screen transforms reality

The camera and microphone can only convey an *impression* of the subject and scene. Whatever the limitations or inaccuracies of these images, they are the only direct information our viewer has available. And, of course, his interpretation must vary with his own experience and foreknowledge. Whether you are aiming to convey an accurate account (newscast) or to conjure an illusion (drama) the screen will transform reality.

You could fill the screen with a shot of a huge aircraft, or with its diminutive model. The pictures would look very similar. Yet neither conveys the subjective essentials; i.e. how you *feel* standing beside the giant plane or handling the tiny model. Introducing a person into shot would establish scale, but it would still not include our characteristic responses to such a situation: the way we would ourselves be overawed by size, or intrigued by minute detail.

Use your camera to select detail from a painting or a photograph and your TV screen puts a frame around it, to make this isolated area into a new complete picture; an arrangement that did not exist in its own right within the original; an arrangement that if sustained in close-up can become detached and dissociated in the audience's minds, from the complete subject.

When you shoot solid sculpture, its three-dimensional form becomes reproduced as a flat pattern on the TV screen. Planes merge and interact as they cannot do when we ourselves examine the real sculpture. Only on the flat screen can a billiard ball become transformed into a flat disc under diffused lighting. In practice, you can continually make use of this falsification of reality. The very principles of scenic design rely on it. But it is as well to appreciate that the camera and microphone do inevitably modify the images they convey; and that these images are easily mistaken for truth.

Interpretative techniques

It is one of those production paradoxes, that although your camera can show what is happening, it will often fail to convey the atmosphere or spirit of the occasion. You can frequently achieve more convincing representative results by deliberately using selective techniques, than by direct reportage.

Straightforward shots of a mountain climb impart none of the thrills and hazards of the situation. But use low camera angles to

281

emphasize the treacherous slope; show threatening overhangs, straining fingers, slipping feet, dislodged stones, laboring breath, slow ascending music . . . and the illusion grows. Even climbing a gentle slope can appear hazardous if strong interpretative techques are used.

Sometimes the audience can be so strongly moved by this subjective treatment, that sympathetic bodily reactions set in when watching such scenes—dizziness, nausea. Even situations outside the viewer's personal experience (the elation of free fall, the horror of quicksands) can be conveyed to some degree by carefully chosen stimulii.

Making the contrived arise 'naturally'

You can introduce techniques *obtrusively* for dramatic effect, or so *unobtrusively* that the effect appears 'natural' and the viewer is quite unaware that the situation *is* contrived:

Obtrusive: The camera suddenly depresses from an eye-level shot to a low-angle viewpoint.
Unobtrusive: The camera shoots a seated actor at eye-level. He stands, and the camera tilts up with him. We now have a low-angle shot.

Where situations seem to occur accidentally or unobtrusively, they are invariably more effective. For example, as an intruder moves towards camera, he becomes menacingly underlit by a nearby table lamp.

Where an effect is blatantly contrived, it often appears to challenge our credulity and we tend to reject it: A sinister figure is reflected in the victim's sunglasses; the camera whip-pans to show him standing nearby.

As conventions become understood and accepted by our audience, even flagrant changes become permissible: *flash cut-ins* (lasting $\frac{1}{2}$–2 sec) to convey 'recognition' or 'recall;' extreme viewpoint changes; close sound on long shots; exaggerated filmic time. But a director who deliberately misuses an established convention, trying to give it a new significance, treads a thorny path.

Many techniques have become so familiar, that we now regard them as the normal and natural way of doing things. But they are really artifices that help us to convey particular concepts:

1. Chipmunk voices (high-pitched through tape speed-up) for small creatures.
2. Echo behind ghostly manifestations.
3. Rim light in 'totally dark' scenes.
4. Background music.
5. Wipes, supers, etc.

Gratuitous techniques

Skilful presentation blends a carefully selected combination of visual and audio techniques to attract and persuade. But you can too easily use techniques for their own sake:

1. Rapid cutting in time with music—that provides a disjointed dissatisfying jumble of images.
2. Video effects—that leave us preoccupied with how it was done.
3. 'Clever' camerawork—focusing on irrelevant foreground objects while main subjects remain unsharp; shooting into lamps or specular reflections.

Such contrived methods have their occasional purpose, but they can quickly degenerate into imposed gimmicks.

Production pressures

Preoccupation with the organization and coordination of production mechanics, leaves most directors little time to meditate on the medium's aesthetics. Rehearsal time is limited. The camera and sound crews are meeting the director's brainchild for the first time and need to be guided in his interpretation. If the treatment is elaborate and exacting, there is greater opportunity for problems that require immediate solution. If the production is shot out of sequence, it becomes that much more difficult to ensure that each segment is coherent, and will provide good continuity when edited together. In such circumstances, there is an understandable temptation to substitute effective known mechanics for creative experiment.

Also, when you are closely involved with a production, it can be quite difficult to estimate accurately the audience impact of pace, timing, tension, etc. Watching a film clip through *several* times, you will see how the stresses, emphasis, speed, and even its significance can change. A word or gesture can have more or less impact with repetition, while heavily pointed action may appear hackneyed or mannered.

15 Production practices

Over the years, various production practices have evolved. Some derive directly from the mechanics of the medium, while others result from established procedures.

Single-camera treatment
Most TV productions are shot with a multicamera set-up. But effective visual treatment does not necessarily involve such complexity. There are normally two situations in which a single camera is used:

1. In the first, we deliberately restrict camera treatment to a single mobile camera; often arranging action and staging to suit this technique.
2. In the second, due to limited resources, we have only one camera available to shoot action.

■ *Discontinuous shooting*   Normal *filming* practice uses a single camera and breaks down all treatment into individually photographed shots; the action and camera set-up (sometimes staging) being arranged to suit each in turn. This technique requires care in maintaining *continuity* (of action, position, expression, etc.), and often involves repeating action (performance) to reshoot it from different viewpoints and so facilitate editing. Exactly the same technique can be followed with the TV camera—providing the extensive videotape editing is practical with available facilities.

■ *Restricted facilities*   With careful organization, it is possible, even with a *single static camera* using a fixed lens angle, to shoot continuously and yet introduce flexibility and shot variety. This is achieved by arranging action to provide mobility; moving people around the setting, transferring attention from one area to another.

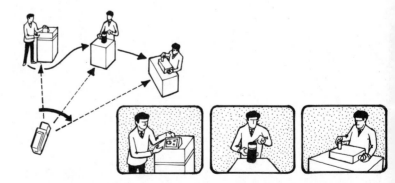

Fig. 15.1 part 1   Subject movement and the single camera—equidistant positions
Items can be arranged equidistant from the camera, to provide a series of similar shots as the talent moves around.

284

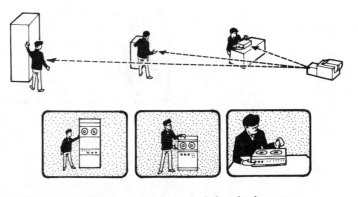

Fig. 15.1 part 2   Subject movement and the single camera—movement in depth
Where subjects of various sizes are involved, the largest can be located furthest away; the smallest closest to the camera.

You obtain variations in shot size by placing subjects at different distances from the camera. Larger items may be located furthest away, while small ones on a foreground table provide close-ups. A turntable helps the camera to see objects from various angles (mirrors are also useful). Occasionally, by pulling focus from one object to another, you can transfer attention.

■ *Variations*   If your single camera is mobile and fitted with a zoom lens, you have further opportunities. To change the camera's viewpoint unobtrusively, you might reposition it during action; for example, arcing round during a walk and perhaps incorporating slight zooming to reduce the amount of dolly movement. As a lecturer turns to point to a wall chart, the camera can zoom in to detail. When he lifts an object we have been studying in close-up, the camera zooms back to a mid-shot. Zooming here is provided shot variety without becoming intrusive.

■ *Inserts*   Use of *insert material* such as slides, film, and video-tape, can give opportunities to reposition the camera or change the shot. Inserts may show information not readily seen by the studio camera (hallmarks on silverware), or may provide supplementary illustrations. Take care, though, not to introduce visual discontinuity.

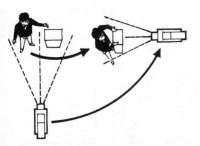

Fig. 15.1 part 3   Subject movement and the single camera—repositioned camera
Here the camera zooms out from a close-up, and arcs round to a new viewpoint as the talent moves to the new subject.

Multicamera treatment

Using two or more cameras you have the opportunity for continuous production: for instant visual change, intercutting different shot sizes and viewpoints of the same subject, or moving from one area to another.

Fig. 15.2   Relocation
During an insert shot, the talent has unexpectedly moved to a new location and is discovered there; often disconcerting the viewer.

So you can immediately provide fresh information, alter emphasis, point new detail, show reactions, shift audience attention, compare relationships, and introduce visual variety. Also you can combine cameras—for supers, inserts, split screen or vignette effects. One camera can be fitted with a visual effect (lens attachment) while others provide normal pictures.

When using only one camera you might have to alter a shot for solely mechanical reasons (e.g. widening it to accommodate an extra person). Two or more cameras help us to overcome this problem.

Fig. 15.3   Continuity
When retaking shots, inattention to detail can lead to poor continuity during editing; due to items having been consumed, moved, action or positions wrongly repeated, etc.

Visual variety

Static shots easily become a bore. So audience interest is encouraged by introducing movement and change. Although, if overdone the pictures can become a strain to watch. Rapid cutting

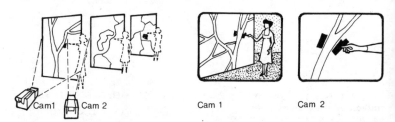

Fig. 15.4   Two-camera treatment
Here shots are divided between two cameras. Cam 1 concentrates on long shots. Cam 2 takes close-ups of maps and speaker.

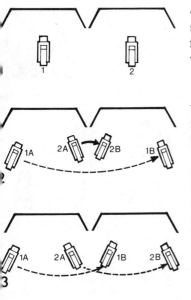

Fig. 15.5 Maximum use of cameras

1. One camera to each set is un-necessarily restrictive. 2. Near the end of a two-camera scene, Cam 2 pans to the next area. 3. Cameras move away successively. Cam 2 moves first (A to B), then Cam 1 joins it.

demands continual viewer concentration . . . that is not readily sustained. Ceaseless action and camera moves, lead to pictures that fidget with movement. As always, the aim is a well-balanced blend with variations in pace, tempo and emphasis.

■ *Performer movement* Most visual variety stems from action within the scene, as people alter positions and regroup. Quite often more meaningful changes are achieved by performer movement than by camera moves or editing.

In a close shot, the performer dominates. As he moves away from camera, the audience becomes more aware of his relationship to his surroundings. In long shots, these surroundings may dominate. So by changing the length of shot, you change emphasis and create visual variation.

These moves are motivated by various ways:

1. By *deliberately arranging* subjects (props, demonstration items) so that a performer is artificially 'motivated' to move to them in the course of the production.
2. By introducing a *gesture* (hand pointing) towards a subject—the camera pans over to see.
3. By a deliberate *turn* of the head or body—to initiate a pan, dolly or zoom to a subject.
4. By a *verbal reference* ('The machine over there') providing a direct introduction to a shot change.

■ *Changes by grouping* Where people are seated (panels, talk shows), you can achieve visual variety by *isolation*; selectively shooting individuals, two-shots, sub-groups.

When people are mobile, you can *regroup* them to provide freshly composed shots by introducing 'natural' moves (sitting, going over to pick up a book), or introduce visually motivated cuts (turning as a guest arrives). By varying emphasis and com-position balance in this way, the movement attracts attention and conceals the deliberate shift of interest.

■ *Shooting static subjects* Visual variety can be introduced into your treatment of non-moving subjects (statuary, pottery, paintings, flowers) by the way you shoot and light them:

Fig. 15.6 Visual variety
Depending on how a picture is arranged, its centre of attention, emphasis, and even the audience's interpretation can change.

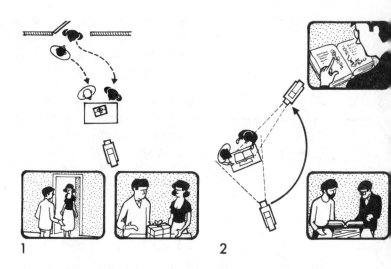

Fig. 15.7 part 1   Subject/camera relationship
1. Here subject and action have been arranged to suit the camera's static view-point (playing to the camera). 2. Here the camera moves, relating to the static subject and action.

1. Camera movements alter the viewpoint.
2. You can selectively pan over the subject, interrelating its various parts.
3. Changing lighting can isolate or emphasize sections or alter the subject's appearance.
4. Small objects can be handled, turned to show different features, or turntable-displayed.

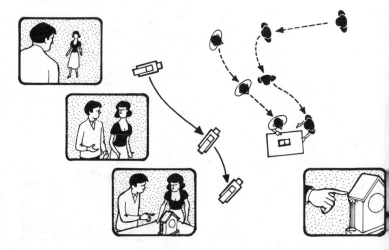

Fig. 15.7 part 2   Subject/camera relationship
Subjects may be arranged to provide shot opportunities, while cameras are moved or intercut to explore these arrangements.

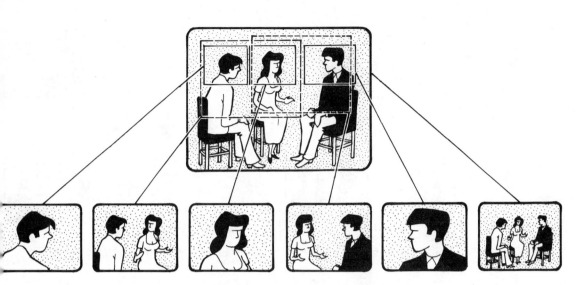

Fig. 15.8 Forming shots by isolation
By intercutting various shots of a group, you can introduce visual variety into a static situation.

■ *Variety by decor* The presentation of certain subjects is necessarily restricted (piano playing, singers). They are relatively static and meaningful shot variations are limited. You can prevent sameness between productions by introducing variations in the decor—whether by scenic changes or lighting. The pictures *look* different, even if the shots and viewpoints follow a recognizable format.

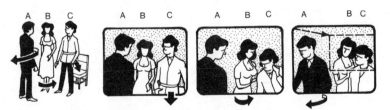

Fig. 15.9 Forming shots by subject movement
By introducing 'natural' moves you can form new groupings. Here *C* sits, *B* turns to him, *A* turns and exits, camera tightens to shot *BC*. A series of compositional changes have moved attention between the people.

Fig. 15.10 Composition change altering significance
By a single change in the picture's composition, you can alter its significance. The girl confronts her father (her frame position gives her strength, even in a rear view). He rises—the strength of his upward move and new position now make him dominant.

■ *Variety by effects* This is achieved by introducing various visual effects such as: combined shots, superimpositions, multi-images, split screens, decorative inserted backgrounds (using chroma key), and synthesized color.

Shot organization

It takes time to experiment, and production time is limited. So directors need a rational plan of campaign. You could distribute

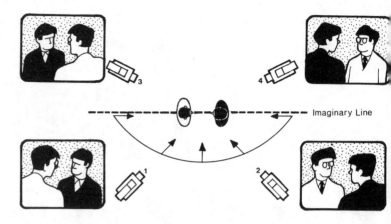

Fig. 15.11 part 1   The imaginary line—intercut facing shots
Shots can be intercut (cross-cut) between cameras located on the same side
of the imaginary line; i.e. between 1 and 2, or 3 and 4. But inter-switching
between cameras on opposite sides of this line, causes jump cuts (1 and 3,
1 and 4, 2 and 3, 2 and 4).

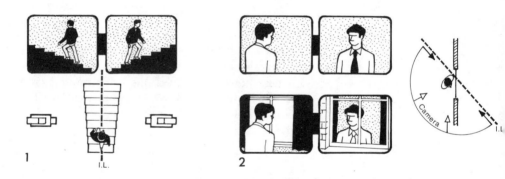

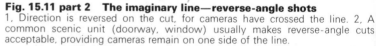

Fig. 15.11 part 2   The imaginary line—reverse-angle shots
1, Direction is reversed on the cut, for cameras have crossed the line. 2, A
common scenic unit (doorway, window) usually makes reverse-angle cuts
acceptable, providing cameras remain on one side of the line.

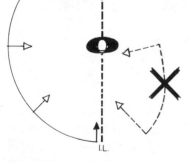

**Fig. 15.11 part 3   The
imaginary line—forward-
facing positions**
Whenever someone faces the camera
(walking, sitting, singing, demonstrating) it is all too easy to interswitch inadvertently across the line
to region X, and reverse direction.

cameras around the scene and choose from shots the cameramen
offered—but there would usually be little relationship between
these pictures. Subject coverage would be patchy. There would
probably be several near-identical shots of the pretty girl guest
and none of the chairman! Perhaps several overall views or
dramatic close-ups—but who could use them?

Shots have to be appropriately chosen and interrelated according to a coordinated plan, and only the *director* is in a position to
do this.

■ *Planned viewpoints*   Using this technique, you begin with the
action's mechanics (where people move, what they do there) and
arrange strategic camera viewpoints from which to shoot it. Each

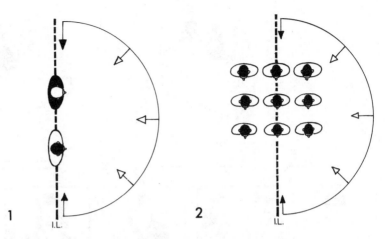

Fig. 15.11 part 4   The imaginary line—aligned positions
Where people are shoulder to shoulder (riding, walking) wide viewpoint changes are acceptable. 2, But for a larger gathering (audience, choir, parade) cameras should keep to one side of the front-rear axis.

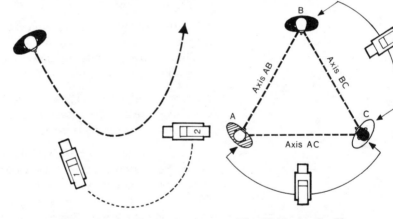

Fig. 15.11 part 5   The imaginary line—curving path
Where action follows a curved path, only a head-on viewpoint is reliable if shooting is discontinuous. With continuous action, viewpoint changes such as 1-2 are practicable.

Fig. 15.11 part 6   The imaginary line—three-way situations
If you assume that the direction of person A's look has established the axis AC, then you can intercut shots along the AC axis. Now, if he turns towards B, the axis is changed to AB (excluding C from shot) and shots can be intercut along the BC axis. Eyelines can be changed (even cheated) to permit inter-axis cutting.

ig. 15.11 part 7   The maginary line—exits and ntrances
during supposedly uninterrupted ction a person exits beside camera-ght 1, a straight cut showing ction continuation should show im entering camera-right also 2.

camera position provides a series of shot opportunities and you select from these available shots as required.

At sports events and other large-scale public occasions, this is the only practical approach. Cameras are often widely dispersed, camera movement is restricted, and the director relies on camera-men to use their initiative to find appropriate shots. The director's knowledge of actual shot opportunities is mostly derived from what his cameras reveal and on-the-spot assistants. He guides

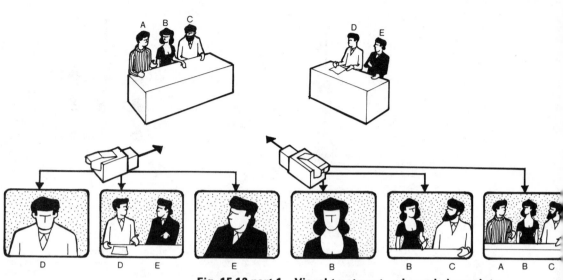

Fig. 15.12 part 1   Visual treatment—planned viewpoints
Cameras are placed at vantage points, from which appropriate shots can b
selected and intercut.

selection, adjusts shot sizes, suggests desirable shots, and choose
from the available material.

You can use a similar strategy in *studio* production, for certai
types of shows with a regular format. For example:

Camera 1—taking wider, *cover shots* (mid to long shots).
Camera 2—primarily close-up shots.
Camera 3—supporting close-ups, close detail, title cards, graphic

This approach can be particularly useful for unscripted, off
the-cuff situations. But it is also valuable for a demonstrator, wh
consequently knows which camera to play to for close-ups at an
time. However, you should resist any temptation to apply such
routine where more exploratory camera treatment is possible.

■ *Storyboard approach*   Where systematic planning is practica
a director may use a storyboard technique; sketching the require
picture treatment for selected *key shots* or for scene-by-scen
treatment. Here we are concerned with *shot significance*; ensurin
that the shots have the appropriate dramatic value, attrac
attention to certain features, emphasize a particular point, an
engender a mood.

Storyboard methods involve analyzing the script, deciding o
composition arrangements, and then working out how to ge
them. You can use such planning as a general or specific guide
Accurate prerehearsal planning can achieve the highest standards
but it must be *realistic*, taking into account how one set-up need
to be developed from another. Most directors dislike the commit
ment of such planning methods and prefer a more flexibl
empirical approach. Although, where complex special effects ar
involved the storyboard may become essential.

Table 15.1   Arranging the shot

	In arranging a shot, you should be able to answer such questions as:
Broad objectives	What is the *purpose* of the shot? What is it aiming to *show*?
	Is it to *emphasize* a particular point or feature?
	Which is the *main* subject?
	Are we *primarily concerned* with: the subject, its relationship to another, or to its background/environment.
The actual picture	Is the shot *too close or distant* for its purpose?
	Is the *attention* reasonably localized—or split or diffused?
	Is the *composition arrangement* appropriate?
	Is the subject suitably *framed*? (Headroom, offset, edge cut-off, overcrowded frame).
	Are subjects *clearly seen*? (Sharp, unobscured, good background contrast?).
	Is there any *ambiguity* or *distraction* in the shot?
The action	Are we aiming to show what a person is *doing*—clearly, forcefully, incidentally, not at all?
	Does action (movement gestures) *pass outside the frame*?
	Are any *important features* or action accidentally excluded?
More specific objectives	Is the presentation to be *straightforward* or *dramatic*?
	Do we want to indicate subject *strength* or *weakness*?
	Do we aim to *reveal, conceal, mislead, intrigue, puzzle*?
	What *effect, mood, atmosphere* are we seeking? (Businesslike, clinical, romantic, sinister) . . .
	Does the shot *relate successfully* to previous and subsequent shots?

Fig. 15.12 part 2   Visual treatment—storyboard approach
Following script analysis, treatment is visualized and broken down into key shots, which are sketched to show required camera treatment.

293

Table 15.2   Selecting camera treatment

	Mechanical purpose	Artistic purpose
Why pan or tilt?	To follow action. To show a series of conjoined parts. To exclude unwanted subjects. To show an area that is too large to be contained in a static shot.	To join separate items. To show spatial relationships. To show cause/effect. To transfer attention. To build up anticipation or suspense As an introductory move.
Why elevate? (ped up)	To see over foreground objects. To see overall action.	To reduce prominence of foreground To look down onto objects. To reduce subject strength.
Why depress? (ped down)	To frame picture with foreground objects. To obscure distant action. To obtain level shots of low subjects.	To increase their prominence. To emphasize subject strength.
Why zoom?	*See tables 12.1 and 12.2*	
Why dolly (track) in?	To see or emphasize detail or action. To exclude nearby subjects. To recompose shot after one of subjects has repositioned or exited. To change emphasis to another subject. To emphasize an advancing subject.	To underline an action or reaction. To emphasize subject importance. To localize attention. To reveal new information. To transfer attention. To follow a receding subject. To provide spatial awareness. To increase tension. To create a subjective effect.
Why dolly back?	To extend field of view. To include more of subject(s). To include widespread action. To reveal new information. To include entry of a new subject. To accommodate advancing action. To withdraw from action.	To reduce emphasis. To show relationship or scale of previous shot, to the wider view (whole of subject, other subjects.) To increase tension, as more significance is gradually revealed. To create surprise (e.g. reveal unsuspected onlooker). To provide spatial awareness. To create a subjective effect.
Why truck (crab)?	To follow subject moving across the scene. To reveal the extent of a subject/ scene, section by section. To examine a long subject or series of subjects.	To emphasize planes in depth (parallactic movement).
Why arc?	To see subject from another viewpoint, without transitions. To exclude/include a foreground/ background subject. To realign or recompose subject(s) when it moves. To reveal new information or a new (extra) subject. To correct for inaccurate subject positions (actor off marks).	To change visual emphasis.
Why follow a moving subject?	In *close shots*—to keep it in frame, while showing reactions or detail information. In *long shots*—to show subject progress through an environment, or its relationship to other subjects.	To spatially interrelate subjects. To avoid transitions or viewpoint changes; maintaining continuity.

294

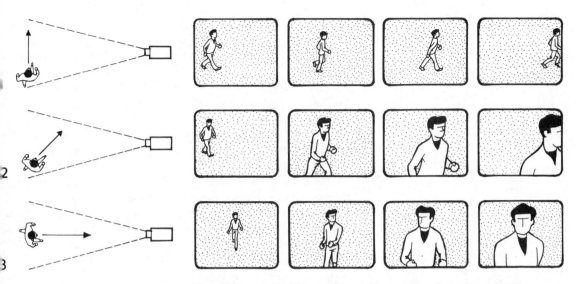

Fig. 15.13 part 1   The moving subject—direction of movement
1, A movement across the screen, quickly passes out of shot. 2, Diagonal moves are preferable, for they are visually more interesting. 3, Moves towards the camera from a distance are sustained longest, but can seem drawn-out.

Program opening

Right from its introduction, you are indicating to your audience the intended character of your presentation. There are a number of opening gambits:

1. *The formal start*   Beginning with a 'Good evening' or 'Hello', introduces the show and gets on with it. Whether the presenter appears casual, reverent, indifferent or enthusiastic, can create an ambience directly influencing the audience's attitude.

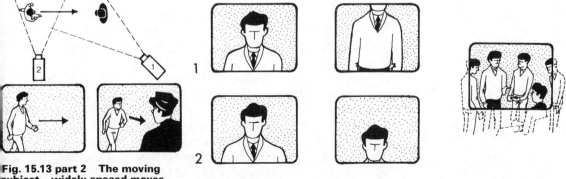

Fig. 15.13 part 2   The moving subject—widely spaced moves
Moves between widely-spaced points are best shot obliquely (Cam 1), rather than with a following pan (Cam 2).

Fig. 15.13 part 3   The moving subject—avoid decapitation
1, The cameraman must be warned of a 'rise' or 'sit', to avoid the bizarre effects of a delayed following tilt. 2, Equally ridiculous effects can arise if framing does not follow action.

2. *The teaser*   Showing dramatic, provocative, intriguing high-lights from the production, before opening titles. Sometimes this inadvertently gives away the plot before the show ever begins!
3. *The crash start*   Takes us straight into the program, which probably appears to have begun already. An automobile screams to a stop outside a store, a figure throws a bomb . . which explodes, an alarm bell shrills, a police siren wails—the show has begun and titles roll over the chasing vehicles.
4. *The character introduction*   A rapid montage of symbols or shots of the hero in various predicaments, provides an intro-duction to the characters we are to meet.
5. *The eavesdropping start*   The camera peers through a house window, sees a group round a TV screen, and moves in to join them.
6. *The coy welcome*   The camera dollies in to someone who is supposedly preoccupied. Realizing that we have arrived, his welcome of 'Oh, there you are!' provides informality—or nauseating artifice.
7. *The slow build-up*   The camera pans slowly round, arousing curiosity or suspense . . . until we reach a climax point (the bloodstained dagger). It is essential to avoid diminishing interest or anticlimax.
8. *The atmosphere introduction*   A series of strongly associative symbols establish the place, period, mood or personality. On a mantleshelf: the brass telescope, ship in bottle, and well-worn uniform cap—the old sea captain is introduced long before we see him.

Subjective and objective approaches

You can use the camera either:

Objectively—in which the viewer becomes an *observer*.
Subjectively—in which he finds himself a *participant* in the action.

In his *objective* role, the viewer becomes an onlooker, eaves-dropper, an invited audience at a vantage point, or a casual bystander. He sees what is going on but is never addressed directly.

Subjective approaches are a regular part of TV, as performers speak straight to camera, using the screen as a communicating opening to our home. The *subjective* role can be an extremely powerful one, when the camera lens moves within the scene as our eyes—or those of a character. It goes up to see an object. It moves amongst a group of dancers. It follows a guide around a museum. The concept can be used forcefully, linking us closely with the action as the camera 'climbs stairs' or jumps out of a plane. Taken to extremes, it all becomes a visual stunt as we find ourselves punched, kissed, fired at, even drowned!

Fig. 15.14   The developing shot
Continuous camera movement may explore the scene where intercutting would disturb the sustained mood, yet you wish to change the focus of attention, reveal new information, show various reactions, etc.

In *developing* (*development*) *shots*, the camera moves from one viewpoint to another; helping us to build spatial impressions and to see varying aspects of the action. This requires carefully controlled camerawork, but it can be most effective in *slow-paced* scenes involving tension, expectancy, solemn or romantic occasions.

Providing the viewer realizes his intended relationship to the scene, you can change techniques between subjective and objective treatment. At times, the distinction may be very nominal.

Focusing audience attention

It is as well not to overrate how much effort the audience is willing to make to evaluate your picture. You can do a great deal to localize and hold their attention, by taking care how you arrange and present subjects.

Audience concentration easily lapses, so you need to continually direct and redirect their attention, to hold interest along particular lines. Such redirection implies *change*. But excess or uncontrolled change can lead to confusion or irritation. Pictorial change must be clearly motivated, so allow the viewer to readjust himself easily to each new situation.

■ *Varying concentration*   You have to strike a balance between *sustained concentration* (where interest soon flags) and free association (where minds wander):

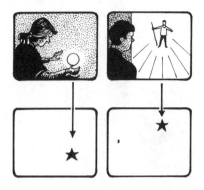

Fig. 15.15   The center of interest
You can attract attention to a specific place in the shot, by the way you arrange and present the subject.

1. In some shots you will aim to concentrate attention on a particular center of interest.
2. In others, you will encourage attention to move around, on ordered detectable lines.
3. While others of a more general nature, encourage free ruminative inspection; concentrating perhaps on the audio (commentary, dialogue, music).

Fig. 15.16   Unifying interest
The eye should normally be led to 1, a single, 2, unambiguous center of interest.

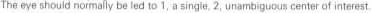

Fig. 15.17   Changing the center of interest
1, The eye-lines and attitudes of performers guide the viewer to a particular center of interest. If there are no visual clues, his attention is not readily transferred to another point. He must look around to see who is speaking. 2, By introducing eye-line and posture clues, the transference becomes obvious. 3, In moving interest around the picture, you must take care not to draw attention to a focal point where there is no real interest or create ambiguous centers. 4, Avoid leading attention out of the frame.

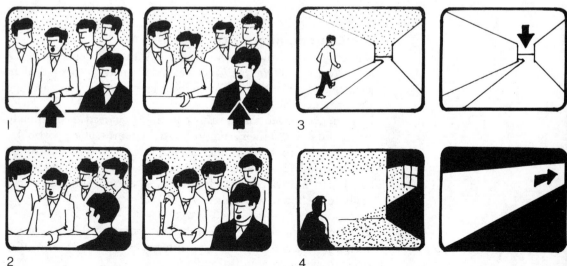

Fig. 15.18 How to focus visual attention
1, Avoiding spurious centers of interest, e.g. prominent lettering, tonal contrast, etc., in supporting subjects. 2, Compositional lines (real or implied). 3, Tonal gradation. The eye naturally follows gradation from dark to light areas. 4, By deliberate unbalance (tonal or linear), with the subject in the heaviest position. By isolating the main subject: 5, In depth. 6, Horizontally. 7, By body position. 8, By height. 9, Using a stronger part of the frame. 10, With lighting. 11, 12, By similarity between the subject and its surroundings, although too great a similarity can pall or 13, 14, lead to confusion.

Although you cannot expect to direct an entire audience's attention with any precision, you can do a great deal to lead them in their selection and influence their associated thoughts.

■ *Shifting visual interest* It is just as necessary to be able to shift the viewer's concentration to another aspect of the picture, as it was to localize his attention originally. You can do this by readjusting any of the influences listed:

1. Having a person stand up within a seated group.
2. Give the first person weaker movements (turn away), strengthening those of the new subject (face the camera).
3. Transfer the original emphasis (contrast, isolation, etc.) to the new subject.
4. Use linking action; first person looks to camera right . . . cut to new subject on screen right.
5. Weaken original subject, having him move to the new stronger subject.

6. Alter the shot size, camera height, etc.
7. Pull focus from old subject to the new.
8. Change the sound source; new person speaks instead.

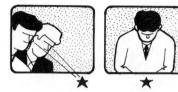

Fig. 15.19 Hidden areas of interest
If an implied center of interest falls outside the frame, or is obscured, you may either intrigue—or frustrate the audience. Subtly introduced, the technique may make a change of viewpoint welcome.

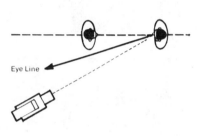

Fig. 15.20 Eyelines
Sometimes a more attractive convincing shot is obtained by having an actor cheat his eyeline. Instead of looking at the other person, he turns his eyes/head along an axis roughly bisecting the centerline and the lens axis (using a distant marker). Cheated eyelines may also be used to intercut across the imaginary line (see Fig. 15.11).

Creating tension

Tension in a dramatic situation derives partly from the dialogue, story line, and interaction between characters; but it can also be considerably influenced by the way in which you present the subject:

1. Using progressively more powerful shots (intercutting closer and closer shots; lower viewpoints, gradual canting).
2. Using suspenseful music and effects.
3. Presenting ambiguous information—is the nocturnal shadow a bush . . .or a prowler?
4. Presenting insufficient information—the audience sees the figure in the doorway . . . is it the villain?
5. Withholding information—have the pursuers arrived yet?
6. The audience knows something a character does not—he runs to escape . . . we know the route is blocked.
7. A character is suddenly confronted with an insurmountable problem—the stairway collapses as he reaches it.
8. We anticipate what is going to happen—surely someone will burst in through that door.

The borderline between tension and bathos can be narrow. An intended climax too easily becomes an *anti*climax. So you have to take care not to emphasize the trivial, or unintentionally to permit an emotional let-down after an emotional peak.

Pace

We might define *pace* as the rate of emotional progression. While a *slow pace* suggests dignity, solemnity, contemplation, deep emotion; a *fast pace* conveys vigor, excitement, confusion, brashness, etc.

A well-balanced show continually readjusts its pace. A constant rapid tempo is exhausting; while a slow sustained one becomes dull. Pace comes from an accumulation of factors:

1. *The script*—Scene length, speech durations, phrasing, word lengths. Sharp snappy exchanges produce a faster pace than lengthy monologue.
2. *The delivery*—Fast, high-pitched, interrupted sounds provide a rapid pace; compared with slow, low-pitched ones.
3. *Production treatment*—The rate of camera movement, switching, performer moves.

The eye can maintain a quicker pace than the ear. While the eye can assess, classify, evaluate almost immediately, the ear has to

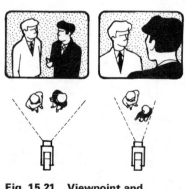

Fig. 15.21 Viewpoint and subject attitude
While frontal viewpoints can suggest alliance, over-shoulder reverse-angle shots can convey conflict.

Table 15.3 Methods of focusing attention

Exclusion	Taking close shots. Excluding unwanted subjects. Using neutral backgrounds.
Visual indication	Indication with a finger, pointer, inserted or superimposed marker (arrow, circle).
Aural indication	A verbal clue or instruction—'Look at the black box.'
Color	Using prominent contrasting hues against neutrals or muted (pastel) colors.
Camera control	Avoid weak viewpoints (side or rear) or weakening angles (high or long shots). Concentrate interest by differential focusing or camera movement.
Composition	Using convergent line or pattern, picture balance, isolation, prominence through scale.
Contrasting the subject with its surroundings	Through differences in relative size, shape, proportions, scale, type of line, movement, association (e.g. elaborate with plain), disposition (e.g. seated with standing), etc. The area of interest might have the lightest tones or maximum contrast in the picture.
By movement	Movement attracts according to its speed, strength, and direction. Change direction during motion, or interrupt and resume, rather than maintain sustained action.
	According to how it is introduced, a movement can attract attention to: *itself* (a moving hand), the *subject* (person whose hand it is), or the *purpose* of the movement (what the hand is pointing at). Remember, when the camera moves (or zooms) it can virtually create the illusion of subject movement, so drawing attention to it.
	Synchronizing a movement *with* dialogue, music, effects (especially the subject's own), gives it strength and draws attention to it. Having a person move on his own dialogue emphasizes both action and speech.
By subject attitude	Having performers use strong movements—upward or diagonal, stand up, play to camera, move in front of others or scenic elements.

piece together consecutive sounds to interpret their overall meaning.

While a fast *visual* pace is readily assimilated, it is usually at the expense of less attention to the accompanying sound; unless visual changes are so rapid that the viewer ignores their information and just listens! Where emphasis is to be on the sound, visual pace generally needs to be relaxed.

Timing

There are two kinds of timing:

Mechanically—timing refers to duration checks that ensure a program keeps to schedule.
Artistically—timing is choosing the right moment and duration for an action—exactly when to cut, the speed of a transition, the pause duration between a comment and a retort.

Inept timing lays wrong emphasis. It can ruin a gag, over-emphasize mechanics, and disrupt continuity.

1   2

Fig. 15.22   Viewpoint and clarity
A good camera viewpoint can mean the difference between obscurity (1) and clear demonstration (2).

Visual clarity

If a picture is to get its message over quickly and unambiguously, your audience must be able to see relevant details easily and clearly. Aim to avoid confusion, ambiguity, obscured information, restricted visibility, distractions and similar visual confusion.

■ *Viewpoint*   A poor viewpoint can make even the commonest subjects look unfamiliar. Occasionally you may want to do this deliberately. But you should generally aim at clear unambiguous presentation; even if, for instance, it becomes necessary for a demonstrator to handle or place items in an unaccustomed way to improve shot clarity.

■ *Distractions*   Poor lighting treatment can distract and confuse, for it can hide true contours, cause planes to merge, and cast misleading or unmotivated shadows (especially from unseen people). It can also unnecessarily create hot spots, lens flares and specular reflections.

Strong tonal contrasts can distract, as can strongly marked detail. Where someone stands in front of a very 'busy' background, the cameraman may inadvertently focus harder on the background (or seem to) than on the subject itself. Close inspection of eyes, necktie, hair, etc., will show us if this is happening.

Slightly defocused details, particularly lettering that we cannot read easily, can sidetrack our attention. For similar reasons, strongly colored defocused background objects are best excluded from shot wherever possible.

Confusing and frustrating techniques

How often, when watching regular TV productions, do you experience total frustration or antagonism at the way the show is being handled? Here is a list of frequent annoyances:

1. Important subjects cannot be clearly seen—they are soft-focused, masked, or shadowed; or merge with the background.
2. Someone points to detail—it is too small or fuzzy to see.
3. An intriguing collection of items fills the table—we are shown only a couple.
4. The demonstrator shows the host a particularly interesting item—the camera never sees it in detail.
5. An item is promised us later in the show—but time runs short and it is eliminated.
6. Titling or graphics are shown—but too briefly for us to read or examine properly.
7. A demonstrator shows us how to prune a tree—but the camera gets a close shot just as he finishes.
8. We are watching interesting action—but it is cut short and now we watch a talking head instead.

Fig. 15.23 Ambiguity
1, In a flat picture, visual ambiguity can arise whenever lines or tones in one plane merge into those of another; as when adjacent patterns or adjacent tones are similar or where scenic lines continue. 2, Take care to avoid items in the set becoming extensions of the performer.

9. We can hear the events taking place off camera—but the camera is still on this voluble announcer.

10. The speaker who says to camera, 'I'll be with you in just a moment'—why isn't he ready?

11. The interviewer who asks questions (quoting from notes), but is not really listening to replies.

12. The interviewer who gives the *guest* information about himself!

13. The interviewer who asks the guest questions which require only 'Yes' or 'No' replies.

14. A too-brief glimpse of a subject—followed by the commentator talking about it instead.

15. Someone speaks in a group—but the camera dwells on the previous person. (They have not yet lined up a shot of the new speaker!)

16. Wrongly cued performers—the action has already started; or action has finished but the shot remains.

17. The shot that leaves us wondering what we are supposed to be looking at or where we are now.

18. The lost opportunities when a film crew goes on location—but mid-shots of a speaker against nondescript backgrounds predominate.

Fig. 15.24 Background contrast
Hand held items are often shot against clothing. Ensure that this background is not similar in hue and tone, or too distracting. If necessary hold it against a more suitable background.

Interest or concentration patterns

Throughout any production, audience interest and concentration will fluctuate; growing and subsiding. Tension and excitement build up and relax. The trick is to arrange these responses so that they come when you need them most.

Directors and writers rely on experience and intuition rather than analysis, to blend the various contributory factors:

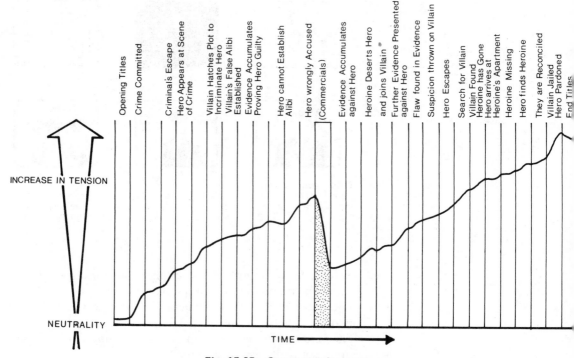

Fig. 15.25 Concentration pattern
Showing audience concentration throughout program.

1. *Script construction*—the order and duration of shots, their dramatic strength, plot outline, etc.
2. *Interpretation*—the performance, its delivery, etc.
3. *Production treatment*—staging, camera treatment, editing, audio treatment, etc.

But by outlining this reaction pattern as a rough graph, you can see how the overall impact changes throughout the production and how the separate factors combine.

In the well-worn story line of Figure 15.25 you can see where one seeks to build tension; where it will be allowed to fall. You aim to grip audience interest at the start of the show, introducing suitable variations, and building to an encouraging peak just before commercial breaks. You can see the bid to recapture interest immediately afterwards . . . and so on to the finale.

Both strengthening and relaxing tension require some care. After a lull, a peak appears greater. But after a long relaxed period, a highly dramatic peak can be intrusive, so that the viewer becomes overaware of his own sudden reaction; responding with laughter rather than shock. A climax may even pass him by, unless he is emotionally preconditioned to receive it.

Preferably, you build interest, concentration, or tension, in a progressive series of peaks. But too slow a development can allow interest to wane. Too many climaxes eventually produce no climax at all.

Relaxation after an interest peak can lead to indifference. Again, sudden tension release can precipitate amusement, after the nervous strain of concentration.

Where action or storyline follow a recognizable pattern, the viewer may grasp the point before the director has time to finish making it! To avoid this happening, you can disguise the obvious through careful treatment; or if this is not possible, at least bypass all superfluous detail (using filmic space and time).

Sometimes attention will wander despite great visual activity. And when quickening pace or high mobility follow a slow episode (e.g. a fast-cutting film clip after slow-speed studio shots), the viewer often finds himself *inspecting* the fast scene rather than stirred by its tempo.

If you present an excess of information or provide facts too quickly, the earnest student may feel frustrated and confused at

Fig. 15.26　Scene analysis
Various factors contribute to the effect of each scene.

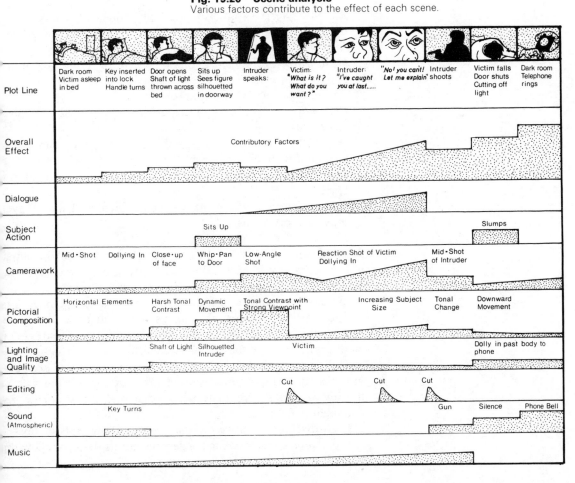

his inability to note them. But a general audience is more likel
to absorb a few random aspects and then relax when they hav
seen enough (not, as some believe, becoming bored when the
have grasped its full import).

Visual padding

Although there are occasions when you can make effective use o
silence, a blank screen is an anathema! Yet many subjects have n
direct visual element. Consider discussions, poetry, music—th
emphasis is on *sound*. What does the viewer gain from *seein*
people perform? He sees expressions, instrumental fingering, etc.
but these are the mechanics. They have nothing intrinsically t
do with the material itself, its significance or purpose (i.e. th
argument or musical effect). In fact, the pictures may actuall
cause *distraction* as he watches the performers—becoming pre
occupied with their appearance, expressions and mannerisms.

■ *The visual problem*   For some programs, directly appropriat
visuals are not obvious:

1. *Abstract subjects*—philosophical, spiritual, social concepts.
2. *General, non-specific subjects*—humanity, transport, weather.
3. *Imaginary events*—hypothetical, fantasy.
4. *Historical events*—before photography, or unphotographe
   events.
5. *Forthcoming events*—future projects.
6. *Filming is not possible*—photography is prohibited or subjec
   inaccessible.
7. *Filming is impracticable*—photography is too dangerous
   meaningful shots not possible.
8. *Concluded events*—event now over and has not been photo
   graphed.
9. *Appropriate visuals too costly*—would involve distant travel
   copyright problems.

■ *Possible solutions*   When the director cannot show the actua
subject being discussed, he often has to provide a suitable alterna
tive picture—a kind of *visual padding* or 'screen filler'.

The most economical solution, and the least compelling, is t
introduce a *commentator*; who *tells* us about what we cannot *see*
as he stands at the now empty location (historical battlefield
site of crime, outside the conference hall).

Inserts in the form of photographs, film clips, VT excerpts
paintings, drawings (typically used for courtroom reports), ca
all provide illustrative material. Occasionally, a dramatic re
enactment is feasible.

When discussing future events, you can show stills or library
shots of a *previous occasion* to suggest the atmosphere, or show th
principles (celebration days, processions).

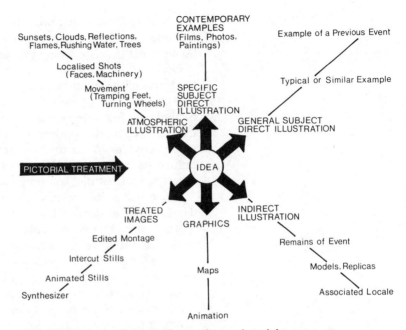

Fig. 15.27 part 1   Subject illustration—pictorial treatment
Where pictures are not readily available, or a more imaginative effect is required
than the direct approach, various substitutes are possible.

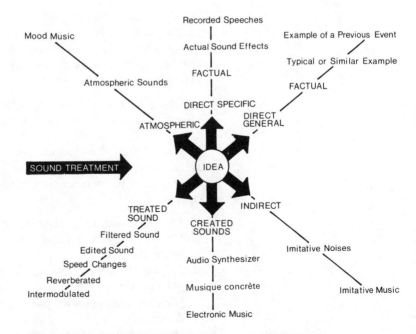

Fig. 15.27 part 2   Subject illustration—sound treatment
Some possible alternatives to direct sound.

You may show a *substitute* subject (not the animal that escaped but one just like it), hoping your audience does not mistakenly assume that they are seeing the actual subject.

Occasionally the camera can show the *absence* of the subject: the frame of the stolen painting; where the castle once stood.

Associated subjects are frequently used. We visit the poet's birthplace, often using stock tourist location shots. However, they may not be strictly applicable (wrong period), inaccurate (wrong part of city), or irrelevant (not architecture, but social conditions influenced his poetry). But, apart from family album snapshots or newspaper cuttings, nothing else is available!

When using library (stock) shots or stills, there is always the danger, that available shots will become overfamiliar through repeated use. This is particularly likely with historic material, or unexpected tragedy (assassination, aircrash).

Some forms of visual padding suit a variety of occasions. The same waving wheatfield can epitomize food crops, daily bread, prosperity, agriculture, the war on insect pests . . .

Abstracts can be pressed into service at almost any time! Atmospheric shots of rippling water, shadows, light reflections, into-sun flares, defocused images, shooting up into a tracery of leaves . . . all have very regular use!

Illusion of time
Motion picture editing has long since accustomed us to concepts of filmic space and time:

Filmic space—intercuts action that is concurrent at different places. As a soldier dies . . . his son is born back home.
Filmic time—omits intermediate action, condensing time and sharpening pace. We cut from the automobile stopping . . . to the driver entering his apartment.

Where all the intervening action has no plot relevance, the viewer would often resent a slow pace, stating the obvious.

■ *Time lapses* You can use several devices to indicate the passage of time. Explanatory titles are direct and unambiguous, but other subtler techniques are generally preferable. For short time lapses:

1. Slow fade-out, new scene slowly fades in.
2. Cutting away from a scene, we assume time has elapsed when returning to it.
3. A time indicator (clock, sundial, burning candle) shows passage of time.
4. Lighting changes with passing time (a sunlit room gradually darkens).
5. Dissolve between before/after shots of a meal, fireplace, etc.
6. Transition between sounds with time association—nocturnal

frogs and owls, to early morning roosters and birdsong.

7. Defocus shot, cut or mix to another defocused shot and then refocus.

For longer time lapses:

1. A calendar sheds leaves or changes date.
2. Seasonal changes—from winter snow to spring flowers.
3. Changes in personal appearance—beard growth, ageing, fashion changes.
4. New to old—dissolving from a fresh newspaper . . . to a yellow crumpled discarded version.

■ *Flashbacks*   A familiar device, the *flashback* turns back time to see events before the present action for reminiscive, explanatory or comparative purposes. Typical methods include reversed time-lapse conventions (e.g. the old becomes new again), or explanatory dialogue during 'mist', edge diffusion, ripple or defocus dissolves. Nowadays, a very brief *flash cut-in* ($\frac{1}{2}$–2 sec long) conveys recognition or moments of memory recall.

Table 15.4   Continuity

	Poor continuity can arise when shooting discontinuously or out of sequence (for later correct order editing). Continuity problems are of several kinds:
Technical continuity	Matching intercut picture quality (color, brightness, key, contrast, tonal balance); light direction; shot heights; perspective.
Pictorial continuity	Maintaining similar atmospheric effect (weather conditions—rain, wind, light quality).
Environmental continuity	Achieving compatible viewpoints within the location.
Spatial continuity	Avoiding the audience losing orientation on reverse cuts.
Physical continuity	People should have identical costume, appearance, etc., in conjoined shots; environment seem identically dressed (same items in the same place).
Performance continuity	Where the director repeats an action sequence (to provide different shot lengths or viewpoints for intercutting), performers should have similar expressions and make similar gestures in each version.
Temporal continuity	There should be appropriate indications of the passage of time during scenes (check variables for continuity: food, cigarettes, drink, candles, clocks, etc.).
Event continuity	Avoid missing part of a continuous action sequence (unless filmic time/space are intended). Avoid showing duplicated action. Avoid rediscovering people in a scene after they have apparently exited; or finding a person entirely relocated after an intermediate shot, although continuity was implied.
Relationship continuity	Avoid mismatching that jumpcuts the subject around the frame.
Attention continuity	Maintain related centers of attention between shots.
Audio continuity (technically)	Ensure that sound volume and audio quality are compatible between conjoined shots.
Audio continuity (artistically)	Maintain similar atmosphere (reverberation, background noise, effects, room tone, audio balance) in an edited scene.

Interscene devices

As well as the usual editing transitions (cut, dissolve/mix, wipe) and their derivations (defocus and ripple dissolves), you can use various interscene devices to link sequences.

1. *Blackout*—a performer moves up to block out the lens. He (or another subject) moves away, revealing a new scene. A balloon is used similarly, being burst to show the new scene.
2. *Pan away*—the camera turns from the action area to a nearby subject. Panning back, we now see different action (or appearance), for time has passed.
3. *Bridging subject*—associative material interconnects two scenes: shots of wheels turning to imply travel, swirling mists to form an atmospheric transition.
4. *Decorative bridge*—abstract forms in light or shadow between scenes. Moving patterns—kaleidoscope, spirals, whip-pan.
5. *Matching shots*—the close-up ending one scene (wrist watch) dissolves to a matching shot (bus station clock) in the next.

Deliberate disruption

The idea of deliberately cutting away from main action, to look at incidental happenings nearby, may seem somewhat incongruous. Yet, if appropriately introduced, this disruption offers valuable production opportunities. Providing the viewer does not feel that he is missing vital information or is robbed of exciting action, the technique can heighten interest and enthusiasm.

■ *Cutaway shots (intercut shots)*   By cutting from the main action to *secondary activity* or *associated subjects* (e.g. spectator reactions), you can:

1. Join shots that are unmatched in continuity or action.
2. Remove unwanted, unsuccessful, dull or excess material.
3. Suggest a time lapse, to compress or expand time.
4. Show additional explanatory information (*detail shots*).
5. Reveal the action's environment.
6. Show who a person is speaking to; how another person is responding (reaction shots).
7. Show what the speaker is seeing, talking about or thinking about.
8. Create tension, to give dramatic emphasis.
9. Make comment on a situation (cutting from a diner to a pig at trough).

Interviews can be shot using a continuous intercut multicamera set-up or as a one-camera treatment. A separate series of cutaway shots (*cutaways*, *nod shots*) are often recorded *afterwards*, in which the interviewer and interviewee are seen in singles or over-shoulder shots, mutely smiling, nodding, 'reacting', looking

Fig. 15.28 part 1   Cheated substitutes—the desired sequence
A fugitive climbs a cliff to escape pursuers. The viewer sees the formidable cliff,
and his climb—but he dislodges a rock, and they look up.

interested. When edited in (to disguise cuts, continuity breaks, or
to add visual variety), the subterfuge appears quite natural.
Without such intermediates, any continuously held shot would
jump frame when edited (dissolves may improve the disruption).

Cut-ins or *inserts* may also be used, reshooting parts of the action
from another camera set-up (different shot size, or angle).

■ *Reaction shots/partials/cut-in shots*   By skilfully concealing
information, you can prime the imagniation and arouse curiosity.
Instead of showing an event, you show its effect.

1. *Reaction shot*—the door opens . . . we see the victim's horror-
   striken face, not the intruder.
2. *Partial shot*—a switch-blade opens . . . is moved out of frame
   . . . we hear the victim's cry . . . then silence.
3. *Cut-in shot*—we watch the victim's cat drinking milk . . . to
   sounds of a fight, and a body falling . . . the victim's hand
   comes into frame, upsetting the milk dish.

This technique can provide maximum impact with minimum
facilities, conveying information by implication rather than direct
statement. It aims to intrigue and tantalize. Ineptly introduced,
though, it can frustrate.

Fig. 15.28 part 2   Cheated subsitutes—the studio procedure
The studio set-up is simple: A gravel-strewn floor, a photograph of a cliff-face,
and a surface-contoured flat over which the fugitive crawls (shot is canted to
suggest vertical climbing).

311

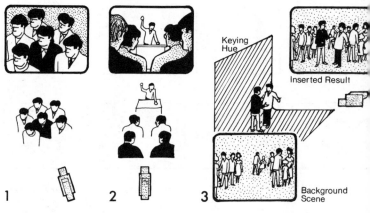

Fig. 15.29 Crowds
Crowds can be simulated by selected viewpoints: 1, Showing part of the crowd
2, Using foreground crowd (real, dummy, cut-outs); or 3, Inserting people
electronically into a recorded scene.

Illusion of spectacle

Budgets do not generally permit lavish staging. Yet occasions
arise when you want to create an impressive illusion of space
grandeur, and spectacle. Several methods are available:

■ *Library (stock) film shots* By intercutting existing filmed
library shots with limited studio action (often close shots) it is
possible to build up extremely convincing illusions. When
highly dramatic spectacle is required, as in battle scenes, hurri-
canes, volcanic eruptions, mobs, earthquakes, floods this is usually
the only solution.

However, available library shots may not be exactly suitable for
your purpose (contents, quality, viewpoint, time of day, unwanted or spurious aspects included). It might even be necessary
to reorganize the studio treatment to suit the library shot!

Fig. 15.30 Staging distractions
Various distractions can appear in shot. Here is a particularly disastrous collection
1, Unexplained, distracting object in shot. 2, Cable in shot. 3, Visible surface
damage on flat. 4, Shadow of someone out of shot. 5, Wall-picture not straight
6, Light reflection. 7, Camera reflection visible in table picture. 8, Multiple
shadows from lamp fitting. 9, Distracting light patch. 10, Wall mirror appears as
'halo.' 11, Wrinkle in background surface. 12, Shooting off the set! 13, Obtrusive
foreground prop.

■ *Direct imitation* A complete or *partial* replica of part of the scene (e.g. a stone wall) is intercut with location material or library shots (the actual castle).

■ *Indirect imitation* Here an environment provided by electronic insertion (chroma key, CSO), rear projection, photo backings, or glass shot, etc., is conjoined with a localized studio setting. For example: an automobile in the studio is inserted into a filmed shot of a highway; an impressive ceiling (a graphic) appears to be part of a studio set.

■ *Cheated substitute* Montage can create an illusion by intercutting small representative parts of the action, to convey a composite impression. Ingenious, imaginative, and most effective if properly handled, you must ensure that the component parts are completely integrated. For instance, you can build up a dramatic impression of a battle simply by rapidly intercutting close shots of details in war photographs, to an audio background of battle noises.

16 Production organization

How any director organizes his show, varies with local procedures, the nature of the production, and his own experience and temperament. Consequently, there are approaches ranging from the 'spontaneous' to the meticulously planned. Much depends, too, on the artistic and operation standards you demand, and the production crew's experience.

Unplanned productions

Total unpredictability is not adventurous. It creates needless stress and haphazard results. Television necessarily involves teamwork. If people do not have a clear idea of what is happening, you must not be surprised when there is no chair for the extra guest, the film insert is not ready, lighting is inappropriate, the camera has a wrong shot, or you cannot hear the speaker.

The amateur often collects material, then considers how to use it. The professional invariably starts with a plan of campaign, however broad, and obtains or arranges material to these concepts; modifying as necessary.

Unrehearsed formats

Even where you cannot rehearse a production, you will usually find that the situation fits some familiar format. So you can arrange suitable staging and camera positions, and allocate

Fig. 16.1 part 1   Regular formats—piano recital
The basic range of pictorially attractive shots is limited when presenting a piano recital.

probable shots. It is possible to precheck potential shots and lighting, using available people. Then, when talent appears, you can quickly preview them to check over voice levels, make-up, lighting, and costume.

If the format is less definite (e.g. a dance troupe arriving just before air time), it is a matter of choosing typical strategic viewpoints (front and cross shooting), then allocating cover shots and close shots to cameras for a defined working area (not forgetting to tell the performers of any necessary movement limits!) The show itself largely involves recognizing effective shots as they are offered; taking action detail and group shots in a balanced ratio.

Regular formats

There are a number of regular formats: interviews, panel discussions, piano and instrumental performances, singers, and newscasts. These are usually familiar to the production team and meaningful shooting variations are limited. What the performer is saying (or playing) is more important than straining to achieve new and original shot development. Consequently, such productions follow recognizable lines, so that the director often starts off with planned viewpoints and then introduces a particular treatment as it becomes desirable.

Planning for regular productions may largely be a matter of coordinating staff and facilities, ensuring that film or VT inserts are available (with known timings and cue points); graphics, titles, etc., are prepared, and any additional material organized. The production itself may be based upon a series of key shots and spontaneous on-the-fly decisions.

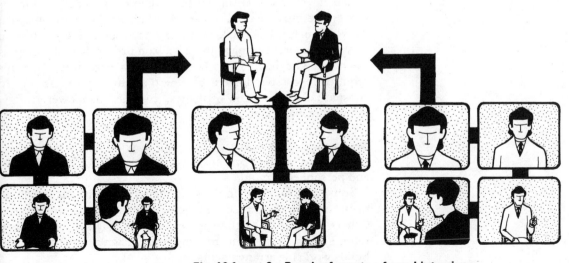

Fig. 16.1 part 2   Regular formats—formal interview
In a formal interview, the number of effective shots is relatively few.

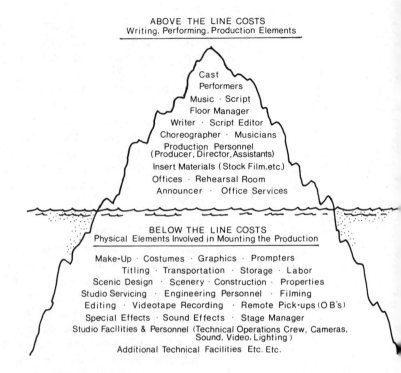

ABOVE THE LINE COSTS
Writing, Performing, Production Elements

Cast
Performers
Music · Script
Floor Manager
Writer · Script Editor
Choreographer · Musicians
Production Personnel
(Producer, Director, Assistants)
Insert Materials (Stock Film, etc.)
Offices · Rehearsal Room
Announcer · Office Services

BELOW THE LINE COSTS
Physical Elements Involved in Mounting the Production

Make-Up · Costumes · Graphics · Prompters
Titling · Transportation · Storage · Labor
Scenic Design · Scenery · Construction · Properties
Studio Servicing · Engineering Personnel · Filming
Editing · Videotape Recording · Remote Pick-ups (OB's)
Special Effects · Sound Effects · Stage Manager
Studio Facilities & Personnel (Technical Operations Crew, Cameras,
Sound, Video, Lighting)
Additional Technical Facilities Etc. Etc.

Fig. 16.2 Budget
Accurate cost estimation and accounting are essential in TV organization, despite the great diversity of the various contributions. Budgets are usually related to the concepts of 'above' and 'below' the line. These differentiate between the 'writing, performing, and production elements', and the various 'back-up' services and physical elements involved in mounting the production.

Complex productions

Whether you are shooting in the studio or on location, time is money and must be used efficiently. While smaller shows may get by with a preliminary talk-over a few days before the production date; for more complex projects, comprehensive planning is essential. The team must be briefed and organized, and this requires discussive agreement; for there are sundry factors to consider, such as time estimates, costing, manpower, safety, union agreements, etc.

■ *Planning preliminaries* Planning begins with *script study*; the director reading and visualizing with his budget, facilities, and personnel in mind.

Some directors rely on the set designer to produce an environment for action and then arrange performers within this setting. Others indicate the strategic features or staging format they require, and from that briefing the designer interprets and builds workable arrangements.

After preliminary discussion, the designer produces a rough

scale plan of the various settings; often on separate pieces of over-lay (tracing) that can be moved around the studio staging plan for optimum layout. He may include sketches, reference photographs, and preliminary elevations.

Those responsible for studio operations including technical organization, cameras, lighting, audio, make-up, costume, etc., may be involved at these early stages.

■ *Treatment breakdown*   Clearly, there are many ways in which the director could set about devising his production treatment. So much depends on individual experience, methods, and the kind of show involved. A typical approach is to closely analyze the script, and within the staging format to visualize potential

Table 16.1   Stages in production procedure

Program concepts	Idea potentials, objectives, potential audience, market, saleability, probable duration.
Executive decisions	Selecting *above-the-line* personnel (production: director, writer, etc.). Budget allocation. Production scheduling (deadlines). Preproduction publicity.
Scripting	Theme, coverage, style, research/advisor. Script editing (evaluating content and practicability). Script conference/rewrites. Director's visualization.
Organizing talent	Contacts, auditions, interviews, casting, contracts. Organization/preparation of performers. Read-through, rehearsals, costume/make-up preliminaries.
Organization of inserts	Any *prestudio* photography, filming, taping. (Arrangements for shooting these: copyright clearances, facility fees, access permissions, permits, insurance, scheduling, staffing, editing, etc.). Selection of illustrative inserts (film clips, VT, stills). Graphics and titles. Musical inserts (disc, tape, performed). Any equipment required for demonstrations (construction or hire).
Staffing	Coordinating services of *below-the-line* personnel (technical, operational, facilities).
Production paperwork	Requisitions. Scripts (draft, rehearsal, and camera script). Rundown sheet, camera cards, etc.
Organizing artistic services	Staging, costume/wardrobe, make-up.
Organizing technical services	Cameras, lighting, audio, telecine, VTR, special effects.
Organizing facilities	Loan/hire of equipment. Storage, transport, catering, accommodations (dressing rooms). Prompters, audience arrangements (tickets, seating, warm-up).
Prestudio rehearsals	Rehearsal of dialogue, moves, action; treatment preparation.
Studio rehearsals	Camera rehearsals (finalization of action, video, and audio treatment).
Videotape recording	Total, discontinuous, or compilation recording.
VT editing	Off-line selection, organization, and editing of video recording. Audio dubbing.
Review and transmission	Evaluation of recording.
Post-production activities	Logs/reports. Publicity/billing/promotion. Correspondence, ratings.

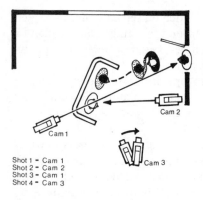

Shot 1 = Cam 1
Shot 2 = Cam 2
Shot 3 = Cam 1
Shot 4 = Cam 3

Fig. 16.3 Action treatment
The plan summarizes the camera treatment for a brief scene (see text).

treatments for each situation. The director then systematically blocks out movement and camera shots, to build up a continuous effective presentation. The same situation can often be tackled in various different but equally successful ways. Let us look at an example for a dramatic production:

Returning husband enters door, wife has unexpected guest.

This could be broken down as:

1. Husband enters door, hangs up hat . . . MLS shows who it is, yet orientates audience.
2. Wife looks up, greeting him . . . Medium 2-shot showing her with guest.
3. Husband turns and sees guest . . . CU of husband's reaction.
4. Guest rises, walks to greet husband . . . From med. 2-shot; pans with guest, dollying in for tight 2-shot of guest and husband.

The interpretation could have been simpler. Its impact would have been different.

1. Husband enters door, hangs up hat . . . Over-shoulder long shot between wife and guest, showing entrance. Watch husband's reactions, tightening shot as he walks to guest.
2. As guest stands . . . Cut to frontal 2-shot of guest and wife greeting him. Zoom in to guest. Intercut CU of guest and husband.

Table 16.2 Typical considerations in technical planning

Talent	Cast discussed, re costume and make-up requirements (styles, fitting etc.).
Staging	Style, treatment of sets.
	Staging plan: details of set structures, scenic changes, storage, audience seating.
	Special visual effects (physical, video, lighting).
	Special props.
	Safety precautions, regulations, etc.
Action	Director indicates possible performer positions, action, business.
Cameras	Number of cameras, types of mountings, main positions for each scene, probable moves.
	Any camera accessories (special lenses, etc.).
	Specialists estimate feasibility of anticipated treatment (checking sufficient working space for cameras/booms, time for moves, cable routing, etc.).
Lighting	Discussion of lighting treatment feasibility, relative to: scenic design, action, sound pickup, time/equipment/manpower availability.
	Discussion of pictorial effects, atmosphere, etc.
	Discussion of picture matching to inserts (film or VT).

Having decided on a particular treatment, the director now arranges camera and audio coverage for these shots. He will often use a 'shot plotter' (protractor or transparent triangle) showing the lens-angle coverage, estimating and marking the camera positions needed (1A, 1B, 2A, etc.), and locating a sound boom (A1). The resultant rough production plan (camera plan), together with his script margin action notes or sketches, form the basis for all rehearsals. Even the biggest productions can be analyzed into shots or sequences in this way.

■ *Production planning meeting* Once a complete provisional floor plan has been prepared by the designer, the specialists involved in studio operations meet with the director and designer. They examine the staging proposals and production treatment; evaluate, discuss, anticipate practical problems, and so on. They outline their own proposed contributions—lighting treatment, audio effects, costume details, etc.

Following this meeting, they then carry through their own organization procedures (documentation, costing, manpower, scheduling, equipment selection), and the designer issues the agreed *staging plan* (*floor plan*, *ground plan*, *setting plan*).

The production planning meeting forms the basis for efficient economic teamwork. Problems anticipated and overcome at this stage, prevent last-minute compromises. Camera cable routing, for example, is a typical potential hazard. Cables get snarled up, can impede other cameras, or drag around noisily. So directors may move around scale card cutouts of cameras (perhaps cord-attached to simulate cables) when devising their camera plans.

Audio	Similar discussion to '*Lighting*'.
	Considering audio pickup methods, potential problems.
	Audio inserts discussed (prerecorded library effects, spot effects, music).
Video effects	Chroma key (CSO), electronic treatment.
Artwork/graphics/titles	Displays, graphics, maps, charts, models, etc.
Further technical facilities	Equipment organization, re filming or prestudio videotaping.
	Technical resources required: telecine, videotape, video disc, slide scanner, picture monitors, prompters, cuing facilities, etc.
Scheduling	Prestudio shooting/recording. Experimental sessions.
Prestudio rehearsals	Read-through, block action, dry run (technical run).
Studio rehearsal/recording	Setting (staging and lighting).
	Camera rehearsal (times and arrangements, meal breaks).
	Technical periods (line-up, relight, reset).
	Recording periods (continuous, discontinuous, rehearse/record).
Editing	Scheduling and facilities.

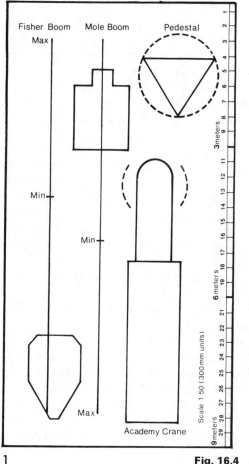

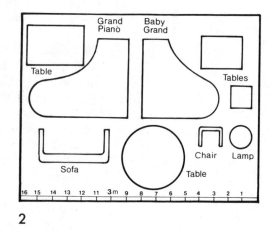

Fig. 16.4   Equipment stencil
Scale stencils are used to mark the floor plan, with equipment positions. 1, Camera/Sound boom stencil. 2, Typical furniture stencil.

Much depends, of course, on the type of show you are considering, how it is to be recorded, elaboration of treatment, any special set-ups (chroma key), rehearsal time, editing facilities, etc.

In a *live* or live-on-tape show, you must allow time for people, cameras, and sound booms to move between situations; time to cover any costume or scene changes, and to reposition equipment or furniture. Discontinuous shooting overcomes such problems; although continuity and editing factors then arise.

The script

■ *Semi-scripted show*   For ad-lib shows, discussions, interviews, variety, compilation shows, and demonstrations, it is normal to use *outline scripts*. This script lists the talent involved, facilities being used, graphics (slides), film and VT inserts (identifying sections, giving durations). Dialogue is only included in detail for introductory announcements, and commentary containing cue-lines (for cuing inserts, editing breaks, or redeploying cameras).

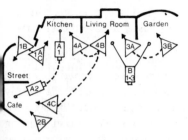

Fig. 16.5   Rough operational sketch
Position of Cameras 1, 2, 3, 4 are successively marked (1A, 1B). Positions of Sound booms A, B, are successively marked (A1, A2).

■ *Fully-scripted show* Here the script develops in several stages:

1. *Preliminary script/draft/outline script/writer's script*—Initial submitted full-page script (dialogue and action) before script editing.
2. *Rehearsal script*—Script prepared for TV and used for pre-studio rehearsal. Script details the locales (settings), characters, action, talent directives, dialogue. (Table 16.3).
3. *Camera script*—A revised script for camera rehearsals, augmented with details of production treatment: cameras and audio, cues, transitions, stage-instructions, set changes. (Table 16.4).

TV scripts follow several 'standard' layouts. The preferred version uses two vertical columns, with picture treatment (cameras, switching) on the *left*, action and dialogue on the *right*, together with stage instructions and lighting/effects cues. Some studios prefer a *single-column* cinematic format, with transitions in a left margin, and all video and audio information in a single main column. Directors often hand-mark this version with their own

Table 16.3   The rehearsal script

SCENE	INT./ EXTR.	LOCATION	TIME OF DAY
3.	INT.	LOUNGE	NIGHT

(GEORGE ENTERS. WALKS TO TABLE. SWITCHES ON LAMP.)

GEORGE: (CALLS) The lights are OK in here.

It must be your lamp.

(GEORGE TAKES GUN FROM DRAWER.)

SLIPS IT INTO HIS POCKET: PULLING OUT TELEGRAM. HOLDS IT UP.

'SORRY CANNOT COME WEEKEND. BRIAN SICK. WRITING . . . JUDY.'

HAND SCREWS IT UP AND THROWS INTO FIRE.)

(DOOR OPENS: EILEEN ENTERS.)

EILEEN: Really, these people are too bad. They promised to be here tonight. Look how late . . .

GEORGE: It's probably the storm that has delayed them. They'll be here all right.

(EILEEN SITS ON COUCH: GEORGE JOINS HER.)

EILEEN: If Judy knows you're here, it'll take wild horses to keep her away.

GEORGE: How many more times do I have to tell you . . .

EILEEN: Why do you keep pretending?

/ LIGHTNING FLASH /

GEORGE: I've warned you. You'll go too far.

instructional symbols indicating transitions and shots. But for more complex productions, this is less easily followed than a two-column format.

The *full script* is not, as some people believe, an artistically inhibiting document that commits everyone concerned to a rigid plan of procedure. It can be modified as the need arises. It simply informs you about what is expected at each moment of the production. Studio rehearsal time is too precious to use up explaining basic mechanics as you go. Far better to have a detailed script that shows the exact moment for the lighting change, to cue the film, or introduce the superimposition. The full script is a plan of

Table 16.4　The camera script

SHOT	CAM.	(POSITION)	SCENE	INT./ EXTR.	LOCATION	TIME OF DAY	F/X
			CAMS. 2A, 3A, 4B.		SOUND BOOM B1.		
			3.	INT.	LOUNGE	NIGHT	
10.	F/U 2.	A			(GEORGE ENTERS. WALKS TO TABLE.		TAPE: WIND RAIN
		LS PAN/ZOOM on GEORGE to MS.			SWITCHES ON LAMP.)		
			GEORGE: (CALLS) The lights are OK in here.				
11.	3.	A	It must be your lamp.				
					GEORGE TAKES GUN FROM DRAWER.)		
12.	2.	A					
		MS			SLIPS IT INTO HIS POCKET: PULLING OUT TELEGRAM. HOLDS IT UP.		
13.	4.	B					
		BCU of telegram.	'SORRY CANNOT COME WEEKEND. BRIAN SICK. WRITING . . . JUDY.'				
		PULL FOCUS on fire as he throws.			(HAND SCREWS IT UP AND THROWS INTO FIRE.)		
14.	3.	A					
		LS			(DOOR OPENS: EILEEN ENTERS.)		
			EILEEN: Really, these people are too bad. They promised to be here tonight. Look how late . . .				
			GEORGE: It's probably the storm that has delayed them. They'll be here all right.				
		DOLLY IN to MS as EILEEN X's to couch.			(EILEEN SITS ON COUCH: GEORGE JOINS HER.)		
			EILEEN: If Judy knows you're here, it'll take wild horses to keep her away.				
			GEORGE: How many more times do I have to tell you . . .				
			EILEEN: Why do you keep pretending?				
				LIGHTNING FLASH			
15.	2.	A					
		CU	GEORGE: I've warned you. You'll go too far.				DISC: THUNDERCLAP

campaign, that has details added to it as the production develops.

For some types of show, the outline script is quite sufficient; in fact, any attempt to include 'pseudo information', such as the entire dialogue for a film insert would usually be pointless. (The duration and 'in and out cues' would suffice.)

Table 16.5 TV script abbreviations

Many organizations duplicate the *rehearsal script* on white paper. Operational information is added to the original stencil to provide the *camera script* on yellow paper.

All *camera shots* are numbered in a fully scripted show. *Inserts* are not numbered (film, slides, VT, titles, graphics), but are identified in the audio column. Where possible, cutting points are marked in the dialogue with a slash mark ⌣ underlining back to the camera number.

Various abbreviations are widely used:

Equipment	CAM	Camera		B	Balop
	MIKE MIC	Microphone		T CP CAP	Telop Caption
	BOOM	Sound boom		G	Graphic
	F/P FISH	Fishpole		ROLLER	Roller caption (crawl)
	VT VTR	Videotape		TC	Title card
	F	Film		PIC PIX	Pictorial graphic (photo caption)
	TC TK	Telecine		RP	Rear projection
				BP	Back projection
	SL	Slide		CPU PROJ	Caption (slide) projection unit
	CS	Caption Scanner			
	TJ	Telejector		VD	Video disc
Position	L/H, R/H; C/S, U/S, D/S; P/S, OP; B/G, F/G:			X	Move across.
	Picture left-hand, right-hand;			POV	From the point of view of person named.
	Centre stage, upstage, downstage; Prompt side (RHS facing the stage), opposite prompt (LHS facing stage); Background, foreground.			O/C	On camera.
Cuing	Q, I/C, O/C: Cue, in cue, out cue.			ROLL, RUN:	Start cue for film (telecine) or VT.
	Pull, flip, Graphic (caption) animate: animation.			C/S:	Change slide.
	S/B: Standby.				
	FX: Cue effects.				
Cameras	BCU, CU, MCU, MS, MLS, LS, VLS. (*See Chapter 12*).			D.I.:	Dolly in.
				H/A, L/A:	High angle, Low angle.
	2-S, 3-S: Two-shot, three shot.			N/A, W/A:	Narrow angle, Wide angle.
	AB: As before, i.e. shot as previously.			P/B, D/B:	Pull back, Dolly back.
	O/S, X/S: Over-shoulder shot.				
	FAV: Favor (make prominent).				

Switching (vision mixing)				
	CUT:	Not marked, but implied unless other transition marked.	LOSE 2:	Cut out Cam. 2 from a super or key.
	MIX (DIS):	Mix.	FI, FO:	Fade in, Fade out.
	WIPE:	Wipe.	T:	Take. (Moment for a transition).
	SUPER (S/I):	Superimpose.	S/S:	Split screen.
	KEY (INSERT):	Electronic insertion (chroma key, CSO).		

Audio				
	F/UP:	Fade up audio.	Spot FX:	Sound effect made in studio.
	FU:	Fade audio under.	STING:	Cue strong musical chord emphasizing action.
	P/B:	Playback.		
	F/B:	Foldback (audience address).	OS; OOV:	Over scene; out of vision (audio heard where source not shown).
	DISC: CART: CASSETTE: S/TAPE: GRAMS:	Recorded audio insert.	REVERB:	Reverberation added.
			B/G:	Background.
			OPT; MAG:	Film audio: Optical; Magnetic.
	ANNCR:	Announcer.	SOF:	Sound on film.
	ATMOS:	Atmosphere (background sounds).	SEP MAG:	Separate magnetic track.
			SOT:	Sound on tape.
			SIL:	Silent.

General				
	TXN:	Transmission.	POS, NEG:	Positive, Negative.
	SEQ.	Sequence.	MON:	Monitor.
	P/V:	Preview.	ID:	Station identification.
	ADD:	An addition (to a story).		
	PBU:	Photo blow-up.	MOS; VOX POP:	Man-on-street; interview.
	PRAC:	Practical.	EXT:	Exterior.
	PROP:	Property.	INT:	Interior.
			LOC:	Location.

When is it necessary to fully script a production?

1. When the dialogue is to follow a prescribed text, that is to be learned, or read from a prompter or script.
2. Where action is detailed, so that people move to certain places at particular times and do specific things there. (This can affect cameras, sound, and lighting treatment.)
3. When there are carefully timed inserts (film, videotape) that have to be cued accurately into the program.
4. When the duration of a section must be kept within an allotted time bracket, yet cover certain agreed subject points. (The speaker might otherwise dwell on one point, and miss another out altogether.) Even where post-production editing is available, it is not always possible to remedy uneven subject-coverage.

5. Where there are spot cues, eg. a 'lightning flash' and an effects disc of thunder at a point in the dialogue.

So you will find fully-scripted approaches in newscasts, drama productions, operas, comedy shows (situation comedy), documentary-type presentations and major commercials. Where dialogue and/or action are spontaneous, there can be no script; only an outline of the routine.

The more fragmentary or disjointed the actual production process is, the more essential is the script. It helps everyone involved to *anticipate*. And in certain forms of production (rehearse/record; total chroma key staging) anticipation is essential for tight scheduling.

In a complex production that is videotaped out of sequence, the production crew may be unable to function meaningfully without a full script, which makes it clear how shots/sequences are interrelated, and reveals continuity. The lighting director may, for example, need to adjust the lighting balance for a scene, so that it will intercut smoothly with a different viewpoint shot yesterday.

So the full script can be a valuable coordinating document, enabling you to see at a glance the interrelationship between dialogue, action, treatment, and mechanics. During planning, of course, it helps the team to estimate how much time there is for a camera move, how long there is for a costume change (perhaps a recording break will be necessary), whether re-arranging shooting order will give the necessary time for a make-up change, the scenes during which the "rain" should be seen outside the windows of the library set (ie. the water spray turned on and the audio effects introduced) . . . and the thousand and one details that interface in a smooth-running show.

The full script is used differently by various members of the team. The director uses it to develop his treatment, control timing, etc., and to demonstrate to the rest what he requires. Thoroughly familiar with it (hopefully!) he refers to his script for certain treatment and cuing details. His assistant(s) follows it carefully, checking dialogue accuracy, noting where retakes are needed, timing (durations) sections, perhaps cuing contributory sources from the script, as well as 'calling out shots' on intercom ("Shot 24 on 2, coming to 3"). The switcher follows the script in detail, preparing for upcoming transitions, superimpositions, effects.

Others such as the cameramen, boom operators, electricians, may use the script as a detailed reference point when needed, but use simplified outlines (breakdown sheet, camera card) as they memorize their operations.

Auxiliary information

In addition to the script, various supplementary summaries may be necessary:

■*Synopsis* An outline summary of characters, action, plo
Appended to a dramatic script; particularly when shot out o
sequence, or to coordinate a series.

■ *Fact sheet/run-down sheet* Summarizes information about
product, or item for a demonstration program; or details of
guest for an interviewer. Provided by a researcher, editor, o
agency for guidance.

■ *Breakdown sheet/show format/running order* Lists the event
or program segments in order, showing allowed duration:
participants' names, cameras and audio pick-up allocate(
setting used, video and audio inserts, etc. (Sometimes ambigu
ously called a run-down sheet.)

Table 16.6   Breakdown sheet/show format/running order

DRAMA PRODUCTION

Page	Scene (Sequence)	Shots	Cameras/Audio (layout across set)	D/N Day/ Night	Cast (Performers)
1	FILM –1 (40") OPENING TITLES (20")		SOF (SEP. MAG.) 4A. S. TAPE (RAIN)		
1	INT. DINING ROOM	2–9	1A. 4A.	N.	GEORGE
3	2. INT. LOUNGE	10–27	4B, 2A, B1, 3A.	N.	GEORGE EILEEN
			R E C O R D I N G B R E A K		
10	3. EXT. GARDEN	28	B2, 3B.	D.	JUDY GEORGE
10	4. EXT. STREET	29	1B. Slung.	D.	BRIAN
11	5. INT. CAFE	30–43	2B, 4C, A2	Eve.	BRIAN JUDY GEORGE
			ETC. . .		

DEMONSTRATION PROGRAM   (TOTAL DURATION: 15 MINS.

Sequence No.	Item (Sequence segment)	Cameras/F/VT	Seq. dur.
1	OPENING TITLES— SIG. TUNE.	4A, 1A,	07"
2	PROG. INTRO.	2A, 3A,	30"
3	TIMBER FELLING.	VT.	24"
4	PREPARING WOOD.	1B, 2B.	2' 30"
5	CARVING DEMO.	1C, 4B, 3B,	8' 00"
6	TIMBER IN BUILDINGS.	FILM.	2' 58"
7	CLOSING LINK.	3A.	22"
8	END TITLES— SIG. TUNE.	4A, 1A,	09"
			15' 00"

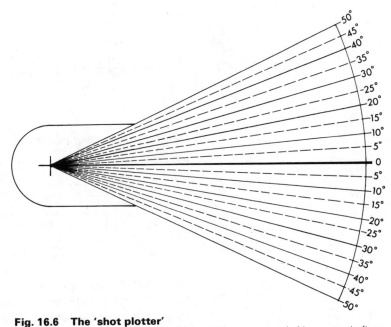

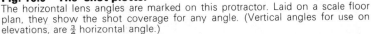

Fig. 16.6   The 'shot plotter'
The horizontal lens angles are marked on this protractor. Laid on a scale floor plan, they show the shot coverage for any angle. (Vertical angles for use on elevations, are ¾ horizontal angle.)

Invaluable for unscripted, semi-scripted, and scripted shows that contain a series of self-contained segments (sequences, scenes). Also as a summary of a complex dramatic script to show at a glance: shot numbers for each scene, operational details, inserts, VT breaks for major resetting, redeployment or costume/ make-up changes, etc.

■ *Camera card*   A sheet provided for individual cameras, showing their successive floor positions (perhaps with a layout sketch), and their allocated shots (shot numbers, type of shots, camera movements, action).

The camera card is not to be confused with a *shot sheet*, which carries brief details of *all* cameras' shots and is used as a summary list when preparing a treatment framework for an unscripted show. (See page 330, Table 16.7)

■ *Requisition forms*   Essential for interdepartmental organization, requisition forms are legion. There are no standard approaches. Each station has its own procedures for booking studios, equipment, facilities, services, manpower, etc. The director may become involved directly in ordering props, graphics, titles, etc., or simply inform specialists. Various requisitions concerning costume, make-up, staging, lighting, video, audio, VTR, filming, editing, are normally handled by the specialists.

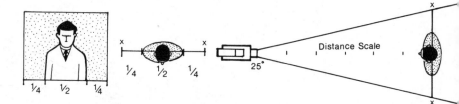

Fig. 16.7 part 1   Shot planning—required camera distance
Instead of finding shots by *trial and error*, you can *measure* the set-up needed on a scale plan as shown here, or *calculate* it from a graph (Fig. 16.8). To measure the camera distance needed to get a particular shot with a given lens angle first check the scale to be used. If 1 cm=0·5 m a person is 1 cm across (on $\frac{1}{4}$ in scale=1 ft, he is $\frac{3}{8}$ in across). Draw the lens angle with a distance scale bisecting it. Decide how much of the screen you want the person to fill e.g. $\frac{1}{3}$ screen width. In this case, draw a line with a scale map $\frac{1}{2}$ its length. Next fit this line (X-X) across the lens angle. On the scale read the distance needed to get your shot.

Calculating shots

Many directors prefer to work empirically, from mental concepts, rough notes, and experience. They make 'guesstimates', or wait until they can guide the cameramen and see which shots are practical. They improvise instinctively and draw inspiration from the occasion. This is fine, as long as time is available to experiment, improvise and coordinate the production team. But as a general practice, it is preferable to plan as fully as possible, modifying, correcting and improving during camera rehearsal and subsequent editing.

■ *Why measure?* Preliminary planning can provide you with the answers to a lot of your camera problems long before even preliminary blocking:

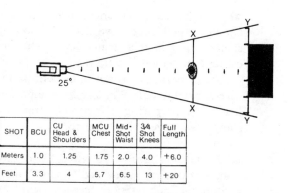

SHOT	BCU	CU Head & Shoulders	MCU Chest	Mid-Shot Waist	3/4 Shot Knees	Full Length
Meters	1.0	1.25	1.75	2.0	4.0	+6.0
Feet	3.3	4	5.7	6.5	13	+20

Fig. 16.7 part 2   Shot planning—shots available from a particular layout
First measure the person's distance. In the example he is 4 m (14 ft) away. (A 25° lens would provide a $\frac{3}{4}$ shot/knee shot—see table.) Draw a line across the lens angle at that point and you see that he occupies about $\frac{1}{5}$ screen width. The object behind him is 1·5 m wide at Y-Y (shot width 2·5 m)=$\frac{3}{5}$ of screen width. So now you have the shot's proportions.

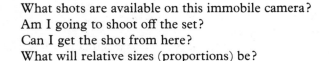

What shots are available on this immobile camera?
Am I going to shoot off the set?
Can I get the shot from here?
What will relative sizes (proportions) be?

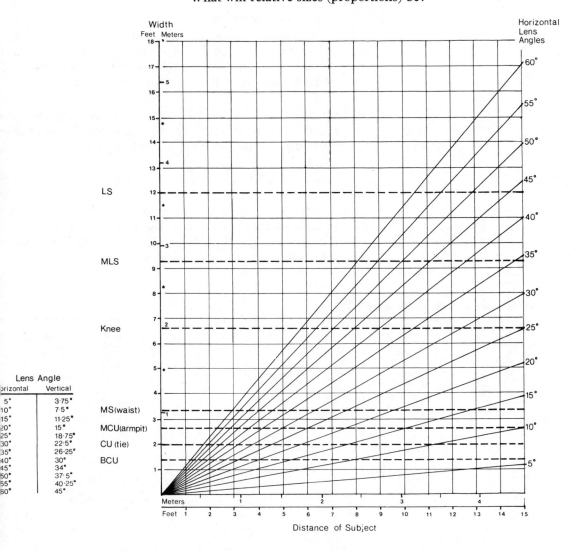

Fig. 16.8 Camera set-up graph
The graph indicates the scene width taken in at any specified distance for any
given lens angle. Height is $\frac{3}{4}$ of the width. 1, To find *camera distance* needed for
a particular shot: select shot type on vertical scale, check to angle used, then
down to distance necessary. 2, To find *lens angle* required for a particular shot:
trace a line across graph from shot type (vertical scale) and a line up from
subject distance. Where they meet, is the lens angle needed. 3, To find the *type
of shot* obtained: trace a line up from distance, to the lens angle used. There
move left onto the vertical scale, to see the nearest shot type. 4, How much
can be seen from a specified camera position? As for 3, but note *scene width
scale*. 5, Different distances giving the same shot size: trace along horizontal
shot-type line (e.g. CU) and note distances where it cuts various lens angles.
6, To make a subject fill e.g. $\frac{1}{3}$ screen width: find subject width then multiply
by 3. Look up this screen width. Trace across to lens angle used, then down to
distance needed. Subject will now fill $\frac{1}{3}$ screen width.

329

How large must we make the graphic in this setting?
Where do I position people for this chroma key effect?

Why *guess* whether it will work on the day, when you can check
it out now?

The tools are simple enough: a scale floor plan and a transparent triangle or protractor representing the horizontal lens
angle. Wherever you place this 'shot plotter' on the floor plan, it
shows what the camera sees from that viewpoint. If the angle falls
outside the marked-out setting, you are shooting off. (Fig. 16.6)

■ *Shot proportions* The edges of the lens angle represent the
left- and right-frame edges. Any object that touches both lines
fills the screen width—whether a small close item, or a large
distant one.

To make a subject fill one-third of the screen width, just mark
on a paper strip a length *three times* the subject's width. At the
distance this marker fits across the angle, the subject will exactly
fill one-third of the screen width. Conversely, if a subject is a
known distance away, measure the scene width there and divide
this by the subject width to find proportions.

Table 16.7 Camera card (See page 327)

CAMERA THREE		'DEATH COMES ONCE'	STUDIO B
SHOT No.	POSN.	(LENS ANGLE*)	SCENE/SETTING
11	A	13°	Sc. 2. LOUNGE CU GEORGE pockets gun from drawer, pulling out telegram.
14	A	35°	LS EILEEN enters. DOLLY IN to MS as she X's to couch.
19	A	20°	BCU GEORGE'S eyes.
23	A	13°	BCU EILEEN screams.
25	A	20°	CU GEORGE puts gun in drawer.
		QUICK MOVE TO GARDEN	
28	B	35°	Sc. 3. GARDEN MLS follow GEORGE carrying body: tightening, depressing to L/A shot: ZOOM IN to hand as he stops.
		MOVE TO RHS LOUNGE	
44	A	50°	Sc. LOUNGE LS DETECTIVE enters. PAN with him into MS as he goes to phone.

*Optional.

330

In Figure 16.7, you have a quick shot-check table, showing the distances for 'standard' shots on a 25° lens. For other lens angles, distances are inversely proportionate:

$$50° = \tfrac{25}{50} = \times \tfrac{1}{2} \text{ table distance.}$$

$$10° = \tfrac{25}{10} = \times 2 \cdot 5 \text{ table distance.}$$

$$40° = \tfrac{25}{40} = \times 0 \cdot 6 \text{ table distance.}$$

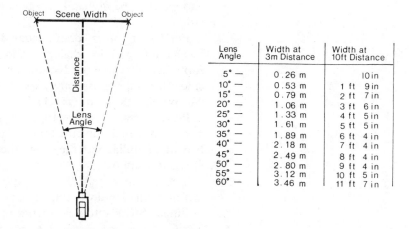

Lens Angle	Width at 3m Distance	Width at 10ft Distance
5° —	0.26 m	10 in
10° —	0.53 m	1 ft 9 in
15° —	0.79 m	2 ft 7 in
20° —	1.06 m	3 ft 6 in
25° —	1.33 m	4 ft 5 in
30° —	1.61 m	5 ft 5 in
35° —	1.89 m	6 ft 4 in
40° —	2.18 m	7 ft 4 in
45° —	2.49 m	8 ft 4 in
50° —	2.80 m	9 ft 4 in
55° —	3.12 m	10 ft 5 in
60° —	3.46 m	11 ft 7 in

Fig. 16.9 Checking your camera's lens angle
1, *To find the angle of your lens:* Place two small objects on a crossline 3m/10 ft away, moving them so that they just appear within the left-right frame edges. Measure how far apart they are and check in the table, the angle that covers this width. 2, *To set the shot-box to a particular lens-angle:* Look up the *required angle* in the table and find the screen-width it takes in at 3m/10 ft away, place them this distance apart (e.g. for 30° they must be 1.61m/5 ft 5 in apart). Adjust your lens to just take in the objects at the frame edges, and it is now at the required angle. Adjust shot-box preset, or note zoom setting. You could make up a regular zoom-lens line-up graphic for adjusting all lens angles at this distance.

17 In production

Any production benefits from some kind of rehearsal prior to recording or live transmission. These rehearsals take several forms.

Prestudio rehearsal

Complicated productions particularly gain from prestudio rehearsal. Even a table demonstration can be discussed, practiced and checked beforehand; perhaps in a *preparation room* near the studio.

For larger shows (drama, comedy), it is essential to work out mechanics, practice dialogue and action, and coordinate performance. So these productions are often rehearsed in halls (outside rehearsal rooms) some two or three weeks before studio date. Meanwhile, the studio is scheduled for other programs.

Prerehearsals begin with a *briefing*, *read through*, or *line rehearsal*, in which the director outlines his interpretation. The cast becomes familiar with the script, the other actors, and the required characterizations.

The rehearsal hall floor is taped or chalked with a full-size layout of the studio setting. Doors, windows, stairways, etc., are outlined. Stock rehearsal furniture substitutes the actual studio items and action props (telephone, tableware) are provided. Rehearsing in this mock-up, actors become accustomed to the scale and features of their surroundings; vertical poles, chairs, dummy doors, marking the main architectural limits.

The director arranges performers' positions, grouping, action to suit his treatment. He rehearses in sections: the cast learning lines, practicing performance, moves and business; until the entire show runs smoothly, ready for its studio debut. Segments and overall timing are checked (and controlled) throughout allowing for subsequent non-action inserts—titles, film, VT graphics, announcements, etc.

During rehearsals, the director scrutinizes action through his viewfinder from potential camera positions, adjusting details as necessary.

A few days before the studio date, the specialists concerned with technical operations watch a complete rehearsal (technical walk-through/technical run) checking and anticipating problems, envisaging camera and boom treatments. Following this meeting, facilities are finalized and the lighting prepared.

Studio rehearsal

Before a studio rehearsal, the staging crew supervised by the set designer, erects and dresses the settings. Lighting units are rigged and guided by the lighting director, are adjusted to cover their plotted areas. Camera and sound equipment are positioned. The

Table 17.1   Initial blocking hints

Timing	A preliminary read-through gives only a rough timing estimate. Allow time for business, moves, recorded inserts, etc. Anticipate potential script cuts if an overrun is evident. Many productions include sequences that can be dropped, reduced, or expanded to trim timing.
Briefing performers	Ensure that performers have a clear idea of the program format, their part in it, and their interrelationship with other contributions.
	Ensure that performers have a good notion of the setting; what it represents, where things are.
Props	Provide reasonable substitutes where real props are not available. Sometimes only the actual item will suffice. Where unfamiliar costume is to be worn (cloak, hoop skirt, space helmet), a rehearsal version is preferable to dummy motions.
Directing performers	Maintain a firm attitude towards punctuality, inattention, background chatter during rehearsals, to avoid time wastage and frustration.
	Check that performers' positions are consistent and meaningful (they do not stand 'on' a wall).
	Avoid excessive revisions—of action, grouping, line cuts, etc. Wrong versions can get remembered, new ones forgotten.
	Warn performers where they are in close shots, to restrict their movements; perhaps indicating which camera to play to.
Shot arrangement	Use a portable viewfinder to arrange shots. Even a card cut-out frame or a hand-formed frame is better than unaided guesses.
	Always think in terms of *shots*, not of theatrical-styled groupings, entrances, exits, business. At script stage, form ideas about shots rather than arranging 'nice groups' and trying to get 'good shots' during rehearsal.
	When setting up shots in a rehearsal hall, do not overlook the scenic background that will be present in the studio. Check shot coverage with plans and elevations.
	Consider depth of field limitations in close shots or deep shots (close and distant people framed together).
	Think in terms of practical studio mechanics. You may rapidly reposition your viewfinder, but the move may be impossible for the studio camera.
Audio and lighting	Try to bear audio and lighting problems in mind when arranging positions and action. For example: where people are widely spaced, a sound boom may need time to swing between them, or have to be supplemented.

performers arrive, seeing the staging for the first time. The show is ready to begin.

Directors organize their studio rehearsals in several ways, according to the complexity of the production, available time, performer experience, and criticality of results. Given the opportunity, you might like the luxury of:

1. *Dry run (walkthrough)*—actors perform, familiarising themselves with the studio settings, etc., while the studio crew watch, learning the format, action, and production treatment.
2. *Camera blocking (stagger through)*—initial camera rehearsal, coordinating technical operations, discovering and correcting problems.

3. *Predress run-through (continuous run-through)*—'polishing' rehearsal, proving corrections.
4. *Dress rehearsal (dress-run)*—hopefully fault-free 'on-air' quality performance.
5. *Videotaping/transmission*—final optimum performance.

In reality, many productions have little more rehearsal time than the show's running duration, and the second and fifth stages have to suffice. So a continual watch on progress, relative to available rehearsal time, is essential.

Rehearsal procedures

■ *Camera blocking/stagger through/first run/stopping run* Here the director controls operations from the production control room, only 'going to the floor' when personal on-the-spot discussion is essential. Otherwise, his eyes and ears are the picture monitors and loudspeaker. All communication with his crew is via the *intercom (production talkback)* system; the *floor manager* cuing and instructing the talent through his intercom guidance.

Many directors use the *whole method*, and go straight through a sequence (segment, scene) in camera rehearsal, continuously guiding until a serious problem arises to require stopping and correction. The director discusses problems, solutions and revisions; then re-runs the sequence with corrections. It should now work.

This method gives a good idea of continuity, timing, transitions and operational difficulties. But, by skimming over the various shortcomings before the 'breakdown point', quite a list of minor corrections may develop.

Other directors, using the *stopping method*, stop action and correct faults as they arise—almost shot-by-shot. This precludes error adding to error, ensuring that everyone knows exactly what is required throughout. For certain situations (chroma key treatment), this may be the only rational approach.

However, this piecemeal method gives the impression of slow progress, and can feel tedious and nit-picking. The continual stopping makes checks on continuity and timing more difficult. Later corrections are given as notes after the run-through.

■ *Floor blocking* Using this method, the director forsakes the production control room and works from a studio picture monitor. Guiding and correcting performers and crew, he calls out required transitions to the switcher in the control room, watching results on his monitor screen.

This technique is often used by directors who find visualizing difficult, feel remote from operations in their orthodox position, and prefer direct contact to relaying instructions.

The director often returns to the control room to run each

floor-rehearsed scene. During videotaping, the show may be shot as segments, scenes, or continuous action.

Where a set-up is particularly complicated or discontinuously recorded, the system may have some advantages. But it has all the weaknesses of the 'stopping method' plus low crew cohesion. When they *all* hear the same intercom instructions, crew members can coordinate more effectively. At worst, this method wastes a great deal of time, and because the director has not been watching channel previews or listening to audio while on the floor, continuity and video/audio relationships may be below optimum.

■ *Rehearse/record* This is really a form of 'floor blocking', in which the director sets up a sequence, rehearses it, then immediately records the corrected result. Each sequence may need to be relit, cameras repositioned and scenery reset or cheated. The separate videotaped sections are later edited together with appropriate transitions and bridging sound.

Ideally, this pseudofilmic technique could provide optimum results for each shot. But in practice, the process can be extremely frustrating and time consuming. Normally the effectiveness of lighting, scenery, make-up, costume, etc., is checked on camera, and improved or corrected while another scene is being rehearsed, or during a studio stand-down. In a *rehearse/record* situation, such changes have to be made immediately, so that the scene can be shot—and this leads to many compromises. To save time, a director may even record the first camera rehearsal with all its imperfections, in case it can be used! A videotape channel is continuously tied to the studio. Themes and variations are shot for later selection, and repeat action may be necessary to ensure editing continuity. Considerable poststudio videotape editing is necessary.

Rehearsal problems

During the course of rehearsal, various problems regularly arise. Here, in summarized form, are typical situations.

■ *The shot*
1. *Subject detail not sufficiently visible*—Move subject/camera closer. Zoom in. Shoot subject in a series of close detail shots or localized insets (magnified view within longer shot). Is greater detail available? (Very close shots of some subjects are confusing or reveal no further information—unsharp photographs.) Is detail defocused? Is subject shadowed or obscured?
2. *Extreme close-ups of details required*—Problems arise due to limited focused depth, unsteady camera, subject moving in/out of frame. Use higher light intensities (levels) to permit stopping down. Use wider lens angle. Locate item on firm surface (use position marks). Use photoenlargement of detail.

3. *Subject's shape does not fit screen aspect ratio*—Take a long shot of total subject, then pan/tilt over detail or intercut closer shots. If a series of items are involved, do similarly or recompose the shot (move them together, arrange in depth, or shoot more obliquely).

4. *Unsatisfactory framing*—Subject cut off by frame edge. Shot too tight or loose. Subject too high in frame or off-center (unbalanced). Unsuitable headroom.

5. *One subject masks (obscures) another*—Reposition subject and mark floor. (Check that this does not spoil earlier shots.) Reposition camera (e.g. arc or truck/crab).

6. *Subjects appear too widely spaced apart*—Move subjects closer together, shoot obliquely, or intercut single shots.

7. *Subject appears too large or small relative to others in shot*—See Table 4.2.

8. *Subject appears too prominent*—Reduce shot size, alter composition, or increase camera height.

9. *Background objects or scenic lines 'grow out' of subject*—Reposition camera or subject. Move, cover or remove background items. Reduce their prominence by lighting adjustment or defocusing.

Table 17.2   Effective Studio rehearsal

Unrehearsed or briefly rehearsed studio production	*Director's assistant* checks that available production information is distributed (breakdown sheet/running order). Also checks contributory graphics (titles, slides, floor captions) and film/VT inserts are correct. Notes word cues for inserts, timings, announcements.
	Director with floor manager, technical director, lighting director, cameras and sound crew, arranges basic performer and camera positions (floor is marked). Even a basic plan aids coordination. He outlines action or moves, and shot coverage. Lighting is set.
	If talent is available, line up shots and explain to them any shot restrictions, critical positions, care needed in demonstrating items, etc. (Otherwise use stand-ins for shot line-up and brief talent on arrival.)
	Check that performers, crew, and contributory services (prompter, film channel, etc.) are ready to start.
	If full rehearsal is impractical, carry out basic production checks: rehearse 'tops and tails' of each segment with intermediate links (e.g. announcer's in-cues/out-cues, to cue film or VT inserts); check any complicated action or treatment. Check any errors or problems.
Intercom and the director	Remember that the production team is interdependent. A quiet, methodical, patient approach is as infectious as an acrimonious one. Be firm but friendly. Avoid critical comments on intercom.
	How detailed intercom instructions need to be, depends on the show's complexity and crew experience. Aim at a balanced information flow, avoiding continuous chat and comment. (A mute key cuts the circuit when necessary.)
	Preferably call cameras by numbers, guiding all camera moves (and zooms) during rehearsal, warning the crew of upcoming action and movements.
	Examine each shot. If necessary, modify positions, action, movement, and composition.
	Consider *shot continuity*. Alterations may affect earlier shots, too.

10. *Unwanted distortion*—Adjust lens angle to nearer 'normal' (25°) and change camera distance accordingly, reposition subject or change viewpoint.

11. *Colors have similar monochromatic values*—Lighten/darken areas concerned. Change hue or materials (surface texture). Put black outlines round areas of similar tone.

■ *Performers*

1. *Out of position*—Check if they have toe marks or location points. Move them and remark floor (instructing performers). Reposition camera.

2. *Performer working to wrong camera*—Floor manager signals. Draw attention to tally (cue) light on camera. Switch shot allocation to the 'wrong' camera.

3. *Dialogue errors (cuts lines, dries/fluffs, or freezes and needs prompt, wrong lines)*—Use prompter, cue cards, etc. Give audible prompt. Is prompter working, script OK? Misunderstanding? Retake sequence, or insert cutaway shot and dub audio retake into audio track of VT.

Remember, the crew and performers are *memorizing*. Their aids are the production paperwork and your intercom reminders/instructions.

Do not be vague. Make sure that your intentions are understood. In correcting errors, explain what was wrong and what is wanted. Not 'Move him left a bit' but 'He is shadowing the map'. Do not assume performers will see and correct any problems.

Avoid excessive changes or revisions, or there will be hesitations and mistakes.

Correct operational and performance errors, even in early camera rehearsal. Misjudgments and inaccuracies should be noted directly they occur (e.g. camera in shot, shooting off, late cue, wrong lines, wrong shot, wrong mike). But avoid an overinterrupted rehearsal, or timing and continuity will be lost.

If a cameraman offers alternative shots (e.g. to overcome a problem), briefly indicate if you accept or disagree, and why.

Where practicable, at the end of each sequence ask if there are any problems, and whether anyone wants to rehearse that section again.

Remember, various staging and lighting defects may be unavoidable in early rehearsal. Certain details (set dressing, light effects) take time to complete. Some aspects need to be seen on camera before they can be corrected or finally adjusted (overbright lights, lens flares). Shot readjustments during rehearsal often necessitate lighting alterations.

Never repeat a segment without indicating whether it is to be *changed* (move faster next time), or is to correct an *error* (late cue), or to *improve* the performance/operations.

Ensure that everyone knows when shots have been deleted or added (e.g. Shot 2A, 2B, 2C).

At the end of rehearsal: check timings, give notes to performers and crew on any errors to be corrected, changes needed, problems to be solved. Check whether they have difficulties that need your aid.

At least one complete uninterrupted rehearsal is essential for reliable transmission.

At the end of videotaping, give details (including shot numbers) and reasons for any retakes.

■ *Senic problems*

1. *Check staging for distracting features*—Overbright surfaces
   reflections. Can relighting cure? Reangle, use dulling spra
   repaint, cover over, scrim over or remove.
2. *Colors or tones unsuitable* (e.g. subject merges with backgroun
   —Modify background, lighting or subject (e.g. change co
   ume).
3. *Background blemishes* (dirty marks, tears, scrapes, wrinkle
   etc.)—Rectify, refurbish, cover over or relight.
4. *Ugly or distracting shadows on background*—Performers
   scenery too close to background? Relight or modify shot.

■ *Operation hazards*

1. *Camera shooting off set*—Modify shot, extend scenery or a
   foreground masking.
2. *Lens flares*—Raise camera viewpoint, improve lens sha
   (hood), shield light off camera or raise lamp height. Alter sh
3. *Unwanted subjects in shot*—Have their positions chang
   since last rehearsal? Alter framing, camera position or le
   angle. Rearrange shot, remove unwanted items. Mask/cov
   items. Leave items unlit.
4. *Boom mike in shot*—Check whether boom operator is awai
   Warn him before cutting from close to distant shot. Tight
   (reduce headroom) or recompose shot. Alter sound pick
   method.
5. *Boom shadows*—Check with audio/lighting personnel if avoi
   able. Consider modifying shot—tighter shot or chan
   viewpoint. Alter sound pickup or lighting.
6. *Cables in shot*—Check for distracting cables in shot—audi
   lighting, monitors, cameras.
7. *Graphics/captions/titles*—Check they are straight, level a
   undistorted. Can all be seen (no cutoff) and read? Time
   read?

■ *Editing/video switching/vision mixing*

1. *Transitions appropriate*? Suitable type? Timing and rate
   transitions OK?
2. *Any distractions*? Any reverse cuts, jump cuts, position jump
   instant changes in size, direction, frame height? Any loss
   orientation or direction on cuts? Any image confusion durir
   supers or mixes?

Floor manager

The *floor manager* (FM) is a person of many parts. On larg
productions, he or she may, with an assistant (AFM), join tl
show at its onset and assist with organization, local liaison, locatio
shooting, and sometimes substitute for the busy director at pr

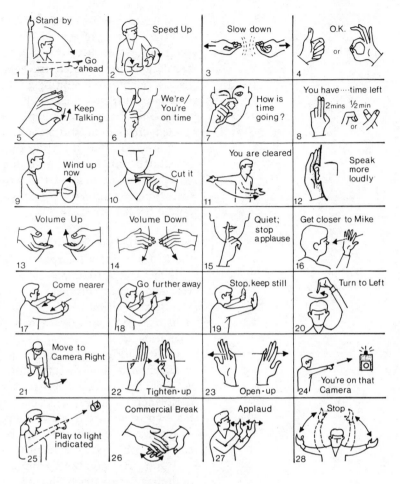

Fig. 17.1 Floor manager signals
1, Stand by . . . go ahead. 2, Speed up, faster pace, quicker tempo (speed according to increase required). 3, Slow down, slower pace, stretch it out (indicated by slow 'stretching' gestures). 4, OK, you're all right now, it's OK (confirmation signal). 5, Keep talking (open/close beak movement). 6, We're/ you're on time, on the nose. 7, FM TO CAMERA. Are we on time? How is the time going? 8, You have . . . time left (illustrated—2 mins, and $\frac{1}{2}$ min). 9, Wind up now, conclude action. 10, Cut it, stop, finish, omit rest of item. 11, You are cleared. You are now off camera, and can move or stop action. 12, Speak more loudly. 13, Volume up, louder. 14, Volume down, quieter (sometimes preceded by 'Quiet' signal). 15, Quiet, stop applause. 16, Get closer to mike. 17, Come nearer, come downstage. 18, Go further away, go upstage. 19, Stop, keep still. 20, Turn round (in the direction indicated). 21, Move to camera right. 22, Tighten up, get closer together. 23, Open up, move further apart. 24, You're on that camera, play to that camera (sometimes with turning head gesture). 25, Play to the light indicated. (When actors are shadowing, point to light source and to area of face shadowed.) 26, Commercial break (brush right hand over left palm—spreading butter on bread). 27, FM TO AUDIENCE: You can applaud now (may be followed by louder sign). 28, Stop. (For applause, widespread action, etc.).

studio rehearsals. Alternatively, the FM may join the production in the studio for camera rehearsals.

During studio preparations for camera rehearsal, the FM checks all *non-technical* aspects of the show, confirming the team's readiness. Progress checks ensure that there are no staging hang-ups (i.e. action props work, scenery and furniture are in planned positions, doors do not stick . . . etc.); that no fire, safety or union regulations are being contravened. He guides the floor crew/stage hands, relative to operating graphics, studio/title cards, cue cards, scenic moves, effects cues; etc.

Around this time, arriving performers may need to be wel-comed, checked in, and accommodated; and told when and where they are required. Ensuring punctuality and the observation of scheduled times are important aspects of the FM's duties; whether for rehearsal starts, *turnrounds* at the end of a rehearsal (returning equipment, props, scenery, to opening positions), or studio breaks (meals).

■ *Rehearsal* As the director's contact man on the studio floor, the FM wearing his earphones, listens to the director's intercom (talkback) and then passes this information on to the talent. He anticipates problems, rearranges action and grouping, furniture positions, etc., as the director indicates, marking the floor where necessary. He supervises staging and property changes.

Only the FM will normally stop floor action, on the director's instructions. (The director only exceptionally uses the studio-address loudspeaker/loudspeaker-talkback, to talk directly to performers). The FM talks back to the director over studio mikes, his own miniature 'reverse-talkback' transmitter, or by gestures in front of preview cameras. Being on the spot, he can often correct problems not evident to the remote director.

A good FM combines calmness, discipline and firmness, with diplomacy and friendliness; putting talent at ease. Always available yet never in shot. He maintains a quiet studio, yet appreciates that various last-minute corrective jobs must progress. He investigates any delay, aiming to clear it; or if rehearsal is likely to be halted, suggests an alternative sequence that is ready instead. Later, the FM may welcome and guide any studio audience, relative to the show format and their contributions.

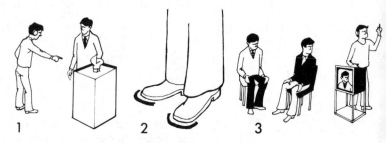

Fig. 17.2 Guiding the performer
Show them where to stand: using (1) a locating point, or (2) toe marks. 3, Keep other action and monitors out of their eyeline.

■ *Taping/transmission* In checking and making ready for 'air time' or videotaping, the FM confirms that all studio access doors are shut, talent and crew standing by in opening positions, the ident slate has accurate details (which he reads during the VT countdown), that studio monitors are showing pictures, etc.

During the show, he is cuing talent, keeping a wary eye on all about him, listening to intercom guidance, anticipating hazards, and generally smoothing proceedings. At the end of taping, he holds talent and crew until results have been checked; announcing and preparing any necessary retakes, finally 'releasing the studio' (performers and crew). He checks on safe storage of any valuable or special props or equipment and ends his duties with a logged account of the production.

Guiding performers

How much guidance performers need, varies with their experience and the complexity of the production. But you must always ensure that talent realizes exactly what you want them to do and where to do it. A preliminary word may be enough, or a painstaking rehearsal may prove essential.

■ *Inexperienced talent* Having welcomed talent, put them at ease, and outlined their contribution, they are best supported by an experienced host who guides and reassures them. Have them talk and demonstrate to the host rather than the camera, for he can steer them by questions and ensure that subject positions suit the shots.

We all feel daunted by unfamiliar conditions, mechanics, and disciplines; self-confidence is essential for good performance. So keep problems to a minimum, with only essential instructions to talent. Avoid elaborate action, discourage improvisation, and have minimal rearrangements.

Even slight distractions may worry inexperienced talent. Sometimes a small cue card or a list of points held beside the camera, may strengthen their confidence. However, few inexperienced people read scripts or prompters 'naturally'; but instead have a stilted, ill-at-ease delivery.

The balance between insufficient and excessive rehearsal, is more crucial with inexperienced talent. Uncertainty or over-familiarity can lead them to omit sections during transmission. Sometimes the solution lies in either taping the item sectionally or in compiling from several takes.

■ *Professional talent* You will meet a very wide range of experienced professional performers in TV production—actors, anchormen, hosts, link men, commentators, introducers, presenters, and MCs (Master of Ceremonies).

Familiar with studio routines, they can respond to the most complicated instructions from the FM, or an earpiece intercom (switched talkback) without the blink of an eye, even under the most trying conditions—and yet maintain cool command of the situation. Comments, feed questions, timing and continuity changes, item cuts, ad-libbing padding, are taken in their stride.

The professional makes full use of a prompter displaying the full script—this means that cameras must always be within reading distance. His performance (action and positions) is accurately repeatable and he will play to particular cameras for specific shots, making allowances for lighting problems (e.g. shadowing) or camera moves.

Cuing

To ensure that action begins and ends at the instant it is required, precise cuing is essential. If you cut to performers and then cue them, you will see them waiting to begin or watch action 'spring into life'. Cue them too early before cutting and action has already begun. So 'cue and cut' is normal practice.

Wrong cuing leaves talent bewildered. If they have finished their contribution and you have not cut away to the next shot, they may stand with 'egg on the face', wondering whether to ad-lib or just grin! Wrong cuing can mean that film or VT runs out to a blank screen, or we see run-up leaders flashing by on transmission. VT editing may compensate for poor cuing and disguise sloppy direction.

■ *Methods of cuing*: (See Fig. 17.1)

1. *Hand cues*—Given by the FM (or relayed by the AFM), these are the standard methods of cuing action to start or stop. Sometimes a performer cannot watch for cues (during an embrace) and a tap on foot or shoulder may be used instead.
2. *Word cues*—An agreed go-ahead word or phrase during dialogue, commentary or discussion, may be used to cue action or to switch to an insert. We note *out cues/out words* (last spoken words) at the end of a recorded insert (VT or film) to ensure accurately timed switching to the next item.
3. *Monitor cues*—Commentators and demonstrators often watch recorded inserts on a nearby picture monitor, taking a go-ahead cue from certain action. Alternatively, they may use a *time cue*, counting down (e.g. from 10 sec) from a cue point. Most inserts have a few seconds without strong action or speech as *run-in* (*top*) and *run-out* (*tail*) to cushion/buffer the cuing.
4. *Light cues*—Performers can take a cue from the *camera tally* (*cue*) *light* (illuminated as the camera is switched to line). For

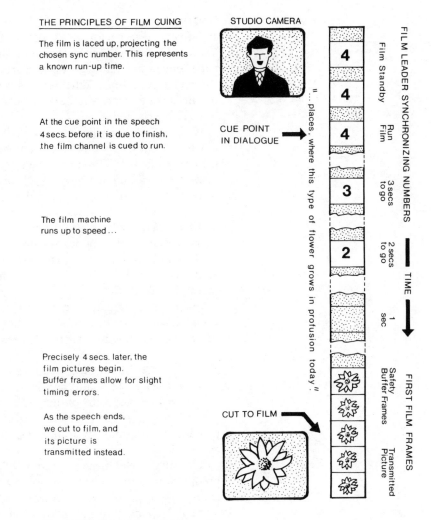

Fig. 17.3 Principles of film cuing

announce booths or when cuing actors behind scenery, small portable *cue lights* are often used; where the standby 'flick' is followed by a steady 'action light'.

5. *Intercom cues*—These are given direct to a performer (news-caster, commentator) wearing an earpiece—interrupted feed-back, program interrupt, switched talkback (Chapter 6).
6. *Clock cues*—A go-ahead at a specific time.
7. *Buzzer cues*—Used in some organizations, for intercom between film/videotape areas and the production control room. (One buzz: 'Yes' or 'Start'. Two buzzes: 'No' or 'Stop'.)
8. *Electronic cue dots*—Used by networks at program change points, these small black/white squares in corner frame cue program out-time.

Prompting

Relatively few performers can be expected to learn a scri
accurately and deliver it at a repeatable speed. Even the mc
experienced are liable to deviate from the written script, 'dr
(forget) or 'cut' lines. Whether this is a problem, largely depen
on the show format, whether live or taped and if it is spontaneo
or strictly scripted.

While a quiet verbal prompt may suffice (perhaps using a mutii
key to temporarily kill the mike), more sophisticated methods a
generally needed. There are generally two types of prompters
reminder notes (*aide mémoires*) that give data, subject heads, or sho
format; and *continual references* giving entire script or lyrics.

■ *Reminders* These have been written on shirt cuffs or near!
scenery, but more orthodox methods range from small hand not
to the clip boards used by many interviewers, commentators, a
anchormen. These boards carry their notes, questions a
research information.

For brief information, cards can be held near the camera I
floor crew (cue card, goof sheet, idiot card), or smaller *flip car*
(*flippers*) suspended under the camera head.

■ *Script* A talking head reading a desk script does not mal
good television. It is dull and lacks 'spontaneity'. But a regul
script ensures accurate timing. Consequently, many shows u
a *roller prompter* (mechanically or electronically operated) th
provides a continuously rolling copy of the script at or near t
camera lens.

The vary-speed display is remotely controlled (by the talent c
an operator) and typically has some 20 words visible in an 8-li
frame. To disguise eye movements or a fixed stare, the perform
'casually' changes his head position while reading. Off-came

Fig. 17.4 part 1   Prompting—direct methods
1, Handwritten prompt cards usually held near the camera, may contain dialog
or reminder notes. (Cue-cards, idiot board, goof sheet). 2, Camera flip car
usually carry brief reminders. 3, A roller-script (speed remotely controlled) m
be attached to the camera, or floor-located.

Fig. 17.4 part 2 Prompting—video prompters
1, Video prompters use a 45° glass sheet at the camera lens, reflecting a picture-tube display of the script. This display tube is fed by a nearby script scanner (a monochrome TV camera shooting a miniature roller-script). 2, The camera lens shoots through the glass sheet at the subject. The reader appears to be looking straight at the lens. (Other systems can result in an off-camera star') 3, The script-scanner video may be fed to floor video prompters.

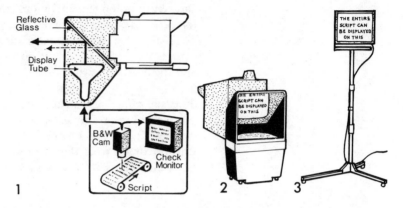

prompters can lead to a 'shifty eyed' presentation, for the speaker has to continually glance away from the lens axis.

Prompters may inhibit shot variations because cameras carrying prompters need to be close enough for comfortable reading. Where only selected cameras have prompters, we must ensure that one of these is always near the speaker.

Production timing

Whereas a *closed-circuit* program needs only approximate timing, a *live transmission* must fit its allocated time slot exactly. Overruns can result in end cut-off or cause scheduling problems. In live *composite productions*, where various contributory sources *'opt-in'* or *'opt-out'* (temporarily join or leave) the main presentation to insert their own material accurate timing is imperative.

Scripts may be roughly timed by reading aloud and allowing for any mute action, business or inserts. Many run at around a minute a full page. During rehearsal each segment or scene is timed, and subsequently adjusted to suit a timed show format/ running order (estimated, permitted or actual).

■ *Intermittent timing* When rehearsal is halted for any reason, the stop-watch time is noted. Rehearsal recommences at a point *before* the fault and the watch is restarted when the fault point is reached. A videotaped production can usually be edited to reduce excess; its associated *time code* providing duration or elapsed time checks throughout.

■ *Shortening inserts* A live transmission needs continual timing surveillance to obviate over- or underrunning—particularly with unscripted material. If one segment *overruns* (*spreads*), another may have to be shortened. Where prerecorded inserts (film or VT) are involved, their durations and *out-words* should be precisely known. However, they offer little time flexibility; particularly if accompanied by recorded dialogue. We can only shorten them on the air by omitting their start (a *late-in*) or end (an *early-out*), or even omitting an insert entirely (*dropping*).

To assist cuing when excerpting from existing material (e.g. a complete speech), we should note both the *run-in* (precue) dialogue and the *in-cue* (in-words).

■ *Back cuing*   To ensure that recorded play-out music finishes precisely at program fade-out, time the music from a recognizable cue point to its conclusion (say 1 min 35 sec). On transmission we start the music from that cue point at 1 min 35 sec before program out-time, fading it up when required.

Table 17.3   On-air intercom

Intercom (talkback) instructions reveal the split-second teamwork involved in efficient TV directing. Although local terminology and personnel vary, the following intercom talk from an actual show certainly reflects the spirit of the occasion.

Director's assistant	Director
Stand-by studio. Going ahead in 1 minute . . .	(FM alerts studio and gives countdown)
Stand-by film. (Acknowledging buzz from film channel).	Start VT. Show slate (identifying board) on 4.
Stand-by announcer.	
Roll film . . . 4–3–2–1–Zero.	Fade up film. Music (on audio tape).
On film for 1 minute . . . Stand-by 1.	Super 1.
Shot 1.	
On 1 . . . Stand-by 2.	Take music under. Cue Announcer. (Cue light in announce booth).
Coming to end of film. Counting down . . . 30 seconds—20–10–5–4–3–2–1–0.	Bring up music. Cue Peter (FM cues talent). Take 2 (cut to Cam. 2). Take music under.
On 2. Shot 2 . . . 1 next.	He's going to sit. Go with him . . . Dolly in slowly to CU . . . Hold it there . . . Super slide title (guest's name) . . . take out super. Steady 1 . . . a bit tighter . . . good . . . take 1.
On 1. Shot 3 . . . 2 to graphic.	
Inserting 2 . . . Stand-by disc (grams).	Wipe in 2 . . . fine . . . take 2 out.
Clear 2 (after combined shot).	Go disc. Stand-by Joan (FM alerts talent).
Stand-by 3 . . . Coming to 3.	Music under. Cue her on and cut. He'll exit left.

On 3. Shot 4 . . . Stand-by 2 . . .
Extra shot 4A.

On 2 . . . We're 15 seconds over
(overrun).

Coming to 2 . . . Mix to 2 . . . She's
going to open the box . . .

On 2 . . . Shot 4A . . . Stand-by VT.

Cue Joan to open it (FM cues) . . .
we're tight for time . . . Drop Shot 5.

Shot 5 is *out* (to be omitted).

Give her a wind-up . . . run VT.

Coming to VT . . . 4–3–2–1–0.

Cut to VT.

On VT for 2 minutes 7 seconds.

1 minute left on VT . . . Coming out
of VT in 30 seconds.

Out words . . . in a pig's eye.

Stand-by 2 for Shot 6.

Stand-by studio. Coming out of VT.

20–10–5–4–3–2–1–0.

Hold it 2. Cut to 2.

On 3. Shot 60 . . . Clear 1 to crawl
(roller titles).

He's going to move to the door and
wave. Give him $\frac{1}{2}$ minute.

Stand-by tape (audio). Stand-by 1
for Shot 61.

(FM signals talent) Wind him up . . . go
crawl, go tape.

On 1 and 3 . . . Off the air.

Super 1. Take out 3 . . . Go to black at
end of phrase.

Stand-by for retakes.

Right everyone. That was great. We'll
just check VT (VTR spot-checked).
OK. Thank you very much. We have a
clear.

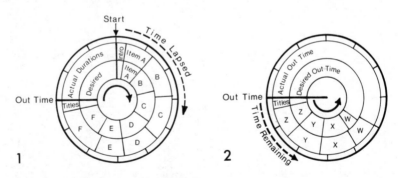

Fig. 17.5 Forward and back timing
1, *Forward timing* (*front timing*): Duration timing (estimated and real) as the
show proceeds.

(Desired durations)	Running stop. watch reads (from prog. start)	Clock start time	Clock end time	Item running times (actual duration)
Intro 30"	30"	19.15.00	19.15.30	30"
Item A 10'	10' 33"	19.15.30	19.25.33	10' 03"
Item B 8'	18' 30"	19.25.33	19.33.30	7' 57"

2, *Back timing*: A 'remaining time' measurement, showing the amount of time
before the production ends.

Item duration	Clock time (item starts)	Remaining time
Item X 1' 0"	20.23.00	2' 0"
Item Y 30"	20.24.00	1' 0"
Item Z 30"	20.24.30	' 30"
End titles, Out time	20.25.00	00.00

Table 17.4   The director's instructions

To cameras	Opening shots please.	Cameras to provide initial shots in the show (or scene).
	Let's line up Shot 25.	Adjust that camera's shot and subjects, for composition required in Shot 25.
	Ready 2: Stand-by 2.	Stand-by cue for Cam. 2.
	Give me a single; 2-shot; 3-shot/group shot/cover shot.	Isolate in the shot the single, two or three persons specified./A shot including a group of people./A wide shot taking in all likely action.
	Zoom in (or out).	Narrow (or widen) zoom lens angle.
	Tighten your shot.	Slightly closer shot (slight zoom in).
	Get a wider shot	Get the next wider standard shot, e.g. from CU change to MS.
	2 match to 3.	Cam. 2, relate to or fit into Cam. 3's shot (using viewfinder 'mixed feeds').
	More (or less) headroom.	Adjust space between top of head and top of frame.
	More (or less) looking room.	Adjust space between angled or profile face and side of frame.
	Center (frame-up).	Arrange subject in picture center.
	Lose the hands.	Compose shot to avoid seeing hands.
	Cut at the hands.	Frame to just keep hands in shot.
	Pan left (or right).	Horizontal pivoting of camera head.
	Tilt up (or down)/pan up (or down).	Vertical pivoting of camera head. (Strictly, 'pan' should apply only to *horizontal* action, but term's misuse is very widespread.)
	Dolly back/pull back/track back.*	·Pull camera mounting away from the subject.
	Dolly in/push in/track in.*	Push camera mounting towards the subject.
	Creep in (or out).	Push in (or pull out) very slowly, imperceptibly.
	Truck left (or right)/crab left (or right).*	Move mounting to the left (or right).
	Arc left (or right).	Move mounting round subject in an arc.
	Ped up (or down).	Raise (or lower) column of pedestal mounting.
	Crane up (or down)/boom up/ (sometimes elevate or depress).	Raise (or lower) crane arm or boom.
	Tongue left (or right)/crane/ slew/jib.	Turn crane boom to left (or right).
	Tongue in (or out)/crane/ boom in.	Crane positioned across the scene, move boom to and from subject.
	Focus up 3.	Warning to a camera 'standing by' that is defocused.
	Lose focus on the boy.	Let the moving boy become defocused, remain sharp on other subject(s); *or* deliberately defocus on the boy.

	Pull focus on the girl.	Change focus deliberately from previous subject, to the girl.
	Follow focus on the boy.	Maintain focus on moving boy.
	Split focus.	Focus equally sharply between main subjects.
	Focus 2/two, you're soft.	Criticism that Cam. 2's shot is not sharply focused.
	You're focused forward (or back).	Focused at a point too near the camera (or too far away—e.g. background sharp while nearer subject is unsharp).
	Rock focus 2. (Focus check).	Focus to and from the subject to check optimum sharpness.
	Stand by for a rise (or sit).	Person is about to stand up (or sit down).
	Follow him over/hold the boy.	Keep subject in shot (well composed) as he moves.
	Let him go.	Do not reframe. Let him pass out of shot.
	Lose him.	Recompose (or tighten shot) to exclude him.
	Cam. 2, you're on lights.	Cam. 2 has lamps or strong reflections in shot (liable to damage the camera tube).
	Clear on 3/you're cleared 3.	A 'clear', usually after a combined shot (super, chroma key); Cam. 3 can move to its next position.
	Release 3 to captions.	A 'clear', directing camera to next position.
To the floor manager	Opening positions please.	Performers (and equipment) in position for start of show (or scene).
	Stand him by/give him a stand-by.	Alert him in readiness for a cue.
	Cue action/cue 'Bob'.	Give sign for action to begin (perhaps name person or character).
	Hold it.	Stop action/performance; to correct an error.
	Freeze there.	Stop movement exactly at that point; hold it quite still.
	Back to the top/take it from the top.	Begin again at the start of the scene; repeat the rehearsal.
	Pick it up from Shot 20.	Recommence rehearsal from Shot 20.
	Would they just walk it?/Go through it?	Without dialogue or performance, actors move through their various positions.
	Clear 2's shot.	Something/someone is obscuring Cam. 2's shot.
	Tighten them up.	Move them closer together.
	Move him downstage (upstage).	Move him toward (away from) the camera.
	Move her camera left.	Move her to *left* of the camera's shot (i.e. stage *right*).
	Give him a mark.	Draw a toe mark or show him a location point.
	Give her an eyeline.	Show her the direction in which to look.
	He'll take a word cue.	He will begin action on hearing certain dialogue.

In production

Give him a light/flash him.	Operate a cue light.
Have him take a beat (or two) before he moves.	He should delay for a silent count of one (or two) before moving.
Have him play to his key.	He should face his main light (key light).
Show David 3/he's on 3.	Indicate to him which camera is shooting him.
He's to work to/play to 3.	He should face Camera 3.
Ask him for more (less) voice.	To speak louder (more quietly).
Show him the mike.	Indicate the mike picking him up.
Speed him up.	Tell him to go more quickly (i.e. time is sh⌐
Stretch him/tell him to spread.	Tell him to go more slowly (i.e. there is time to spare).
Tell him to pad/keep talking.	Tell him to improvise until next item is ready. (Often a hand signal, thumb and extended fingers touch and part.)
Give him 2 minutes . . . 1 minute . . . 30 seconds.	He has 2 minutes left (followed by countdown on fingers).
Give him a wind/wind him up.	Signal him to finish.
Give him a hard wind/quick wind.	It is essential for him to stop immediately. (Faster signal.)
Kill him/cut him.	He is to stop immediately.
Give him a clear.	We have left him. He is free to move away (or relax).
Lose the flowers.	Remove (kill) the flowers (for the shot or scene).

To audio (sound mixer)	Fade up sound/fade up boom.	Fade up from zero to full (general or specific source).
	Stand by music (disc/grams, tape, effects, etc.).	Warning before cue.
	Cue music/go music.	Go-ahead for music (live, disc or audio tape).
	Creep in music/sneak in music.	Begin very quietly, gradually fading up audio.
	Down music.	Reduce volume (level) of audio.
	Music under/music to background.	Keep music volume low, relative to other sources.
	Hit the music.	Begin music, at full volume.
	Up music.	Increase volume of music, usually to full.
	Sound up (or down).	General instruction to increase (or decrease) overall volume.
	Kill (cut) the music.	Stop the music.
	Fade sound.	Fade out all program sound.
	Fade tape (disc, film sound, etc.).	Fade out specified source. Usually leaving other sources.
	Cross fade; mix.	Fade out present source(s), while fading in the next.
	Segue (pronounced 'seg-way').	One piece of music immediately follows another without a break.

To switcher (vision mixer)	Take 1/cut to 1/cut 1.	Switch to Camera 1's picture.

	Fade up 2.	Bring Cam. 2's video-fader from zero to full.
	Stand by to fade 2/ready to fade 2.	Prepare to fade Cam. 2's picture out.
	Fade/go to black/fade out.	Fade selected camera's video fader down to zero.
	Stand-by to mix (dissolve) to 3.	Warning before mixing cue.
	Mix (dissolve) to 3.	Mix from present camera(s) to Cam. 3's picture.
	Supering 3/ready to super 3.	Warning before super cue.
	Super 3/superimpose/add 3.	Fade Cam. 3's picture up, adding to existing sources.
	Take 2 out/lose 2.	Remove (usually fade) Cam. 2's picture, leaving the rest.
	Wiping 3 over 2.	Stand-by for wipe.
	Wipe to 3.	Using prearranged pattern, Cam. 3's picture obliterates present picture.
	Ready black . . . go to black.	Stand-by . . . fade picture to black.
To captions (graphics)	Flip, animate (de-animate). Change caption. Go crawl (roller).	Cuing caption changes. ('De-animate' indicates 'return to preanimation position'.)
To other source	Cue telecine/roll film/run TK.	Start film projector (similarly for *videotape*, although some organizations ROLL film and RUN VT.).
	Go ahead PARIS.	Start cue for remote source in PARIS.

*British/European terms.

18 Imaginative production

Pictorial function

Most pictures are factual, showing subjects in a familiar un
distorted fashion, without any predominating emotional appea
Yet, by judiciously arranging these same subjects, by caref
composition, and a selective viewpoint, you can modify their enti
impact and give them quite a different implied *significance*. Yo
interpret the scene.

You can deliberately distort and select reality, so that you
presentation bears little direct relationship to the actual situatio
you may do this to create a dramatic illusion . . . or to produce a
influential force (advertising, propaganda).

Abstracting further, you can stimulate emotions and idea
simply by the use of movement, line, form, etc., which the view
personally interprets.

Most TV production aims at the direct reportage of events. Bu
there are occasions when we seek to stimulate the audience'
imagination; to evoke ideas that are not conveyed directly b
camera and microphone. In this chapter we shall examine thes
concepts and how they can be used.

■ *Picture usage*   The purposes for which pictures are employe
are, of course, extremely varied. Some of the more common use
are:

1. *Factual*—Conveying information directly.
2. *Environmental*—Establishing location. (Big Ben suggest
   London.)
3. *Interpretative*—Visual associations, conveying ideas, thoughts
   feelings. (Plodding feet suggest weariness.)
4. *Symbolic*—Symbols associated with people, events, etc. (Ol
   Glory to symbolize USA.)
5. *Imitative*—Imitating appearance or action. (The camer
   'staggers,' as a drunk reels.)
6. *Identifying*—Features associated with particular people, events
   etc. (Trademarks, logos.)
7. *Recapitulative*—Recalling subjects met earlier.
8. *Coupling*—Linking events, themes, etc. (Pan from toy boat . .
   to water . . . to ship at sea.)
9. *Montage*—An interplay of successive images, or a juxta
   position of subjects.

Production rhetoric

Rhetoric is the art of persuasive or impressive speech and writing.
Unlike everyday conversation, it stimulates our imagination
through style and technique, by inference and allusion, instead of
direct pronouncement, by appealing to our inward ear and eye.
The rhetoric of the screen has similar roots, that film makers such

as Hitchcock have explored over the years to great effect.

The analysis that follows is more than a concise list of facts. It is a distillation of production techniques with exciting persuasive potentials, techniques that have moved audiences to tears, of laughter and of pity; that have held them in tense expectation. Translate these principles into living illustrations and we see how the camera can, without a word of dialogue, convey the whole gamut of human responses.

An example: A veteran performer ends his brave but pathetic vaudeville act amid gibes and cat-calls. He bows, defeated . . . we hear hands clapping . . . the camera turns from the sad lone figure . . . past derisive faces . . . to where his aged wife sits applauding.

Summary of devices

Regard this analysis as a series of fingerposts, pointing the opportunities to build up original situations of your own. The summary outlines the principles involved. The examples that follow show those principles in action.

1. Making a *direct visual contrast*.
   (a) Of the picture quality—brightness, clarity, tonal contrasts.
   (b) By editing, contrasting the shot duration, transitions used, cutting rhythm.
   (c) Contrasting camera treatment—shot sizes, viewpoint height, camera movement, composition.
   (d) Through the subject itself. Creating a change that reveals new information, altering the picture's significance, e.g. by introducing a subject movement (the blonde turns . . . to reveal it is a man in drag); by cutting to a new viewpoint (the formally dressed butler . . . is wearing sneakers); by lighting. Contrasting the movement of one subject with another (contrasting energy with langour). Contrasting subject associations, i.e. the mood, qualities, properties, state, of one subject with another (new with old). Contrasting subject form (building construction with its subsequent demolition).
2. Making a *direct audio contrast*. Contrasting the sounds' relative volume, pitch, quality, reverberation, speed, rhythm, duration, methods of transition, composition, sound movement, association.
3. We can *directly contrast the picture and its sound*. By comparing any aspect of (1) with another in (2) (e.g. A picture of starving children singing *Pennies from Heaven.*)
4. Similarly, we can make a *direct comparison* between two ideas, situations, etc. for visual comparison, audio comparison, picture and sound compared.

5. Showing identical subjects with *different associations*.
   (a) Identical or similar subjects having different purposes, values, significances, etc.
   (b) The original purpose (or associations, etc.) of a subject has become changed.
6. Linking a variety of subjects, through *common association*.
7. *Juxtaposing apparent incongruities*.
8. *Implication*. Hinting at a situation without actually demonstrating it. Examples range from filmic time to censorable innuendo.
9. *Unexpected outcome*.
   (a) Climactic build-up to an unexpected outcome.
   (b) Anticlimax, following a build-up.
10. *Bathos*. A fall in significance; from the sublime to the ridiculous.
11. *Deliberate falsification or distortion*. 'Accidentally' causing the audience to misinterpret.
12. *Imitative interpretation*.
    (a) Between subjects.
    (b) Between mechanics and subjects.
13. *Associative selection*.
    (a) Direct, using part of a subject to represent the whole.
    (b) Recalling a subject by referring to something closely associated with it.
    (c) Symbolism—using a symbol to represent a subject.
14. *Deliberate overstatement*. Excessive emphasis on size, effort, etc., for dramatic strength.
15. *Deliberate understatement*. Preliminary underemphasis, to strengthen the eventual impact of size, effort, etc.
16. An *unreal effect* seeming to evolve naturally.
17. A *natural effect* introduced through *obviously contrived* means.
18. *Repetition*.
    (a) Of sound.   (b) Of picture.
19. *Sequential repetition*.
    (a) A series of sequences, all beginning with the same shot, associations, etc.
    (b) A succession of similar circumstances.
20. *Successive comparison*.
    (a) Showing the same subject in different circumstances.
    (b) Showing the same subject in different manifestations.
21. *Pun*. Play on a subject's dual significance.
22. *Irony*. A comment with an inner sardonic meaning, often by stating the opposite.
23. *Modified irony*. An ironic modification of the real significance of a subject, situation, etc.
24. *Dramatic irony*. The audiences perceiving a fact that the character involved is unaware of.
25. *Personification*. Representing an inanimate object as having human characteristics.

26. *Metaphorical transfer*. Transferring the properties of one subjet to another.
27. *Flashback*. Jumping back in time, to a point earlier than the narration has reached.
28. Referring to *future events* as if *already past or present*.
29. Referring to *the absent as if present*. Usually relating concurrent events by montage.
30. Referring to *the past as if still present*.
31. *Cutaway*. Deliberately interrupting events to show concurrent action elsewhere.
32. *Fade-out on climax*. Fading at the crucial moment in action:
    (a) To leave the audience in suspense.
    (b) To prevent the climactic peak being modified by subsequent action.
33. *Double take*. Passing by a subject casually . . . then returning to it quickly, having suddenly realized its significance.
34. *Sudden revelation*. Suddenly revealing new information that we were not previously aware of . . . immediately making the situation meaningful.
35. *Incongruity*. Where a character:
    (a) Accepts an incongruous situation as normal.
    (b) Exerts disproportionate effort to achieve something.
    (c) Displays disproportionate facility (i.e. exaggerated speed, etc.)
    (d) Is unable to perform a simple act.
    (e) Imitates unsuccessfully.
    (f) Does the right thing—wrongly.
    (g) Caricature.

Examples of production rhethoric

These are, of course, only the bare bones of opportunity. You can take any of the devices listed, and apply them in many quite different ways. For example, let us take (10) BATHOS. Here is the basis of the 'banana-skin' and 'custard-pie' jokes, and our reactions when the tattered hobo dusts off the park-bench before sitting. In the following examples you will see how readily these ideas can be introduced.

1. (a) A sudden mood change, by switching from bright gaiety . . . to macabre gloom.
        Contrasting a soft-focused dream-like atmosphere . . . with the hard clear-cut state of harsh reality.
        Contrasting the airiness of a high-key scene . . . with the restrictiveness of heavy chiaroscuro treatment.
   (b) Contrasting the leisurely pace of prolonged shots . . . with fast-moving short-duration shots.
        Contrasting the peaceful effect of a series of fades . . . with the sudden shock of a cut.

Contrasting the jerky staccato of rapid cutting . . . with the deliberation of a slower cutting rhythm.

(c) Contrasting the size of a giant aircraft . . . with its diminutive pilot.

Contrasting an individual's dominance in close-up . . . with his relative insignificance within his surroundings.

Contrasting the subject strength from a low-angle shot . . . with its inferiority from a high-angle viewpoint.

Contrasting a forward aggressive move . . . with a backward recessive move.

Contrasting the resctriction of limited depth of field . . . with a spacious deep-focus shot.

Contrasting the normality of a straight-on shot . . . with the instability of a canted shot.

(d) The only refuge on a storm-swept moorland is revealed by a lightning flash . . . as a prison.

2. Contrasting the busy crowds' noise by day . . . with the hush of the empty street at night.

Contrasting realistic . . . with unreal sounds.

3. A hippopotamus lumbers along, in step with delicate ballet music.

4. A pair of lovers embrace . . . and the camera tilts up to show a pair of lovebirds.

A soprano's high C . . . merges into the scream of a factory siren.

A helicopter hovers . . . to the sound of a bee buzzing.

5. A favorite record which has been played at a party, is later used to cover the sounds of a murder.

6. Shots of sandcastles, rock pools, beach ball; sounds of children's laughter, sea-wash, suggest the seaside.

7. Shots of a massive French locomotive . . . end with the shrill, effeminate toot of its whistle.

8. Suspecting that he is followed, a fugitive leaves a café . . . the sound of feet joins his own in the empty street.

9. A thief grabs a valuable necklace . . . it breaks . . . the pearls scatter.

It is spring. Migrant ducks arrive and land on a lake . . . but skid on its still-frozen surface.

10. After ceremonial orders, a massive gun is fired . . . producing a wisp of smoke and a pop.

11. South American music, a striped blanket on sun-drenched stone, a bright straw hat, cactus . . . but only a sunbather in a suburban garden, listening to the radio.

12. Someone is speaking to a deaf person . . . we see his lips in close-up, but without sound.

An upward movement . . . accompanied by rising-pitch sounds.

13. (a) The name on a ship's life-belt identifies the vessel—
TITANIC.

    (b) The guillotine used to epitomize the French Revolution.

    (c) A shot of the Golden Gate bridge to represent San Francisco.

14. A close shot of an auctioneer's gavel descending.

15. A smoker casually throws aside a cigar-butt. A fire starts . . . which develops into a devastating forest blaze.

16. As we watch a moving picture . . . a large hand appears and turns it over like the page of a book.

17. Watching a street fight as a distorted reflection in the chromium wheel-trim of a nearby automobile.

18. A searcher shouts the lost person's name . . . it echoes and re-echoes.

    An angry crowd closes round a central figure. Close shots cut alternately between the accused and individuals in the crowd.

19. A successful concert tour is shown by the artist taking a series of similar curtain calls.

20. A succession of shots show the same policeman in a variety of situations; directing traffic, guiding a sightseer, rescuing a would-be suicide, making an arrest . . .

21. A parrot at an open window whistles for food . . . a passing girl turns at the 'wolf whistle'.

22. It is a power blackout. The lost traveler strikes a match . . . and sees a poster—'Save Energy'.

23. A newspaper advertisement shows extraordinary bargains . . . then we see it is an old copy, used to line a drawer.

24. A mountaineer climbs . . . unaware that his rope is fraying.

25. An animated coffee-pot describes the great coffee it makes.

26. An elephant that flies ('Dumbo').

27. An old woman tells of her childhood . . . brief shots showing her life as a girl.

28. Looking at the projected plans of a ship . . . we hear the launching festivities.

29. A superimposed montage showing a missing man surrounded by headlines; radio announcers telling of his disappearance.

30. A derelict ballroom echoing to the sounds of bygone dances.

31. The fugitive escapes down a side alley. But just as we see that his way is blocked, the shot changes to show his pursuers.

32. A little man arrogantly challenges a person sitting nearby . . . who stands and towers over the challenger. The picture cuts to another scene.

33. Walking past a poster of a wanted robber, a man suddenly stops, reacts, and returns to it. He realizes that its picture looks like himself.

34. Entering a room, we see someone reading . . . a close viewpoint reveals the dagger-hilt protruding from his back.

35. A man takes off his hat . . . and eats it.

Imaginative sound

Although the aural memory is less retentive it is generally more imaginative than the eye. We are more perceptive and discriminating towards what we see. Consequently, our ears accept the unfamiliar and unrealistic more readily than our eyes, and are more tolerant of repetition. A sound-effects recording can be reused many times, but a costume or drapes design may become familiar after a couple of viewings.

In many TV shows the audio is taken for granted, while attention is concentrated on the visual treatment. Yet, without audio the presentations can become meaningless (talks, discussions, interviews, newscasts, music, game shows, etc.); whereas without video the production would still communicate.

Audio can explain or augment the picture, enriching its impact or appeal. Music or effects can suggest locale (seashore sounds) or a situation (pursuing police heard), or conjure a mood (gaiety, foreboding, comedy, horror).

A non-specific picture can be given a definite significance through associated sound. Depending on accompanying music, a display of flowers may suggest springtime, a funeral, a wedding, or a ballroom.

Sound elements

■ *Voice* The most obvious sound element, the *human voice* can be introduced into the presentation in several different ways

1. A single person addressing the camera, formally or informally
2. An off-screen commentator (voice over) providing a formalized narrative (e.g. travelogues); or the spontaneous commentary for a sports event.
3. We may 'hear the thoughts' of a character (reminiscive or explanatory narration) while watching his silent face, or the subject of his thoughts.
4. Dialogue—the informal natural talk between people (actual or simulated), with all its hesitance, interruptions, breaking off, overlapping; and the more regulated exchanges of formal discussion.

■ *Effects* The characteristic sound picture that conjures a particular place or atmosphere, comes from a blend of stimulii: from action sounds (e.g. footsteps, gunfire), from environmental noises (e.g. wind, crowd, traffic), and from the subtle ways in which sound quality is modified by its surroundings (reverberation, coloration, distortion).

■ *Music* Background music has become near-obligatory for many programs. It can range from purely melodic accompaniment to music that imitates, or gives evocative or abstract support. You can even use musical instruments to create audio effects (creaks, clicks, whines, etc.).

■ *Silence* The powerful dramatic value of *silence* should never be underestimated. However, silence must be used with care, for it may too easily seem to be just a loss of audio.

Continued silence can suggest such diverse concepts as: death, desolation, despair, stillness, hope, peace, extreme tension (we listen intently to hear if the marauders are still around).

Sudden silence after noise can be almost unbearable:

A festival in an Alpine village . . . happy laughter and music . . . the tumultuous noise of an unexpected avalanche engulfing the holiday makers . . . then silence.

Sudden noise during silence creates an immediate peak of tension:

The silently escaping prisoner knocks over a chair and awakens the guards (or did they hear him after all?)

Silent streets at night . . . then a sudden scream.

Dead silence when the audience has been following action that would logically lead to a tremendous noise can give a scene a taut unreal quality:

To a crescendo of sound, intercut shots of two locomotives traveling towards each other at speed on the same track . . . they crash in silent slow-motion.

The explosive charge has been set . . . the detonator is switched . . . nothing happens . . . silence.

Sound emphasis

You can manipulate the relative volumes of sounds for dramatic effect; emphasizing particular sources, cheating loudness to suit the situation. A whisper may be amplified to make it clearly audible, a loud sound held in check.

You may establish the background noise of a vehicle and then gradually reduce it, taking it under to improve audibility of conversation. Or you could deliberately increase its loudness so that the noise drowns the voices. Occasionally you may take out all environmental sounds to provide a silent background—for a thoughts sequence, perhaps.

You can modify the aesthetic appeal and significance of sound in a number of ways.

For *factual sound* you can use:

1. *Random* natural pickup (e.g. overhead street conversations).
2. *Selective pickup* of particular sources.

For *atmospheric sound*:

1. By choosing certain natural associated sounds, you can develop a *realistic* illusion. (Cockcrow suggests it is dawn.)
2. By deliberately distorting reality, you create *fantasy* to stimulate the imagination. (A Swannee whistle's note suggests flight through the air.)
3. By *abstraction*, the pitch and rhythm of sounds can evoke ideas and emotions without direct reference to naturalistic phenomena. (Film music, cartoon sound-tracks, musique concrète.)

Sound usage

As with picture usage, this can be extremely varied:

1. *Factual*—Conveying information directly (normal speech).
2. *Environmental*—Establishing location (traffic noises imply a street scene).
3. *Interpretative*—Of ideas, thoughts, feelings (a slurred trombone note as a derisive comment).
4. *Symbolic*—Of places, moods, events (air-raid siren denoting an attack).
5. *Imitative*—Of a subject's sound, character, movements (music imitating a cuckoo's call).
6. *Identifying*—Associated with particular people or events (signature tunes, leitmotif).
7. *Recapitulative*—Recalling sounds met earlier.
8. *Coupling*—Linking scenes, events, etc. (Musical bridges between scenes.)
9. *Montage*—A succession or mixture of sounds, arranged for dramatic or comic effect. (A bassoon and piccolo duet.)

Off-screen sound

When someone speaks or something makes a sound, it might seem logical to show the source as a matter of course. But it can be singularly dull if we do this repeatedly: she starts talking, so we cut and watch her.

You can use *off-screen sound* in many ways, to enhance program impact:

1. Having established a shot of someone talking, you might cut to see the person they are speaking to and watch their reactions, or cut to show what they are talking about. The original dialogue continues, but we no longer see the speaker. So you can establish relationships, even where the two subjects have not been seen together in the same shot.

2. Background sounds can help to establish location. Although a mid-shot of two people occupies the screen, the audience interprets that they are near the seashore, a highway, a sawmill.

3. Off-screen sounds may be chosen to intrigue us, or arouse our curiosity.

4. A background sound may introduce us to a subject before we actually see it, informing us about what is going on nearby or is going to appear (e.g. the wheezy spluttering of an approaching jalopy).

5. Tension can build as a character recognizes and reacts to a sound that the audience cannot interpret. Again, tension grows when a character hears a sound (that we also hear) but cannot understand its significance. Alternatively, we may realize the significance of a sound that the character has not heard or understood.

6. Off-screen sounds can exaggerate or emphasize our impressions of a scene (a crowd, traffic); perhaps indicating them even where none exists.

7. By deliberately *overlapping* sound you can create a linkage between scenes; introducing the next scene's audio before you cut to it. For example, an old man reminiscing at dinner: 'In those days, Vienna was a city of wonderful music.' While watching him, we hear a waltz in the background . . . the picture dissolves to show a soirée.

8. The background sound may create audio continuity, although the shots switch rapidly. Two people walk through buildings, down a street . . . their voices are heard clearly throughout at a constant level.

9. An audio montage of several different sources, may be used to suggest thoughts, dreams, etc.

Substituted sound

Surprisingly often, instead of reproducing the *original* sound, we shall deliberately devise a substitute to accompany the picture. There are several reasons for this approach:

No sound exists—as with sculpture, painting, architecture, inaudible insects, prehistoric monsters.

Sometimes the actual sounds are *not available*, *not recorded* (mute shooting), or *not suitable*. For example: absence of birdsong when shooting a country scene; location sounds were obtrusive, unimpressive, or inappropriate to use; a location camera may obtain a close shot of a subject (using a narrow-angle lens) that is too distant for effective sound pickup.

The sounds you introduce may be just *replacements* (using another lion's voice instead of the missing roar), or *artificial substitutes* in the form of effects, music, synthesized or treated audio.

Background music and effects should be added cautiously
They are easily:

1. Disproportionate (too loud or soft).
2. Hackneyed (too familiar).
3. Overobvious (imitating every action—'Mickey Mousing').
4. Obtrusive (surging into slight gaps of silence).
5. Out of scale (overscored music).
6. Inappropriate (have wrong or misleading associations).

Controlling sound equipment

Various working principles are generally accepted in sound
treatment:

1. The scale and quality of audio should match the picture
   (appropriate volumes, balance, audio perspective, acoustics
   etc.).
2. Where audio directly relates to picture action, it should be
   synchronized (lip movements, footsteps, hammering, other
   transient sounds.)
3. Video and audio should normally be switched together. No
   audio advance or hangover on a cut.
4. Video cutting should be on the beat of music, rather than
   against it; preferably at the end of a phrase. Continual cutting
   in time with music becomes tedious.
5. Video and audio should usually begin together at the start of
   show; finishing together at its conclusion, fading out as
   musical phrase ends.

Audio analysis

We all recognize that some sounds seem exciting, martial, happy
. . . while others are melancholy, soporific, wistful. Is this entirely
fortuitous, or are there working principles to guide our audio
selection. Experience suggests that there are.

If you analyze a series of sounds creating a particular emotional
impact, you will find they have many common features. The
Table 18.1 shows some 42 sound characteristics and typical
associated responses. These effects can, of course, combine in
various complex ways. For example, from Table 18.1, it is obvious
that a sound containing the features—1, 4, 9, 10, 13, 15, 19, 20,
23, 24, 26, 27, 31, 33a, 35, 39, 42 must necessarily produce an
exciting, vigorous impact. On the other hand, one containing
features—2, 5, 8, 11, 14, 16, 18, 21, 22, 28, 32, 34b, 36, 41 must be
a sad, peaceful, sound.

■ *The effect of combining sounds*   When we hear two or more

362

sounds together, we shall often find that they *interrelate* to provide an emotional effect, that changes according to their relative loudness, speed, complexity, etc.

Overall *harmony*—Conveys completeness, beauty, accord, organization.

Overall *discord*—Conveys imbalance, uncertainty, incompletion, unrest, ugliness, irritation.

Marked *differences* in relative volume, rhythm, etc.—create variety, complication, breadth of effect, individual emphasis.

Marked *similarities*—Result in sameness, homogeneity, mass, strength of effect.

Table 18.1   Analysis of sounds and their effect

	Sound characteristics	*Associated with*
Volume	1. Loud sounds	Big, strong, assertive, powerful, energetic, rousing, earnest.
	2. Soft sounds	Small, soothing, peaceful, gentle, subdued, delicate, little energy.
	Against a quiet background	Alerting, persuasive.
Pitch	3. Pitch	Often suggests physical height.
	4. High-pitched sounds	Exciting, light, brittle, stirring, invigorating, elating, attractive, distinct, sprightly, weak.
	5. Low-pitched sounds	Powerful, heavy, deep solemn, sinister, undercurrent, depression.
Key	6. Major	Vigour, brightness.
	7. Minor	Melancholy, wistful, apprehensive.
Tonal quality	8. Pure, thin (e.g. flutes, pure string-tone)	Purity, weakness, simplicity, sweetness, ethereal, daintiness, forthright, persuasiveness.
	9. Rich (possessing strong overtones, harmonics)	Richness, grandeur, fullness, complexity, confusion, boisterous, worldly, vitality, strength.
	10. Edgy, brassy, metallic	Cold, shrill, gay, bitter, snarling, vicious, forceful, hard, martial.
	11. Full, round tone (e.g. horn, saxophone, bowed basses)	Warm, rich, mellow.
	12. Reedy (e.g. oboe, clarinet)	Sweetness, nostalgic, delicate, melancholy, wistful.
	13. Sharp transients (a) High-pitched (e.g. xylophone, breaking glass)	Thrilling, exciting, horrifying.
	(b) Low-pitched (e.g. timpani, thunder)	Dramatic, powerful, significant.
Speed and rhythm	14. Slow	Serious, important, dignified, deliberate, ponderous, stately, somber, mournful.
	15. Fast	Exciting, gay, hopeful, fierce, trivial, agile.

	16. Simple	Uncomplicated, deliberate, regulation, dignity.
	17. Complex	Complication, excitement, elaboration.
	18. Constant	Uniformity, forceful, monotonous, depressing.
	19. Changing	Vigorous, erratic, uncertainty, elation, wild.
	20. Increasing (Accelerando)	Increasing vigour, excitement, energy or force; progressive development.
	21. Decreasing (Rallentando)	Decreasing vigour, excitement, energy or force; concluding development.
Phrasing	Repetition of sets of sounds:	
	22. Regular repetition	Pleasurable recognition, insistence, monotony, regulation, co-ordination.
	23. Irregular repetition	Distinctiveness, personality, disorder.
	24. Strongly-marked accents	Strong, forceful, emphatic, rhythmical.
	25. Un-accentuated sounds	Continuity, lack of vitality.
	26. Interrupted rhythm (Syncopation)	Character, vigour, uncertainty, unexpectedness.
Duration	27. Brief, fragmentary	Awakening interest, excitement, forceful, dissatisfaction.
	28. Sustained	Persistence, monotony, stability, tiredness.
	29. Staccato	Nervous vitality, excitement.
Movement	30. Movement pattern	Movement pattern of sound suggests corresponding physical movement, e.g. upward—downward—upward— glissando pitch changes suggesting swinging movement.
	31. Upwards	Elation, rising importance, expectation, awakening interest, anticipation, doubt, forceful, powerful.
	32. Downwards	Decline, falling interest, decision, conclusion, imminence, climatic movement.
Pitch changes	33. Sudden changes (a) rise (b) fall	Increasing interest, excitement, uplift. Force, strength, decision, momentary unbalance.
	34. Slow changes (a) rise (b) fall	Increasing tension, aspiration, rising motion. Saddening, depression, falling motion, reduced tension.
	35. Well-defined pitch-changes	Decision, effort, brightness, vitality.
	36. Indefinite pitch-changes (e.g. slurs, glissando)	Lack of energy, indecision, sadness.
	37. Vibrato	Instability, unsteadiness, ornamentation.
Volume changes	38. Tremelo	Uncertainty, timidity, imminent action.
	39. Crescendo	Increasing force, power, nearness, etc.
	40. Diminuendo	Decreasing force, power, nearness, etc.

Reverberation	41. Dead acoustics	Restriction, intimacy, closeness, confinement, compression.
	42. Live acoustics	Openness, liveliness, spacious, magnitude, distance, uncertainty, the infinite.

Focusing attention

Audience attention is seldom divided equally between picture and sound. One aspect usually dominates. However, you can transfer concentration between ear and eye. For example, a movement will emphasize a remark made immediately afterwards. Dialogue before a move gives it emphasis.

The ear is particularly drawn to certain types of sound:

1. Loud sounds, increasing volume.
2. High-pitched sounds (around 1000 to 4000 c/s).
3. Sounds rich in overtones (harmonics); edgy, metallic sounds; transients.
4. Fast sounds, increasing speed or rhythm.
5. Complex rhythms.
6. Briefly repeated phrases, syncopation, strong accents.
7. Short-duration or staccato sounds.
8. Aural movement.
    (a) Especially increases in volume, pitch, etc.
    (b) Clear-cut, unexpected, violent, changes.
    (c) Interruption, vibrato, tremolo.
9. Reverberant acoustics.
10. Marked contrast.
    (a) Between the principal and background sounds.
    (b) Between the sound and the picture (i.e. their associations, composition, etc.).
11. Marked similarity.
    (a) Between sounds (e.g. one source echoing another).
    (b) Between sound and picture (e.g. simultaneous upward movements in both).

We can transfer aural attention to another subject by:

1. Giving the original subject's sound pattern (rhythm, movement, etc.) to the new source.
2. Weakening the original subject's attraction and strengthening the new source.
3. Linking action e.g. having the pattern of the original sound change to that of the new subject, before stopping it.
4. Transferring aural movement through e.g. by carrying over a solo sound, while changing its background.
5. Cutting to a shot of the new source alone.

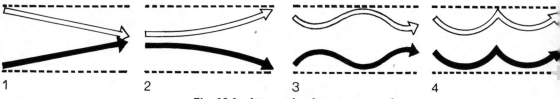

Fig. 18.1 Interaction between sounds
1, Towards a common focal point—suggesting conflict, concentration. 2, Away
from a common focal point—suggesting divergence, broadening. 3, Contrasting
sound movement—suggesting diversity, variety, interdependence. 4, Parallel
sound movement—suggesting similarity, unanimity.

6. Changing the original composition lines e.g. whereas upward
   sounds lead attention towards high notes, downward sounds
   lead attention towards lower notes.
7. Dialogue attracting attention either to its source or to its
   subject.

Selective sound

In recreating the atmosphere of a particular environment, the
trick is to use sound *selectively* if you want the scene to carry
conviction, rather than try to include all typical background
noises. You may deliberately emphasize, reduce, modify or omit
sounds that would normally be present; or introduce others to
convey a convincing sense of location.

Table 18.2 Selective sound treatment

Sound selection	*Interpretation*
Scene: After a long hopeless day seeking work, a cripple returns through emptying streets.	
All sounds audible—of subject and background.	His footsteps sound amidst traffic and crowd noise.
The subject alone is heard.	His stumbling footsteps echo through quiet streets.
The subject plus selected background sounds.	His slow tread contrasts with the brisk steps of passers-by.
General background sounds alone.	Traffic noises. Passers-by.
Significant background sounds alone.	The laughter and gaiety from groups he passes; contrasting with his abject misery.
Interpretative sounds, not directly originating from the scene.	His echoing footsteps become increasingly louder and distorted. By progressively filtering out the higher audio-frequencies, his labored tread becomes emphasized.
Significant selected sounds from another *scene* (providing explanation or comment).	Voices of people refusing him work, echo in his brain.

The selection and blend of environmental sounds can strongly influence the interpretation of a scene. Imagine, for example:

The slow even toll of a cathedral bell accompanied by the rapid footsteps of approaching churchgoers.

In developing this scene, you could reproduce random typical sounds. Or, more persuasively, you might deliberately use audio emphasis:

1. Loud busy footsteps with a quiet insignificant bell in the background.
2. The bell's slow dignity contrasted with restless footsteps.
3. The bell's echoing notes contrasted with the staccato impatience of footsteps.
4. The booming bell overwhelming all other sounds.

So you can use the same sounds either *environmentally* or *atmospherically* ; to suggest hope, dignity, community, domination . . . simply through selection, balance, and quality adjustment.

As you saw earlier (*substituted sound*), instead of modifying a scene's natural sounds, you might augment them or replace them by entirely fresh ones.

Audio-visual relationship
The picture and its audio can interrelate in several distinct ways:

1. *The picture's impact may be due to its accompanying audio.*
   A close-shot of a man crossing a busy highway . . .
   (a) Cheerful music—suggests that he is in lighthearted mood.
   (b) But automobile horns and squealing tires—suggest that he is jaywalking dangerously.
2. *The audio impact may be due to the picture.*
   (a) A long shot of a wagon bumping over a rough road . . . and the accompanying sound is accepted as a natural audio effect.
   (b) But take continuous close-ups of a wheel . . . and every jolt suggests impending breakdown!
3. *The effect of picture and audio may be cumulative.*
   A wave crashes against rocks . . . to a loud crescendo in the music.
4. *Sound and picture together may imply a further idea.*
   Wind-blown daffodils . . . birdsong, lambs bleating . . . can suggest Spring.

19 Visual effects

We use 'effects' to create an illusion. It may be simple and unobtrusive, or elaborately impressive. In more extreme applica tions, our audience will realize that there must have been some subterfuge, but many effects are subtle and their methods undetectable.

Pictorial effects are achieved by various means:
1. *Special effects* such as fire, explosions, fog, are a specialist craft and best left to the expert. In amateur hands, the results can be uncertain, unconvincing or dangerous.
2. *Staging effects* include scenic illusions such as false perspective scale changes (Chapter 8).
3. *Lighting effects* range from environmental impressions such as sunsets, to illusory effects—passing vehicles, water ripples firelight (Chapter 7).
4. *Mirror effects* to reposition the camera viewpoint or multiply images.
5. *Projection effects* using front- or rear-projection images.
6. *Camera effects* using optical attachments.
7. *Foreground effects* in which the camera shoots through a foreground device.
8. *Electronic effects* resulting from manipulating or controlling the video signal.
9. *Temporal effects* involving the speeding, slowing or freezing of time.

Mirror effects

Mirrors are a simple straightforward way of creating certain illusions. Although vulnerable to marking, surface-silvered mirrors are preferable to rear-silvered types, which produce a tinted less-sharp image. Mirror-plastic, ferrotype or chromium sheeting, are successful for less critical applications where distortions can be tolerated or are required.

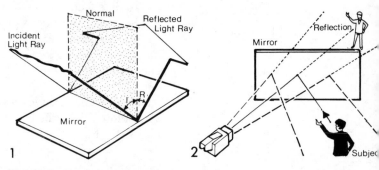

Fig. 19.1   Basic laws of reflection
1, Light is reflected at an angle (R) equal to the angle of incidence (I). The *normal* is a line perpendicular to the mirror surface and starting from where the beams meet. The *incident* and *reflected* rays and the *normal* lie in the same plane. 2, The reflected image appears laterally reversed and as far behind the mirror as the subject is in front. The camera focuses on the subject image, not the mirror surface.

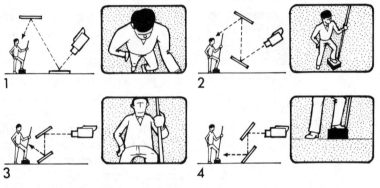

Fig. 19.2 part 1   Changing the viewpoint with mirrors—using a single mirror
1, Overhead (top) shot—image inverted and laterally reversed. 2, Reverse-angle top shot. 3, Very low angle shot.

Fig. 19.2 part 2   Changing the viewpoint with mirrors—using two mirrors
1, Overhead shot. 2, High-angle shot. 3, Low-angle shot. 4, Low level shot (mirrors may be mounted separately, fixed in a portable periscope stand, or attached to the camera mounting).

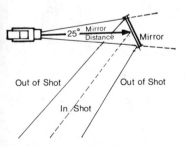

Fig. 19.3   Area seen in a mirror
The easiest method of planning a mirror position and judging what it reflects, is to cut out a paper triangle of the lens angle (e.g. 25°), folding it at the anticipated mirror position and placing this on a scale floor plan. This indicates mirror size and distance required. (The smaller the mirror, the closer it must be to the camera.)

Mirrors are mostly used for overhead shots (demonstrations, piano, dancers) and for low-angle (periscope) shots. They can be time consuming to adjust and usually permit little camera movement. The mirror size needed increases with the camera/mirror distance and when using oblique mirror angles. Large mirrors are heavy and expensive items.

Having adjusted any mirror set-up during rehearsal; ensure that camera, mirror, and performer positions are clearly floor marked.

Where the lateral reversal (mirroring) is unacceptable, a further intermediate mirror or reversing prism can be introduced to correct the effect. While monochrome cameras can be either laterally or vertically reversed electronically, only the latter is possible directly with most color cameras. (Digital video effects systems where available, may provide lateral or horizontal reversal.)

■ *Pepper's ghost*   An old theatrical device, the camera shoots through an angled sheet of clear glass or semi-transparent mirror (pellicle), which reveals the direct scene and/or a reflected scene, according to their relative lighting intensities. (See Fig. 19.9)

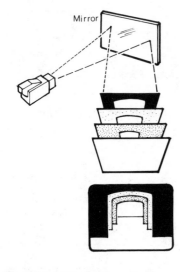

Fig. 19.4 Mirrors to extend space
The camera's effective distance can be increased, to obtain a longer shot than studio space permits.

The idea is widely used with display dioramas in museums (comparing before and after conditions). In television, we find it in *shadow boxes* which present a series of graphics simultaneously to one camera; cuts, superimpositions, dissolves between them being achieved by lighting changes.

The device is a useful aid to shadowless lighting for multiplane animated graphics. It also provides the background image for *reflex projection* systems. (See Fig. 19.14)

Rear (back) projection (See Fig. 19.10)

Here the camera shoots action against a translucent screen of 'frosted' plastic; a film or slide being projected onto its rear side.

Once used extensively in motion pictures and monochrome TV, rear projection has been largely replaced in film by process work (matting during optical printing), or reflex projection; and in color TV by electronic insertion (chroma key, CSO).

However, rear projection still has some useful applications in TV studios. As well as being used for pictorial rear projection, the translucent screen can be used to display silhouettes, shadows or pattern effects.

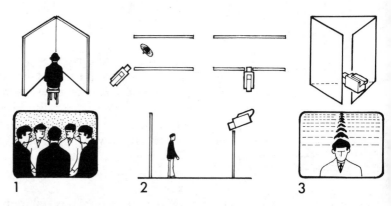

Fig. 19.5 Multiple images
For multi-reflections, you can arrange mirrors: 1, Angled—the smaller the mirror-angle, the more the images (Number = $\frac{360}{\text{Angle}}$).
2, Parallel mirrors. 3, Triple mirrors.

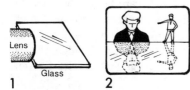

Fig. 19.6 Split screen reflection
1, A horizontal reflector (foil, mirror, clear glass) across the lens can provide 2, reflected images, mirage or water-reflection effects.

Front projection

We can project still or moving images directly onto a light-toned background, or even onto action for decorative or scenic effect.

Basic light patterns (from metal stencils) are most effective, but under typical full-lighting conditions, subtle half-tones and low-key images are disappointing. Geometric patterns and shapes, clouds, etc., are regularly projected onto cycloramas; and used to simulate natural light and shadow effects (window shadows). Scenic projectors can provide certain moving images (flames, sea-swell, snow) that are effective enough under favorable conditions.

Given *sufficiently sensitive* cameras and no spill light to dilute the image (so reducing its clarity and contrast), further interesting effects are possible, e.g. an announcer with newsprint projected onto his face, dancers against black drapes who are illuminated only by striped lighting.

Reflex projection

In this system (also called *front axial projection* or simply *front projection*) the background is projected directly along the TV camera's lens axis, via a half-silvered mirror, onto a special background screen material. The secret lies in this *highly directional* surface, comprising millions of minute glass beads, each reflecting nearly 92% of the projected background image back to the camera. (The principle of this Scotchlite sheeting is identical to that used in many reflective traffic signs and night-safety clothing strips.)

Normal subject lighting cannot dilute the background image (unless placed near the camera) thanks to the screen's directionality. Where the front projected *background* image falls onto the performer or foreground subjects, it is swamped by action lighting and so is not visible on camera. Shadows from it are cast directly behind subjects and so are unseen by the camera, providing the system is accurately aligned, and the lens angles of the camera and projector are comparable.

Although not a foolproof system, it can provide near-miraculous effects for both film and TV cameras. Its limitations include: mirror light losses; lighter subject tones can be reflected in dark screen areas, causing edge haloes; the system must be aligned to avoid dark subject edges; and the studio camera must be *static* in most applications.

If pieces of reflex screen material are cut out and positioned to match objects or planes in the background picture (e.g. wall, tree, couch, archway), the studio performer can seemingly walk 'behind' them very convincingly.

Camera filters

Filters of various kinds can be clipped over the front of the lens or inserted in a *filter-wheel* holder within the camera.

■ *Neutral density filters* These have various light transmissions (Chapter 3) and are used to prevent gross overexposure when working under strong sunlight.

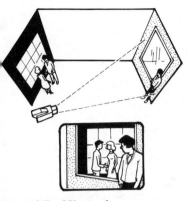

Fig. 19.7 Mirror shot
Here a subject is shot together with a background mirror. The result is a combination of the direct and reflected images.

Fig. 19.8 One-way mirrors
The thin reflective coating on this one-way mirror makes it appear transparent (with a considerable light loss) from the rear unlit side (Cam 1); but as a normal mirror in the brightly-lit scene (Cam 2).

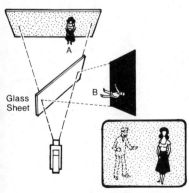

Fig. 19.9 Pepper's ghost
The out-of-shot person (B) is reflected into the lens as a transparent ghost within the main subject picture. Depending on the relative illumination, you can see A or B, or both (B ghostly).

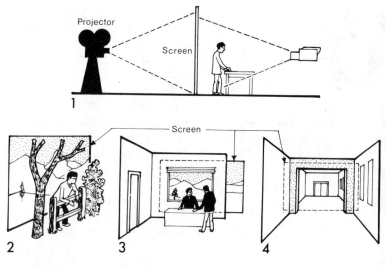

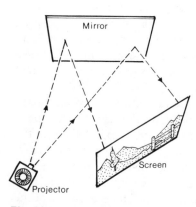

Fig. 19.10   Rear projection
1, A slide or film image is projected onto one side of a matte translucent screen. On the reverse side, a camera shoots action for which the screen image provides a background. 2, The rear-projected image can be used as a complete scenic background with foreground set-pieces, furniture, etc.; or 3, used to supplement a built set; (e.g. window backing); or 4, to extend a built set.

■ *Corrective filters*   When there are changes in the color temperature of light sources these filters can be used to compensate e.g. when moving from daylight to tungsten-lit areas.

■ *Star filters*   These are clear discs with closely scribed grid patterns. Diffraction effects produce multi-ray patterns (typically

Fig. 19.11   Folding the light path
By folding the light-path with a mirror, floor space can be saved. Using a second mirror even greater compactness is possible, but with a further loss of clarity.

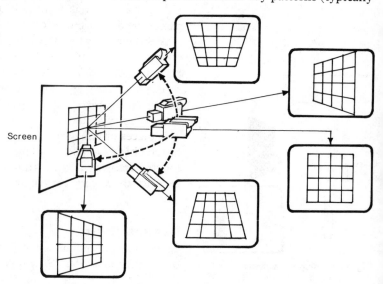

Fig. 19.12   Background distortion
Unless the studio camera is perpendicular to the flat two-dimensional background picture, distortion occurs. Similarly, the projector must be dead square to the screen to avoid distortion.

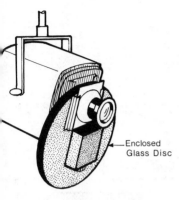

—Enclosed
Glass Disc

ig. 19.13   Scenic projector
n enclosed glass disc with painted
photographed patterns is rotated
an adjustable speed by a small
otor. This device is attached to a
ojector spotlight and focused onto
background to simulate moving
ouds, snow, water patterns, etc.
ne low image-intensity usually re-
ricts the usable image-size.

4, 6, 8) around highlights from flames, specular reflections, and lamps. (Sunburst versions may have some sixty rays!) The rays' directions change as the filter is turned.

■ *Diffusion discs*   Available in various densities, these filters provide general image softening through fine surface scratches or dimpling on a clear disc. Nylon or muslin net or hose may be similarly used. Sharp detail is reduced and highlights develop glowing haloes. To produce localized softening or distortion, a clear glass disc can be lightly smeared with grease or oil where required—usually round the edges. The effect can be ethereal, or woolly and irritating.

■ *Low-contrast filters*   When used, these desaturate and mute colors, while softening contrast.

■ *Fog filters*   Made in different densities, these create image diffusion from slight mist to dense fog, with strongly haloed highlights.

■ *UV (haze) filters*   When used on daylight exteriors, these help to reduce haze blur due to ultra-violet light.

■ *Night filters*   These are graded neutral density or blue-tinted filters (denser at the top) used to simulate night or evening scenes.

Table 19.1   Rear Projection/BP

Advantages	The system is adaptable and simple to use.
	The background image is visible to performers.
	Compact screens can provide backgrounds outside small scenic openings (e.g. automobile windows).
	Rear projection can provide display panels in settings to show graphics, titles, microscope shots, film clips, etc.
Disadvantages	Studio space is wasted due to the projector-to-screen distance (*throw*) required, especially for large screens.
	Projectors occupy floor space.
	Projector noise may be audible.
	Subject illumination may spill onto the screen, diluting its image. (A black scrim layer over the screen may improve matters.)
	The projected image may be: insufficiently bright, have a central hot spot or edge darkening.
	Only slides or film can normally be projected. (Film is susceptible to vertical *hop* and horizontal *weaving*.) Projected TV images are seldom of suitable quality when shot by the studio camera.
	The rear projected picture's apparent sharpness will vary with the studio camera's focusing and depth of field.
	Matching of the projected background and foreground subject/scene may pose problems: perspective, scale, light direction, contrast, color quality, etc.
	Problems can arise in matching audio perspective and quality to varying projected backgrounds.

Fig. 19.14 Reflex projection

1, In Reflex (axial) Front Projection, the background scene is projected along th lens axis, via a half-silvered mirror, onto a highly directional glass-beaded screer Actors' lighting swamps the image falling on them. 2, People appear *in front* c the background scene. 3, If people move behind a surface covered with beade material, they will appear *behind* corresponding areas of the background scen (walls, trees, etc.).

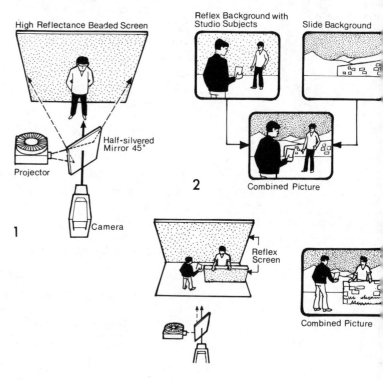

By heavy filtering (particularly the sky areas) and considerably reducing overall exposure, it is possible to shoot night scenes daylight ('day for night' shooting)—thereby saving money and avoiding lighting problems.

■ *Color filters*   These are rarely used on *color* TV cameras, a: hue changes can be introduced electronically. In monochrome they facilitate lightening/darkening the tonal reproduction o: parts of the spectrum.

■ *Polarizing filters*   These are occasionally used to reduce strong reflections or flares from smooth or shiny surfaces such as glass or water. (They have little effect for rough materials.) Polarizing filters can be used to darken an overbright sky without affecting overall color quality; although light loss is inevitable (e.g. 75% loss). The filter's properties derive from its crystalline structure, which enables it to discriminate against light reflected at around 35°. By rotating the filter, you can selectively reduce or suppress particular reflections.

Fig. 19.15 Prismatic lenses
The number and position of images is determined by the facets on the lens. The peripheral images move round as the lens is turned but remain upright.

Fig. 19.16 The Kaleidoscope
1, Shooting through a three- or four-sided mirror-tube, produces an upright central image surrounded by angled reflections, that move round as the tube turns. 2, Two angled mirrors provide multiple converging images. (Number of images equals 360 divided by angle of mirrors.) Surface-silvered mirrors are needed to avoid spurious ghost images.

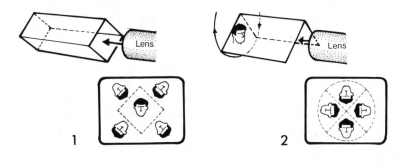

Lens attachments
A variety of add-on devices can be attached to the lens to modify the image:

■ *Multiple images* These can be produced by prismatic or multifaceted lenses. One form splits the screen into a series of repeat-image strips. Another has a static central image facet surrounded by a series of repeat versions, while a further form provides a ring of merging images. Their image patterns can be repositioned or rotated about the center by a small handle (or motor control), the images remaining upright; unlike those from the simple *kaleidoscope* mirror tube.

■ *Image distortion* Methods available for distorting the image include shooting through molded ripple or patterned glass, via a flexible mirror, or water dish, or a foil tube giving peripheral effects. Fisheye lenses cover a wide angle of view (e.g. 100°–360°), providing extreme geometrical (curvature) distortion.

■ *Image rotation* Several mirror or prismatic attachments enable you to rotate, invert, or tilt the picture. Apart from ceiling walking, you can simulate floating in free-space, flying, cliff climbing and other feats—the easy way. The most frequent use, however, is for *canted (dutch angle)* shots in which the picture is horizontally tilted for dramatic effect. If the image is to be

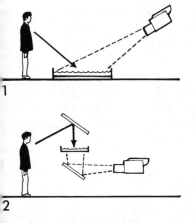

Fig. 19.17 Ripple distortion
1, The camera shoots into a shallow mirror-bottomed water tank, giving an inverted ripple image. 2, The camera shoots into a mirror periscope containing a water-filled glass tray. Ripples can be made to flow across the picture, finally breaking it up altogether. If an opaque liquid is poured into the tray, the picture blots out with a fluid wipe.

Fig. 19.18 part 1   Tilting the picture—using mirrors
Tilted shots can be obtained from (1) elevated or (2) depressed angles using canted mirrors. Shooting via a 45° mirror (3) attached to the lens system, causes the scene to cant as the camera tilts.

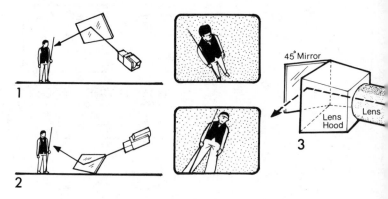

rotated continuously, this must be done smoothly and at a suitable speed.

■ *Split-field close-up lenses*   These are used to overcome extreme depth-of-field problems, for they can provide sharp images of both close, *and* distant subjects simultaneously.

Camera mattes (gobos)

For many years, solid or transparent foreground devices have been used in motion picture making to create composite visual effects. Although often replaced in TV by electronic insertion methods, they still remain a useful facility. These *mattes* (masks) or *gobos* ('go-between') take several forms. (The term 'gobo' is also used for a black surface that masks off lights or cameras appearing in shot.)

■ *Foreground matte*   At its simplest, this is just a plain surface with a hole in it, through which distant studio action is seen. Painted black, it provides a soft-edged isolating *vignette* (stencil) selecting part of a scene, for example:

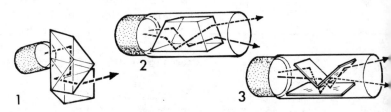

Fig. 19.18 part 2   Tilting the picture—using prisms
1, The Porra prism (a pair of reversing prisms). Rotating one of the prisms causes the image to rotate. A similar device can be constructed from two sets of double mirrors. 2, Image inverter prism (straight-through reversing prism). Rotating the assembly causes the picture to revolve. 3, A similar device to 2 uses a three-mirror assembly. Lateral reversal takes place.

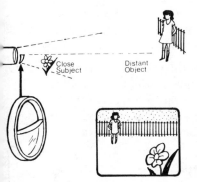

Fig. 19.19 Split-field close-up lens
A half-lens ($\frac{1}{2}$ to 3 diopters) provides a sharp image of a close object, although the main lens is sharply focused on a distant subject; so simulating a considerable depth of field. The broad band of confused focus at the border, is arranged in a subject-free area.

To superimpose or insert somebody's head into another shot.

To create a shaped border, simulating a telescope, binoculars, keyhole.

To provide localized *chroma key insertion*, a colored foreground matte board (e.g. blue) is used.

The foreground matte may be used to introduce a *decorative border*, or a *framed opening* (picture frame), or a *realistic* scenic effect (a window). Often the camera dollies forward and 'enters' the scene as it loses the matte. *Cartoon gobos* are 'funnies', where the performer appears behind the cut-out section of a comic drawing, or pokes his head through a hole.

■ *Insertion matte*　Sometimes the foreground matte is intended to merge imperceptibly with the background scene, e.g. a foreground photograph with a cut-out area, or a painted *glass-shot* putting a matching ceiling 'onto' a studio setting. The scale, perspective, lighting, colors, tones, etc., of 'foreground' and background must be matched accurately to be really convincing and action must not normally move behind the matte opening (or it will disappear). When using insertion mattes, we must have considerable depth of field or overall sharpness may not be feasible.

Foreground miniatures (models) are occasionally used to extend the scene, but these are usually rudimentary—foliage, saloon swing doors, drapes, pillars, arches. *Transparent mattes* can provide a decorated foreground plane showing titling or graphics (a world map, shop window sign, patterns, etc.); the camera pulling focus from the foreground to the background scene.

Electronic effects

The pictorial opportunities that electronic effects can provide are endless. The more you use them, the more will you discover fresh applications for both realistic and abstract picture treatment. They combine practical economies with a fascinating magic that never palls.

■ *Superimpositions*　As we saw in Chapter 6, when two picture sources are faded up together, their pictures intermix. The relative strengths of their images are *additive* and can be adjusted, but:

1. Where *black* appears in either shot, the other shot's detail appears firm and solid.
2. Where *white* appears in either shot, the other shot's detail appears diluted, paled out. A peak white can burn through another shot's detail.
3. Where different hues appear in both shots, they form a new color mixture (e.g. $R + G = Y$).

■ *Black level adjustment*   By electronically darkening the lowest picture tones until they crush to a solid overall black (batting down on blacks, sitting down) you can:

1. Ensure that 'black' materials (black drapes, black title cards) reproduce as a firm black without wrinkles or shading.
2. Prevent slight tonal variations and construction being seen in hand-animated graphics—e.g. to make a black strip (with white titling) merge with its black background card.

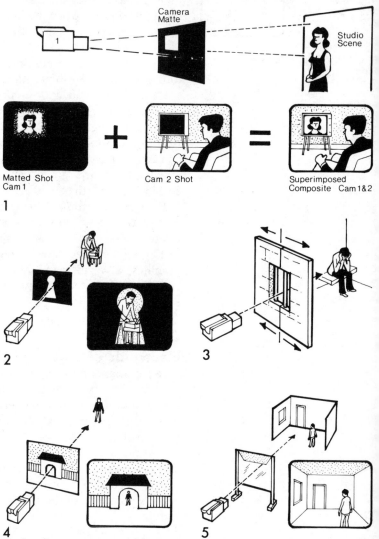

Fig. 19.20   Foreground mattes
This is a foreground surface (plain or decorated), through part of which the distant scene is visible. It has several purposes: 1, Selectivity isolating an item e.g. for localized superimposition. 2, A pattern vignette e.g. keyhole effect. 3, Scenic foreground e.g. a window through which camera shoots. A break-through frame parts to enable the camera to dolly 'through' the foreground. 4, Insertion e.g. a studio subject is apparently inserted into an area within a foreground graphic. 5, A glass-shot e.g. positioning a foreground painted ceiling 'on' a studio setting.

Normal   Weave

Fig. 19.21 Electronic weave/ ripple effects
An S-distortion (of variable strength and speed) is created electronically by varying the line-trigger pulse.

3. Make parts of a subject disappear—e.g. a black-gloved hand against black drapes will disappear while light-toned objects that are held, remain visible.

To make a near-*white* tone merge with a white area, you increase exposure so that lightest tones merge (burn-out, clip off) at white-clipper video limits.

■ *Picture weave/electronic ripple* You can introduce a sideways ripple of adjustable rate and strength into a picture, by time displacement of the camera line-scan process. Straightforward picture weave is used for flashback and transformation situations. Combined with wipe patterns, you obtain a variety of decoratively shaped undulating transitions.

■ *Negative picture* Electronic switching enables you to reverse picture *tones* (*reverse polarity*) so that they become *negative*. White reproduces as black, black as white. This is a useful facility for changing standard white-on-black titling to black-on-white (to improve clarity on light backgrounds). It can also be used to produce negative images or to reproduce negative monochrome film.

■ *Complementary picture* Color can be transformed into its complementary hues—red, green, and blue become cyan, magenta, yellow. This system enables us to directly transmit negative color film. Reversal can be used dramatically, too: transforming smoke, clouds, steam; creating black highlights glistening in a white sea; black lightning in a white sky.

■ *Camera channel adjustments* The color TV camera channel requires careful adjustments for optimum picture quality. But, by making certain changes from optimum line-up, you can obtain some interesting visual effects:

1. Reduction of the camera tubes' beam current (*beam limiting/ debeaming*) causes light tones to lose sharpness and intensity, and wash out (*puddle*).
2. The *video gain* or the *black level* controls of the RGB channels can be altered to give or correct a color cast in light or dark tones, or overall. So you can provide color effects that would otherwise require extensive colored lighting (suggesting moonlight, firelight, candlelight).

■ *Video feedback* TV cameramen soon discover that by shooting a picture monitor showing their own output, they can obtain interesting repeat-image effects. When you mix this camera's repeat-video with that from *another source* (in the studio switcher) and feed this combination back to that picture monitor, all manner of interesting multi-effects develop. (Two or more cameras can be used.) This technique can produce a variety of dynamic patterns, transformations and surround images.

Interesting effects are obtainable, too, if you take a defocused shot of a subject on a picture monitor and superimpose it on the original picture. The resultant halo creates a convincing aura around 'supernatural' beings.

If positive and negative images of a subject are intermixed, a strangely 'solarized' image results, that can be further processed by video feedback.

Keyed insertion (electronic matting)

Electronic insertion processes can provide in an instant, illusions that in film making involve costly expertise, laboratory apparatus, and time. Nowadays TV production switchers have many of these associated effects facilities ready at the touch of a button! They have become an everyday tool, offering challenging opportunities.

■ *How it works* Electronic insertion is an ingenious inter-switching process, used to selectively combine different picture sources. It is virtually a process of 'cutting a hole' in one picture and inserting exactly a corresponding piece of another picture. (Seldom putting one picture *over* another.)

As you know, a TV picture is normally formed by scanning the scene's image in a series of horizontal lines. The switcher selects which source is sent to line. However, suppose we had an instant-aneous electronic switch available, so that the system switches itself between sources *during* this scanning process. We might,

Fig. 19.22   Inter-source switching
1, If an electronic switch changes from Cam 1's picture to Cam 2's output halfway through each line period, the resultant *split screen* combines those parts of the shots. 2, If the switching moment is earlier in the line, the division is displaced left. 3, Occurring later in the line, the division is displaced right. The shots may be composed to suit the division being used; or the division located to suit the composition. 4, If the apparatus is arranged to select the top half of Cam 1's picture, switching to Cam 2 for the remainder of the frame, the result with these particular shots is nonsensical.

for instance, begin *each* scanning line by showing Cam. 1's picture but half way through that line switch to Cam. 2's shot instead. (Switching back to Cam. 1 at the end of each line.) Because of this interswitching, the resultant transmitted composite would show the left half of Cam. 1's picture and the right half of Cam. 2's picture. (You will recognize this particular example as a split-screen effect.)

Altering the exact switching moment, repositions whereabouts the division appears in the frame. If we *gradually alter* the switchover moment, a *moving* division, a *wipe* is produced; in this example, a horizontal wipe movement.

This intersource switching is initiated automatically by a *keying* or *matting* signal we apply to the system. This signal can be produced in several ways and may be:

1. An *external key or matte*—from a separate specially generated waveform (SEG, camera matte, rostrum camera, prepared film inlay).

2. An *internal key or matte* (self-keying)—derived from a selected tone or color in one of the actual pictures involved (chroma key, CSO, overlay).

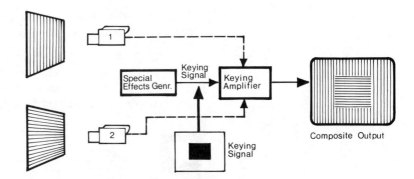

Fig. 19.23　Special effects generator (SEG)
The SEG provides voltages of variable waveforms (square, parabolic, triangular, etc.). These control the exact switching pattern of the keying amplifier, so producing a series of geometrical matte shapes.

If sources are instantaneously *interswitched*, there will be a sharp hard-edged division between their pictures. If the insertion circuits rapidly *mix* (dissolve) between them, the result is a soft edged (diffused) transition with one source merging into the next at the borders.

■ *Tonal and color mattes*　Depending on the equipment used, you can actuate any keying process by:

1. A *tonal matte*—The system switches from the background picture to the foreground (subject, master) shot, wherever it sees a black (or white) matte—however this matte is derived.

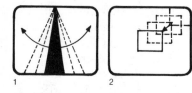 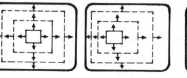

Fig. 19.24 Wipes
Wipes can be controlled in various ways: 1, direction; 2, position; 3, symmetry; 4, edge sharpness. Changes can be made at any speed.

2. A *color* matte (as in chroma key or CSO) switches only where it sees a particular hue (usually blue).

External key

■ *Special effects generator (SEG)* This produces a series of electronic waveforms. Its output is used as an external keying signal to interswitch sources. Depending on the shape and frequency of the waveform used, so the picture divisions will differ. If its switching sequence continually repeats, the pattern appears stationary. If it gradually alters, the divisions will move across the frame and/or change the divisions' shape. The effects produced are entirely independent of the pictures involved.

The interswitching divisions produced by the SEG are invariably *geometric*. They have the following features:

1. They can be controlled in *speed* (fader lever or *autowipe*).
2. They can be *directed* (expand/contract, up/down, left/right).
3. Their *symmetry* and *position* are controllable (wipe center selection by a joystick positioner control).

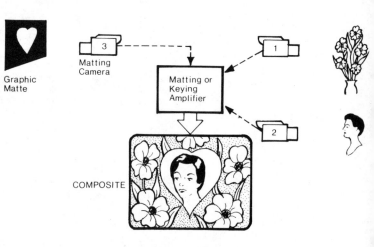

Fig. 19.25 Camera matte
The graphic in front of the matting camera (3) results in a video signal which inter-switches Cam 1 (flowers) with Cam 2 (girl) in a heart-shaped mask form. Wherever complex, non-geometrical matte shapes are required (e.g. tree-branch patterns, flower shapes, decorative screens, etc.), camera mattes are used.

Fig. 19.26 Electronic rostrum camera (inlay desk)

Using this external keying apparatus, even elaborate masks are practical. 1, TV camera—the overhead monochrome TV camera shoots a back-lit white panel. Cut-out card, inked or painted shapes, provide a matte for the keying signal. 2, Picture tube—the matte is placed or drawn on a *cel* (clear plastic sheet) or glass, over a picture tube displaying a plain white picture (raster)—a flying-spot scanner. The phototube picks up a constant scanning output (i.e. peak white video) except where interrupted by the matte. Adjustment of the scanning width or height alters the effective matte shape. Zooming produces matte size changes. 3, Keying signal—this is fed to the montage amplifier, where it inter-switches between the cameras' outputs.

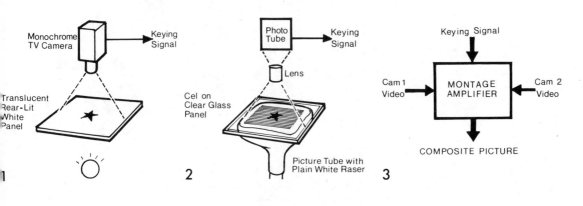

4. The pattern can be adjusted and left as a static *insert* (*inset*).
5. The *edges* may be hard (sharp) or soft (diffused).
6. A border (black, white, color) can be placed around the insert.
7. The pattern may be made to *weave* (ripple) at an adjustable speed and strength.

■ *Physical mattes* Instead of using electronically operated waveforms (SEG), intersource switching can also be derived from a physical matte (mask)—'separate key mode'—provided by:

1. A *camera matte*—a graphic placed in front of a TV camera.
2. An *electronic rostrum camera* (masking scanner, inlay desk). (Fig. 19.26).
3. *Prepared film or VT masks.* (See vignette/surround matte).

■ *Camera matte* Here a studio camera shoots a graphic with the required matte shape (mask, key) e.g. a black silhouette on a white background (or vice versa). The video from this masking camera is used solely to operate intersource switching (via a *keying amplifier*). Its picture as such is not used. Even if the matte is of a complicated shape (a map silhouette), it will key a corresponding area from one chosen picture into another.

The key could equally well be derived from a static matte in a *slide scanner*:

Fig. 19.27 Self-keying/self-matting
Wherever a selected *tone* (or *hue* if you use *chroma-key/CSO*) appears in the *master shot* (the subject or foreground shot), a corresponding part of the background shot is seen instead. The system switches back to the master shot for all other tones (or hues) in that scene. *Background* scene tones or hues do not affect keying in either system.

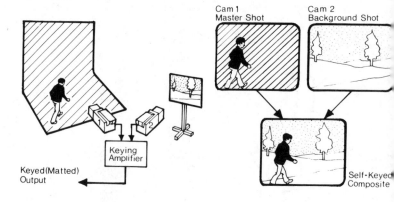

Cam. 1 shooting a brick wall photograph.
Cam. 2 shooting a striped surface.
Matting slide—Title slide.
Matted result—Lettering in striped paint, on a brick wall.

Moving camera mattes are also possible:

Cam. 1—A cloud scene.
Cam. 2—Glittering reflections (defocused tinsel).
Cam. 3—Silhouette of dancer in studio, used as a matte.
Matted result—A glittering shape dancing amongst clouds.

Self-keying (self-matting internal key)

The most sophisticated method of keying requires no *prepared* matte/mask or waveform. Instead, it continually examines the video from the master or foreground shot and wherever the system detects a specified tone or hue (depending on the process used), it switches instead to another picture—the background. In this *common-key mode*, therefore, the matting signals originate from the master picture itself.

If you are using a *tonal-keying* process (brightness separation), wherever the chosen tone (black or white) appears in the foreground shot, a corresponding part of a background is seen instead in the composite picture. So areas of the keying *tone* must only appear in the *foreground* scene where you require insertion; otherwise you will see spurious breakthrough. The background would be visible through a black necktie! Any tone may appear in the *background* scene, however. This *tonal keying* process can be used in monochrome *and* color TV systems.

Using a *color keying process* (chroma key, CSO color separation overlay), wherever the selected keying/switching *hue* appears in

Fig. 19.28 Choice of system

External key does not normally allow foreground subjects to move *in front of* background features. The subject disappears when it moves outside the matted area. 1, This is appropriate when people are to move behind walls, trees, etc. 3, It is incongruous if they were to get out of a vehicle matted in by external key. *Any* tone or color may be used in foreground and background scenes. (Switching is due to a separate matte.)

Self-keying relies on hues (or tones) within the master (subject/foreground) shot. So subjects cannot normally move *behind* background features. 2, The effect can be incongruous in certain situations (people may appear to stand in space). 4, Where appropriate, the technique is convincing. The subject itself should not contain (or reflect) the keying hue (or tone). The background may contain any colors or tones.

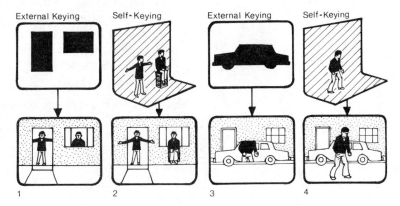

the master scene, the background is switched in instead. Again, any color or tone may appear in the background scene. The *color* keying process can only be used in a color system.

Using keyed insertion

Some applications for keyed insertion are obvious; others are extremely subtle. Certain effects can be achieved by either external or self-keying methods. Others are only possible using specifically designed equipment. In the following section, we shall look at the most frequently used processes.

■ *Inserted titles* Printed information has often to be introduced into a picture to supply identification or to add data. This titling can be provided by camera title cards, a photoslide, or an electronic alphanumeric character generator, and inserted in several ways:

1. *Superimposition* may result in 'transparent' lettering; tone and color varying with the background.
2. Where the switcher has a *non-additive mix* (NAM) facility, the source with the *lightest* tones (usually titling) inserts 'solidly' into the associated picture.
3. *Keyed insertion* permits lettering of any tone or hue to be solidly punched into a background picture, without interaction effects.

However, even an inserted letter becomes indistinguishable i
its tone or color exactly match its background. To improv
clarity you can:

1. Use a *border generator*. Edge-keying electronically creates a
   edge shadow outline around lettering. This can be blac
   (black edge), white, or of selectable color; the width of thi
   surround being constant throughout, or adjustable to produc
   contouring (drop shadow). It is even possible to use th
   original lettering as a matte and transmit its outline instead.
2. Change the lettering tone or color by *infilling* it with a differen
   one; e.g. white lettering is used to produce a matte, which key
   in any chosen color; the result being colored titling.
3. Using *double re-entry switching*, you can introduce texture o
   patterning into titles.

Table 19.2   Keying methods compared

	Advantages	Disadvantages
External key (Using SEG, camera matte, slide matte)	Simple to operate and adjust. Reliable and definite in action. No spurious breakthrough between pictures. Any tones/colors usable in foreground subjects *and* background shots. Only the matted area is inserted into the composite. No mutual degradation of picture quality (unlike *superimpositions*). Studio area required for insertion may be confined. Sharp matte edges keep edge ragging to a minimum. Subjects disappear behind matted areas. Appropriate when moving *behind* walls, objects . . .	Complex scenic mattes (ceiling insert) may require accuracy in construction and adjustment. *The camera set-ups for recording must usually be identical to rehearsed version. *Normally, foreground and background cameras should remain static. Any film must be free from hop or weaving in static scenes. *In pictorial insertions, foreground and background shots should match in perspective, proportions, scale, lighting, etc. *The *whole* of one picture source cannot normally be inserted into *part* of another shot (*see* 'Digital techniques'). But inappropriate or accidental cut-off or disappearances can arise.
Self-keying/self-matting (using chroma key, CSO)	Operation is automatic. No mattes/masks have to be prepared. Subject movement is not confined (but may need restriction, to suit background scene). Various tones or hues can be used, for differential keying. Foreground subjects always move *in front* of background and cannot be cut off by matte or inadvertently disappear . . .	Similar to *external key* disadvantages marked*. The keying/matting tone or color must only be present where 'background' is required. Shadows or reflections may cause spurious breakthrough. Edge ragging, color fringing, arise. When inserting into long shots, a large studio area may require chroma key background and floor unless keying foreground matte/gobo is used. But the effect may not be appropriate and foreground matting flat or external matte may be necessary also.

Fig. 19.29   Title insertion

1, Titles introduced by *superimposition* are satisfactory against plain backgrounds (preferably white letters against black), but otherwise, mutual interaction reduces evenness and clarity. 2, *Inserted* letters cut through all picture tones; but for titling against similar tones (white on white, gray on gray) black-edging is desirable. 3, Sometimes an *inset area* (mid-gray presenting black or white titling) is more distinctive. It does not require a border. But the overall effect may be intrusive.

1   2   3

colorize background and/or titling independently. *Cycling* can cause this color to self-change throughout the spectral range. In addition, lettering may be rhythmically blinked or flashed if required.

A *background generator* (field generator, colorizer, color synthesizer) produces color electronically, that may be used to

---

Table 19.3   Chroma key;   internal key/self-keying

Check for *unwanted areas of keying color* in the master scene, that could cause spurious switching and background breakthrough, eg: blue clothing or articles, nearby chroma key surface color reflecting onto subject, blue reflections in eye-glasses, shiny surfaces, etc..

Watch for *deep shadows* (particularly subject's) falling on chroma key surfaces, patchy lighting, unevenly painted or draped chroma key surface. These can all lead to spurious breakthrough, ragged edges, etc..

To give a performer an accurate *eyeline*, have him look towards a distant colored lamp.

In a total chroma key situation (blue background and floor), place discs of chroma key hue on the floor to give talent *location marks*.

Where a person (or object) has to be keyed in *correct scale/ perspective* into a background scene, at various distances, use the following technique: Lock off the *background* camera, and watch its picture displayed on a picture monitor. Mark reasonably accurately scaled vertical lines (wax pencil, felt pen) on this scene, that are proportionate to a person standing in those positions. Using a normal lens angle, take a long shot of the chroma key area so that the foreground performer matches the nearest market (in location and height). Then move him to the farthest position.

If the perspective and camera height are correct, he will match the marker. If he looks shorter than the distant marker, zoom in and dolly back a little, and repeat near and far matching again for optimum results.

If the subject appears to move upwards disproportionately as he goes to the further position, lower the subject camera's viewpoint slightly.

If the subject camera overshoots or sees cameras, boom, lights, obscure them with a chroma key camera matte or a localised wipe.

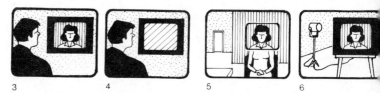

1   2 Subject Cut Off   3   4   5   6

Fig. 19.30   Display screen
1. A 'wall display' screen can be simulated by self-matting (chroma key).
2. External keying can only be used when nothing moves into this area. 3. A 'screen' may be used to include a distant guest in a studio interview. 4. This is achieved by using a corresponding chroma key area in the main shot. 5. Care must be taken that the person to be inset is in proportion and position for the display screen. 6. To insert the whole of one picture into part of another, a picture monitor must be used (or a digital effects unit). Monitors should be shaded from reflected or spill light, color-adjusted (yellow bias), preferably using a dark screen (black matrix) tube. Quality unavoidably deteriorates.

■ *Display screen*   Keyed insertion can simulate a wall-screen in a setting, to display illustrations, photoenlargements, maps, etc.

1. *External matte* (using a box wipe inset or a camera matte) is easily introduced into any shot. While avoiding chroma-key problems, this method does restrict performers, who *must not* move in front of the matted area (*cut-off* problems).
2. *Self-keying* requires a chroma-key panel in the setting (a blue panel or blind, a blue light patch), this keying hue being excluded from the rest of the master scene.

Main Scene   Detail Shot   Matte   Composite

Fig. 19.31   Inset
To introduce close-up detail while retaining the main shot, an inset can be used (created by any method).

■ *Inset/Insert*   You normally introduce a *detail insert* into a wider shot by wiping in and holding the required shape (e.g. to show a close-up soloist within a full orchestra long shot, a box or iris wipe is positioned in the selected part of the screen). A hard-edged inset is usually preferable, to prevent image confusion— perhaps with a colored border to emphasize demarcations. Occasionally a soft (diffused) edge is unobtrusively effective— for a dream montage or a 'thinks' bubble.

■ *Vignette/surround matte*   You can use keyed insertion to provide decorative surrounds or borders to an insert subject, or to form symbolic shapes (keyhole, telescope, binoculars, heart). Geometrical patterns are available on the SEG.

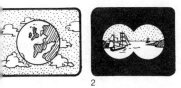

ig. 19.32 Surround matte
Mattes can be used: 1, to mask off
nwanted material in the shot, or
, to simulate a vignette.

By blanking out parts of the shot, you can exclude unwanted subjects. The support stand of a world globe could be matted out, so that it appeared in 'free space'—perhaps inserting a skyscape into the now black surround.

Regular opening titles sequences (film or VT) can include chroma key color patches in the pictures, into which live studio shots of 'today's guest' can be inserted.

■ *Indicators* Many switchers include facilities for inserting indicators, that draw attention to specific information—in maps, charts, diagrams, photographs. Arrows, surround-rings, spotlight effects (a lightened disc on darkened background) are widely used, and located by joystick positioning; often with a 'blink' option.

Fig. 19.33 Indicators
Visual indicators in the form of an arrow, ring, spotlight (devised electronically or by any matting method), can identify specific detail.

■ *Appearance/Disappearance* This is simply achieved, by switching or fading the insertion in and out.

■ *Multisplit screen* This is a method of showing information from several shots simultaneously, by dividing the screen into 2, 3, 4 or more sections (with or without border lines).

Individual segments may show:

1. *Different views* of the same situation (reactions and prize giving at a game show).
2. A *montage* of concurrent events (different rescuers moving to the disaster area).
3. *Sequences* (steps in the development of a product).
4. *Groupings* of subjects with a common significance (all members of the team).

The multisplit screen can be achieved by: (See Fig. 19.34)

1. *Sequential switching*—A rapid autoswitching process in which a section of each source is selected in turn (a combined 'multi-wipe' process).
2. *Sequential insertion* (Quad overlay)—The switcher treats pictures in turn; matting out a chroma-key area, passing the matted result on to combine with the next source (with its

chroma-key area), matting from this combination, and so on
(Each insertion could use a different keying hue if required.)
3. *VT build-up*—Recording a split screen pattern; then repro
ducing the tape and matting in a further segment(s).

Chroma-key insertion techniques

This is the most exacting application of keyed insertion and the
one with potentially the most far-reaching impact. Chroma-key
is increasingly used to substitute built staging in the studio and is
often combined with other processes (e.g. external key).

Table 19.34   Electronic insertion terms

Background	The picture into which the foreground subject is electronically inserted. Can usually contain any tone or color.
Blink, flasher	Blinks a keyed insert (titling, symbol) at an adjustable rate.
Character generator, alphanumeric generator	Equipment synthetically generating lettering, numerals, signs.
Chroma keyer	1. Selects hue of keying color. 2. Selects input for chroma key treatment. 3. Selects edge-switching (hard, soft, color).
Color synthesizer, colorizer, color field generator	Equipment generating color synthetically, by applying selected voltages to the RGB channels.
Cycling	Synthesizer cycles through spectral colors (or flashes the luminance/brightness level).
Downstream key	Provides titling insertion while the main effects system is already in use (e.g. for chroma key). It inserts titles or any external key into the switcher's *line output*. May be used to fade program to black.
Double re-entry, cascade re-entry,	The output of one mix/effects-bus amplifier (A/B) can be re-entered into another (C/D) for successive additional treatment. So we can mix to or from combined shots (keyed-in titles, split screen, supers, chroma key). Also, a key from an external matte source may be used to create a matted area in one shot, that is filled in by another camera's background.
Edge key, border generator, black-edge	Introduces a black, white, or color border; or derives an outline on any matted-in subject.
External key	Interswitching between two sources, actuated by a *separate* video source (SEG, camera matte).
Foreground	The subject inserted into the background shot. In chroma key, contains the keying color.
Hue control	Adjusts the actual color used for keying matte (usually blue, yellow or green).
Internal key, self-keying, self-matting	Interswitching between two sources, using a tone or hue within the subject picture itself to provide the key (fed into a mix/effects bus).
Key edge	Controlling keying amplifier gain for the chroma key insertion of *transparent materials*. (*Linear keying*).
Key fill	Selects the source or color of infill for insert titling.
Key level, clipper, clipping level, threshold level	Adjusts the exact brightness (luminance) in the foreground matte, at which the background shot is inserted.

Although used primarily to insert people into backgrounds from any video source (photographs, film, VT, video disc, models, artwork), it can provide endless naturalistic or stylized effects.

■ *Compatibility* Foreground and background usually need to match in *pictorial* inserts: scale, perspective (lens angle, height, eyeline), color quality, contrast, light direction, etc. Incongruities can arise too, if the depth-of-field is restricted in foreground subjects, yet the inserted background appears quite sharp.

Linear keying	A facility enabling transparent subjects to be inserted, while permitting background to be seen 'through' them.
Luminance	Control of brightness level (in monochrome) or saturation (in color).
Matte, key	A shape (mask) supplying a signal to a keying amplifier, causing it to switch from one source to another in the matted area, integrating the respective images. Can be derived from a camera-picture, or specially generated electronic waveforms.
Mode	Selects the function of the *bank fader lever*. The same lever may be used to: 1. MIX (LAP DISSOLVE) between buses (banks). 2. MASK KEY for wipe-pattern movement. 3. MIX KEY to fade in a key (matte) signal. 4. To provide a pattern transition between A and B buses.
Non-additive mix (NAM), peak mix mode, white insert mode	A circuit facility in which the lightest tones of one picture are inserted solid (opaquely) into another source's picture. Used for title insertion.
Pattern direction	Selects direction of movement with operation of fader control.
Pattern edge	Adjusts pattern's edge sharpness (hard, soft-edged) or width.
Pattern generator	Selection of interswitching patterns used for wipes and insets.
Pattern symmetry	Adjusts pattern shape and proportions.
Pattern weave	Adjusting a weave or ripple effect—speed, amplitude (extent), frequency (rate), direction, position; pattern shape.
Polarity	Determines whether white or black is used for keying.
Position	Joystick positioning a pattern in the frame.
P/V (preview) bus key	The *preview bus* signal mattes a hole in the A bus picture and fills it with the B bus video.
Quad split	Simultaneous display, dividing the screen into four separate pictures, putting pictures into the four corners of the frame. (SPLIT or VAR adjusts the division positions.)
Sequential effects amplifier	Enables a picture to be matted in successive steps to produce a complex composite.
Special effects generator (SEG)	Apparatus generating a series of electrical waveforms (square, sawtooth, triangular, parabolic), at line and field rate, from which switching patterns are derived; for wipes and insets.
Spotlight	Background brightness is reduced around a circular disc matte (joystick positioned).

Fig. 19.34 part 1 Type of split screen

Several forms of divided screen are widely used: 1, split screen; 2, triple split 3, quad split; 4, multi-split.

1   2   3   4

Fig. 19.34 part 2 Methods of creating split screen effects

Several methods are used: *sequential switching*—the system selecting each source in turn for a certain period. *Sequential insertion*—after the first picture has been treated, part of a further picture is matted in. The next source is inserted into that composite . . . and so on. Other techniques include VT build-up (see 19.42), or digital image-processing.

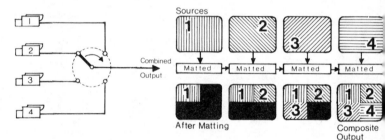

If the inserted foreground camera pans or tilts, the subject will appear to slide around the background. If you dolly or zoom the foreground camera, scale will be upset as the subject grows and shrinks. Unless you want such effects, therefore, foreground and background must normally remain stationary.

However, if foreground and background cameras move in synchronism, realistic pan/tilt/zoom changes are often possible. In simpler situations, skilled cameramen may actually coordinate

Fig. 19.35 Background insertion

1, Substituted background—Using a self-matting/self-keying (usually chroma-key) surface behind the subject, a different background can be substituted of plain tone or hue (perhaps color synthesized), decorative or pictorial. 2, Camera matte—This selects part of the studio scene, matting it into a background scene (graphic, photo, film, VT).

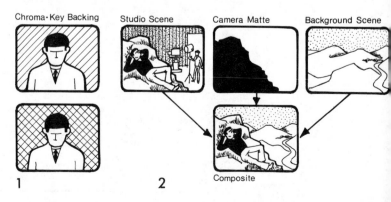

Chroma-Key Backing   Studio Scene   Camera Matte   Background Scene

1   2   Composite

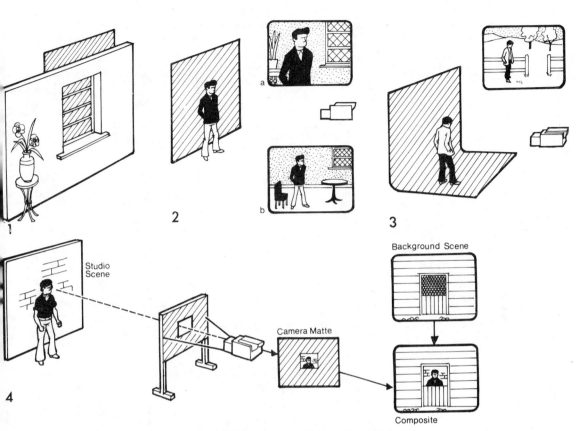

Fig. 19.36 Typical chroma-key set-ups
1, A chroma-key backing permits an 'exterior' to be inserted from a photograph.
2, Where (a) the floor is not in shot or (b) the studio floor area is required and foreground objects are to be included, a simple chroma-key flat is sufficient.
3, For longer shots, or where people are to appear higher in the frame, or the studio floor is to be substituted, a continuous chroma-key background and floor covering is necessary. 4, A camera matte in chroma-key hue masks off most of the studio, and inserts the visible area into the background scene. This method also avoids the need for large keying areas in the studio when inserting into long shots.

their cameras' movements and achieve effective results. But the most reliable synchronism becomes possible if foreground and background cameras are ganged by electronic servo devices (as in 'Magicam' process). Then the foreground camera can tilt/pan/zoom and the background camera shooting a photograph, graphic, or model will move proportionately. In an alternative arrangement, the background derives from a static camera shooting a servo-controlled graphics stand or a rear-projection device (slide or film)—'Scene-Sync'.

Results can be extremely convincing as the foreground camera follows action and adjusts shot length. However, the background camera must turn on a correctly adjusted pivot center and its displacement rate match the relative lens angles used. Where possible, you should avoid visually critical situations; scenes involving perspective and parallactic movement (relative displacement of planes at different distances).

Fig. 19.37 part 1 Foreground mattes—matting flat

A flat of chroma-key hue can be positioned within the chroma-key set-up to coincide with a wall in the photo background, so enabling a performer to 'move behind' it.

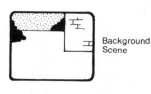

Background Scene

Chroma·Key Backing and Floor

Walks in front of Wall

Foreground Matte Matches Wall in scene

Walks behind Wall

Fig. 19.37 part 2 Foreground mattes—camera matte

Normally, when subjects are chroma-keyed into a scene, they appear in *front* of the background. But by introducing an *additional matte* (e.g. a silhouette camera matte as an external key), to suppress the chroma-key insertion in that area, you can make subjects 'move behind' parts of the background scene. Such matting can be switched in and out to allow moves in front and behind background objects.

Background Scene | Normal Insertion | Camera Matte | Insertion + Camera Matte

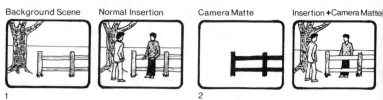

1   2

Fig. 19.37 part 3 Foreground mattes—relating to background objects

By appropriately positioning suitable items in the chroma-key set-up (blocks, parallels, stairs, etc., painted in the keying hue), people can apparently sit, lean, climb, or put objects on features in the background shot.

Photo Background   Chroma·Key Backing Floor Matching Areas in Studio   Composite Result

■ *Color fringing and edge ragging*   Color fringes (of the keying hue) can develop round the borders of keyed-in subjects for several reasons: colored spill light, reflected keying hue, optical-flare problems, and inherently poor color detail (restricted band width). The often recommended 'remedy' of using yellow back-light only alleviates the problem, and is likely to tint the subject yellow! Some chroma-key systems are less susceptible to color

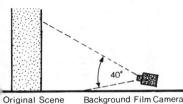 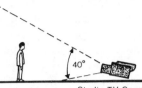 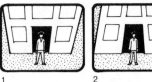

Original Scene   Background Film Camera   Studio TV Camera

Fig. 19.38   Matching camera height
The height, angle of tilt, and lens angle, of the studio camera should be identical
to that used to photograph the background scene, if distortion is to be avoided.
Often the errors are not obvious. 1, Unmatched viewpoints; 2, matched viewpoints.

fringing than others. Soft-edge keying can diminish the effect,
but for total suppression special circuits (hue suppressors/fringe
eliminators, exclusive hue matrix) may prove necessary. Unfortunately, these may modify the reproduction of certain color
mixtures.

When inserting fine detail (hair, lace, fur, feathers) or vague,
unsharp subjects (smoke, steam) you will encounter edge ragging
or break-up due to indecisive switching action, without such
electronic compensation.

■ *Keying surface and hue*   Strictly speaking, the actual color of
the matting surface in the foreground scene should exactly match
the keying hue chosen for the system—whether a colored surface
or colored lighting is used. And it should present an even surface
brightness overall. In practice, set-ups deviate considerably from
this ideal and work successfully. Most studios use colored cyc
cloths, flats, stretched cloth, painted floor cloths, and their brightness and spectral distribution can vary noticeably. The matte can,
of course, originate equally well from graphics, film, videotape,
or any other video source.

Although any *keying hue* may be used, this color must not be
present in the foreground picture, or you will see spurious break-
through and the background will appear instead. If you were to
dress someone in clothes of the keying hue, they would progressively disappear. So total or localized 'invisibility' is readily
achieved!

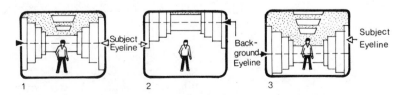

Fig. 19.39   Matching eyelines
1, The foreground subject matches its background's perspective only when the
television camera's eyeline coincides with the eyeline of the background.
2, Background eyeline *above* subject eyeline. 3, Background eyeline *below*
subject eyeline.

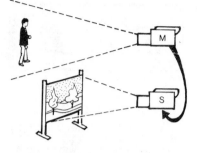

Fig. 19.40 Ganged backgrounds
By using a servo-controlled inter-connection ('ganging') between the subject (master) camera and the background (slave) camera, the latter pans, tilts, and zooms automatically as the subject camera follows the action. Alternatively, the background graphic itself may be moved vertically/horizontally by servo systems (SCENE-SYNC SYSTEM) using a static slave camera.

A highly saturated *blue* (approximately 'cobalt') is widely used for the keying hue, because it is chromatically most distinguishable from flesh colors and least likely to arise in foreground scenes. But where blue subject matter is unavoidable, yellow or green keying hues are often used instead. Today's sophisticated circuitry can be so color specific (*exclusive hue*), that it operates on the selected hue alone and even close colors or color mixtures do not trigger spurious switching.

■ *Shadow insertion* The chroma-key system extracts everything from the foreground scene that is not *blue* (if that is the keying hue) and inserts it into the background scene. So a person's shadow falling onto some foreground scenery will be keyed-in but not their floor shadow (a shade of blue). Performers' shadows falling onto a keying surface can actually cause irregular switching (reduced key matte strength), so we often avoid or illuminate them to prevent this.

Without floor shadows, however, inserted people may appear to 'float' and the illusion is spoiled—particularly where shadows are evident in the background scene. With appropriate circuitry a natural shadow can be *simulated*. It detects the lower brightness (luminance) of the shadow on the keying surface, deriving a matte from it of adjustable sharpness and density, and introduces it transparently over the background scene. Extremely convincing, even its position can be relocated to match the background by including keying surfaces in the foreground scene.

■ *Transparent/translucent insertion* Basic chroma-key insertion tends to block out detail beyond transparent materials (glass, smoke) or show spurious triggering (tearing, ragging). By using suitable circuitry (*linear keying*, hue suppression) it becomes possible to 'see through' a glass surface in the foreground shot and key it into the background scene without spurious effects.

■ *Moving backgrounds* Where a photostill is used for the background scene, you should avoid any 'frozen' features (waves, smoke, people) drawing attention to the falsity. When keying into moving backgrounds from film or VT, you should make sure that sequences last long enough for the foreground action and where necessary are suitably synchronized (e.g. a 'moving' automobile's background).

■ *Foreground pieces* With care, you can introduce subjects in the foreground scene, so that they blend with the background shot into which they are inserted. Furniture, scenic units (pillars, walls), foliage, properties, can be used by performers to achieve a totally convincing integrated effect. Also, by carefully positioning a foreground piece, you can even cover an obtrusive or unwanted feature in the background scene picture.

Fig. 19.41 Shadows

1, Chroma-key insertion relies on color to actuate switching, but does not normally differentiate between different intensities of the keying-hue (i.e. shadows on the chroma-key surface). So subject shadows are not inserted. 2, Special extra circuitry permits 'shadows' to be produced (actually luminance-derived mattes), but the shadow falls along the floor and may be inappropriate. 3, If a carefully-positioned matting flat is used in the chroma-key set-up, so as to co-incide with planes in the background picture, the subject shadows falling on it seem to be on the photographed wall. 4, The inserted shadow's sharpness must be comparable with that of the photo background. 5, Given compatible shadow direction, contouring and hardness, a totally convincing illusion is possible.

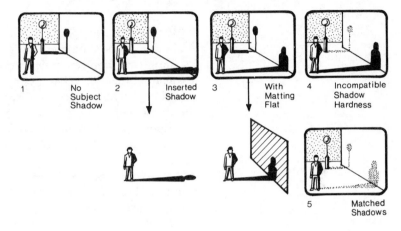

| 1 | No Subject Shadow | 2 | Inserted Shadow | 3 | With Matting Flat | 4 | Incompatible Shadow Hardness |

5   Matched Shadows

Complex insertion

■ *Moving 'behind' background subjects*   When you want people in the foreground scene to 'move behind' subjects in the background picture, you can use either *keying areas* or *external mattes*. If you exactly align an area of keying hue (e.g. blue) in the foreground scene, with a subject in the background shot, you effectively 'detach' that item in the background picture. Make a blue matching flat in the studio correspond with a wall in the background photograph and a performer can paste a bill on it, or walk behind it in the chroma-keyed composite! Using blue-covered risers (blocks) parallels (rostra), treads (stairs), a person can apparently sit, lean, climb, place objects on the background items!

Where your performer is to move behind fine detail (tracery, open screens, foliage, fences) in a background photograph, you can use an *external matte silhouette* that inhibits the chroma-key insertion, letting the background through instead. A suitable silhouette can be traced from the background picture or the projected image of a background slide. As this inhibiting matting signal can be switched in or out, a person can walk behind (matte in) or in front of (matte out) the background feature at choice.

■ *Video build-up*   You can key a person into a background, replay the composite, and then key him into that. So far, he has become twins! You could then continue inserting into composites (with quality deteriorating at each dubbing) to provide any number of different insertions of the same performer; thus

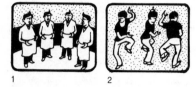

Fig. 19.42 VT build-up
1, Using chroma-key, extra subjects can be inserted into videotape to provide action build-up e.g. a series of separate recordings (passes) using the same person in different costumes. 2, Multi-camera shooting can be integrated to produce simultaneous multi-images of the same subject.

generating a barber-shop quartet, choir, or orchestra! One ingenious example showed a man sitting beside himself, taking a tiny figure (of himself) out of his pocket . . . and so on!

Solid superimpositions can be used (supers against black background), but chroma-key insertion offers most flexibility. Using video disc, varispeed VT, or digital processors, the movement of inserted subjects can be adjusted (fast, slow, freeze, reverse motion).

Color synthesis

Any color can be simulated in a TV system, without a TV camera, by supplying appropriately proportioned voltages to the red, green and blue video channels.

The color synthesizer (colorizer, color field generator) produces these voltages either for switch selection or joystick control (paintpot). The latter is a hemispherical device with a central joystick and provides rapidly variable changes in the R-G-B balance.

This synthesized color can be used as a source for electronic insertion: lettering infill, title backgrounds, background insertion, and decorative effects.

■ *Multilevel synthesis* This keying process switches in a chosen synthesized color wherever a particular tonal level appears in the treated picture. So for example, in a two-tone system, a white on black slide can be reproduced as a red on yellow picture. Both economical and adaptable, this is a regularly-used facility in many studios; titling being prepared in monochrome and then video processed.

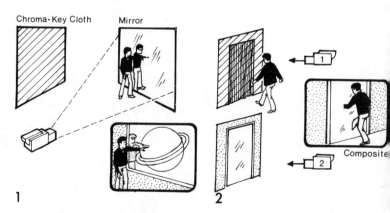

Fig. 19.43. Mirror effects
1, A demonstrator stands before a 'magic mirror' displaying a series of changing images. In fact, the mirror reflects a cloth of keying hue, so turning it into a self-matting area, into which subjects are inserted (from slides projected on obliquely-angled screens). 2, A person walks through a mirror. In fact, an opening in a blue flat is hung with a curtain of thin blue strips and corresponds to another camera's mirror flat.

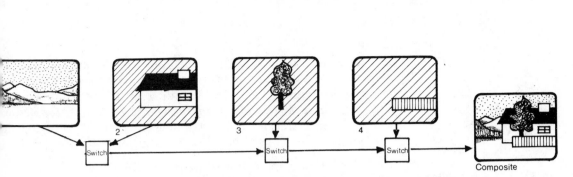

Fig. 19.44 Sequential keying (sequential insertion, cascade insertion, quad overlay)
Each picture is keyed in 'on top of' the previous combination, to build up an overlapping composite.

Using *multilevel digital synthesis*, greater elaboration is possible. The picture is analyzed into perhaps eight tonal levels, each of which can be colorized to a different hue or tint. So a multitone graphic becomes transformed into a multihue display. A normal varitoned picture (monochrome or color) becomes strangely posterized in synthetic colors; akin to a colored contour map. The same device may be used to produce *tonal coarsening*, reducing half-tones to eight or less gray scale levels, or to solid areas of black and white.

Digital systems

The current TV camera produces a fluctuating voltage (the video signal) which is a direct *analogue* corresponding to the scene's light and shade. However, there are many technical advantages in transforming the video into a different form, through digital analysis. To the engineer, this provides exciting opportunities using digital picture processors to improve picture quality (image enhancement), reduce picture noise (snow removers), improve apparent definition (edge enhancement), stabilize picture synchronism from mobile recorders, build switchers (vision mixers) that accept non-sync and synchronous sources, permit multigeneration dubbing without quality deterioration. Also to provide more compact higher quality equipment such as standards converters, teletext facilities, electronic still stores, time-base correction, etc., etc.

Digital analysis is remarkably simple in principle. It checks the strength of the video signal at regular intervals and stores away these samplings in numerical form (binary digits—i.e. an on-off pulse code representing numbers). Now, instead of fluctuating video, we have a store of numbers representing the brightness and color of each tiny area (element) of the picture. These stored digitally coded signals can be recalled (or replaced) at will and reconstituted into the original picture.

The particular advantage of what otherwise seems a tortuous transformation, is that we now have random access to each stored

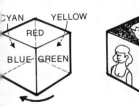

Fig. 19.45 Multi-hue chroma-key
Several keying hues are used in turn, each keying in a particular video source. So different pictures appear on this rotating multi-colored block.

Fig. 19.46 Outline effects
The outline of a girl dancer transforms into the outline of a tree. To
obtain this: Camera 1 shoots a white
on black silhouette of the dancer,
gradually zooming in until the white
shape fills the screen; while Camera
2 is close on the white tree silhouette,
gradually zooming out. The combined
effect is fed through a border generator and the outline transformation
develops.

element of the picture! Normally, picture information would be
taken from the store in its original order. But the equipment can
now manipulate the video according to the rate and order in
which it reads this information: repeating parts of the picture,
sampling in reverse order, skipping information, changing the
sampling rate . . . and here lies the magic of digital effects!

Video image processing provides many facilities:

1. Squeezing or expanding the picture vertically and/or horizontally, from full frame to zero. So you can change the aspect
   ratio, alter the picture size overall, 'zoom' in or out (e.g. × 4)
   and have the frame shrink to a spot.
2. The total picture can be moved vertically/horizontally.
3. Areas of the picture can be selected, processed, repositioned
   leaving the rest unchanged—or eliminated.
4. Pictures can be laterally reversed, inverted.
5. Pictures can flip contract . . . switch . . . expanding to a new shot.
6. The picture(s) can be reduced to $\frac{1}{4}$ frame and repositioned
   (singly, half, quad frame).
7. A multisegment montage can be built up, with independently
   switchable parts.
8. Push-over (push-off) wipes can slide the picture off the screen.
9. Split wipes can cut the picture and move the segments apart.
10. Inserted titles can be zoomed and/or repositioned.
11. Frame grab ability enables a frame to be frozen, strobed for
    effect, interated freeze (successive freeze frames).
12. Dynamic spiralling wipes (rotary pattern generator).
13. Autokey tracking maintains the position and scale of a chroma-
    key inset as the total combined picture is controlled.
14. Audio-activated patterns can be created.
15. A display in which a multiple image built from a succession of
    freeze frames, can show the movement pattern of a subject
    ('Action Track'—AT) and selective strobe-action effects.
16. 'Electronic Palette'—a computerized system for simulating
    sketching or painting directly on to the screen. Digital graphics
    aid conversion of artwork to video animation.
17. Non-linear shape changes . . . teardrop, hourglass etc.
18. Ooze function . . . one shot fluidly merges into the next.

Temporal effects

Once *real time* has been recorded, you can control the speed at
which this action is reproduced; using film, VT, video disc, or
digital processing.

■ *Fast motion*   Fast motion effects can be introduced for various
reasons:

1. Speeding up subject movement and time scale.

2. To exaggerate energy for comic effect, pixillation, or dramatic force (fast-moving clouds).
3. To shorten an action's normal duration.
4. To reveal the development of a slow process—time lapse shots speeding plant growth.

In film this is achieved by shooting at less than 24 fps and reproducing at normal projection rate of 24 fps; or by regularly omitting frames.

■ *Slow motion*   The uses of slow motion include:
1. Slowing subject movement and time scale.
2. To exaggerate movement, form or pattern.
3. To allow rapid movement to be discerned (humming birds' wing-beats).
4. To emphasize movement, giving it force and importance (emphatic body movements).

In film, slow motion is achieved by shooting at more than 24 fps and reproducing at normal speed, or repeat printing alternate frames (stretch/skip frame). In television, slow motion is achieved by video disc or by slow replay of helical VTRs.

■ *Freeze frame*   Stopping action during its course, to show development or prevent completion. Achieved by video disc, digital processing, or by repeat-printing the required film frame.

■ *Reverse motion*   Reproducing action backwards for comic, magical, or explanatory effect—a demolished building reforms, articles wrap themselves up, scattered lettering arranges itself.
Video disc is generally used. Film can be reversed by: feeding it into telecine tail first and electronically inverting, running film in reverse, or reverse printing.

■ *Animation (single-frame, stop frame)*   Achieved on film or VTR, by recording a single frame at a time and changing the subject a little for each. When reproduced in quick succession, the shots develop a group significance—usually creating apparent movement. Thus you can animate cartoons, puppets, still-life.
Photoanimation uses a series of photographic prints as the basis for treatment, rephotographing them after progressively adding titling, retouching, sectional movement, etc. Thus various process effects can be achieved economically, including complex wipes, spins, phantom sections and explosions.
Video rostrum cameras are increasingly used for animation treatment (often combined with digital video effects) for promotion material, advertisements, graphics.

■ *Cyclic motion*   Recurrent repetition of an action; achieved by a film loop or conjoining a series of identical copies of a sequence.

■ *Extended and contracted time/space*   By editing-in a repeated part of an action sequence (from another viewpoint, to avoid an obvious jump cut), we create the impression that space or time are extended. Similarly, by editing-out action (with a viewpoint change or cutaway shot) we contract space and time.

■ *Transformations*   By animation; or by stopping action during recording (with locked-off cameras), then modifying the scene and continuing recording. When the sections are joined together, we can, for example, make trees burst into bloom and a man turn into a mouse!

20 On location

Until the advent of relatively lightweight TV equipment, video programming originated in television studios, larger mobile control rooms, and lightweight mobile vans that covered smaller scale field events. All used standard broadcast-quality TV cameras. Film crews with 16 mm cameras covered most news reporting, documentaries, and high-mobility situations.

Then, in the early 1970s, competitive broadcasters turned to new lightweight TV cameras and VTRs (mostly designed for closed-circuit and industrial use). This move greatly increased production flexibility. Highly mobile units now set fresh standards in newscast topicality. Spurred by events, rapid technological advances provided increasingly higher quality and lightweight TV cameras began to supplant film-shooting for various program purposes.

Initials proliferated, to designate applications:

EJ—Electronic journalism.
ENG—Electronic news gathering.
EFP—Electronic field production.
ESG—Electronic sports gathering.

The facilities we need, depend largely on the type of program material involved. A large-scale remote may require a formally organized set-up to coordinate the production on site (e.g. major sports coverage). For improvised snatch shots in a disaster area, high mobility and a low profile are essential. So broadcasting organizations often have a comprehensive systems range.

Large mobile control room (location production unit)

■ The biggest vehicles used for remotes are usually mounted in large truck or trailer form. The unit provides a full broadcast standard production control center with complete video and audio facilities (switcher, audio and lighting boards, audio and video recorders, caption scanner, etc.). Occasionally a small mobile studio is attached. The sizable production and operation team controls up to five broadcast-standard cameras dispersed about the site.

The unit may be used too, as a *drive-in control room* for a temporary studio; or alternatively, equipment may be removed from the vehicle and installed within a site building (e.g. for conferences). Program material may be relayed to base by microwave links for direct program insertion, taping, or provide a complete live production. For large-scale telecasts in which on-site program-control is essential, this traditional *remotes (outside broadcast)* approach has considerable merits.

Remotes van

A more compact system uses a small control room in a lightweight van. This contains full broadcast-standard equipment for one or two studio-type or lightweight cameras. Basic switching, monitoring, and audio facilities are provided. The program material may be videotaped (2 in quad or 1 in helical) for subsequent editing at base; or microwave transmitted to base.

Typical applications include EFP (for documentary and drama inserts), ESG, and some ENG.

Lightweight truck/small van units

Fitted with one or two *studio-type lightweight cameras* and a small format VTR (1 in helical), this type of mobile vehicle has proved highly adaptable for ENG, EFP, ESG, and commercials. Versatile designs often include arrangements for roof-mounted cameras, a cab hatch, and/or a rear platform for flexible camera viewpoints.

Program material is recorded on the internal VTR, and/or microwave transmitted to base.

The vehicle may also be used as a *satellite unit*, transmitting back to a mobile control room, e.g. traveling at speed around a horserace track, cycle-race circuit, etc.

Station wagon/small van units

Designed to meet the demands of rapid news-gathering techniques, this unit carries one or two 'broadcast ENG' cameras and a small-format helical recorder. The cameras may feed audio/video to the support vehicle by cable or small microwave transmitter. The program is videotaped and/or transmitted to a central relay station (at a high vantage point) from which it is retransmitted to the news center. There the signals are processed (timebase correction, image enhanced, color corrected, noise reduction) and videotaped.

Portable units

Here a single small industrial-quality hand-held color camera feeds a lightweight VTR (e.g. $\frac{1}{4}$ in shoulder-slung). After correction, the collected material is transferred to a high-grade VTR before editing or airing. Although below broadcast standards, news-gathering opportunities may outweigh sub-standard picture quality.

Program handling

Program material (audio and video) can be handled in several ways for field production:

1. Videotaped in the field and delivered to base by courier or

Fig. 20.1 Field transmission
The illustration shows one typical method of handling program material. 1,
Camera may be hand-held or tripod mounted, its video being fed to a 'window
unit'. Local mike connects to camera's audio pick-up socket. 2, *Window-unit*
is a lightweight battery-operated transmitter. Range 1–8 kms (1–5 miles) in
line-of-sight contact with support vehicle (13 GHz). 3, Coaxial cable from
camera to support vehicle. 4, Portable VTR. 5, Portable dish reflector of 60–120 cm
(2–4 ft) diameter, transmits audio/video from support vehicle to central relay
station. Range is 16–24 kms (10–15 miles). (Transmitter 10–20 W, 2–7 GHz,
circular polarization.) (In certain situations a helicopter may serve as a relay
point.) 6, Remote signal is picked up on 4-horn remotely controlled 360° dish
antennae system on a distant high building. The signal from the uplink receiver
is up-covered to 7 GHz., for transmission to news center. 7, At news center,
signal correction is introduced (digital timebase correction, image enhancer).
8, Two-way communication system from news center (program and engineering
coordination) to field unit (450 MHz).

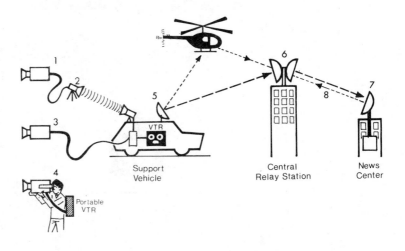

Support Vehicle   Central Relay Station   News Center

when the unit returns. A growing technique involves single
or two-camera shooting with a VTR per camera and without
a switcher; all editing being a postproduction operation.

2. Videotaped on site and transmitted to base via a suitable relay
   station. In remote areas, where local topography prevents a
   line-of-sight transmission path (dead spots due to hills or
   ridges), it may be necessary to drive to a known good transmission point to replay to base.

3. Transmitted live to base over microwave links (occasionally via
   helicopter or satellite stations) for live broadcast.

It is normal procedure in many organizations, to videotape the
incoming material (after timebase correction and image processing) on quad or 1 in helical VTRs. Then the material is edited
onto quad to avoid further quality losses in multigeneration dubs.

While some mobile units are essentially self-contained, others
such as ENG teams maintain continuous two-way VHF communication with base for engineering and production control.
When the team is remote from their vehicle, it serves them as a
repeater station (receiving and retransmitting) for their program
audio/video and control.

Fig. 20.2 Typical ENG organization

1, Incoming program material scrutinized, controlled by News Coordinator. 2, Video recordings prepared with time-code: helical work-print for editing decisions and quad master copy. 3, Decision booth where editor (writers/reporters) examines incoming material; log in-out times (time-code) of required excerpts, audio cues for edit points, etc. 4, Dub editing by assembler from editor's instructions (commentary, graphics, titles added), on to 2 in quad tape cart (cartridge) player using computer-controlled selection. 5, Cart player with edited version of program; one cart on air, other cued up ready.

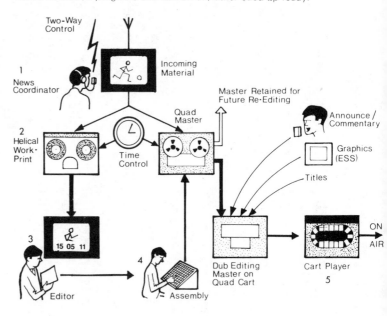

ENG editing

The production demands of ENG are considerable and special organization has evolved to meet its particular needs. In some systems, the mobile crew (reporter/director, engineer/operator) is closely coordinated from the news center through two-way radio communication. The center provides news evaluation, late-breaking information, updating, talent cues, production timing, and interrupted feedback. The editorial coordinator serves as field producer, continually examining story progress and giving editorial guidance.

The program material is transmitted from site to the base news center (via a relay point), where it may be continuously taped; both as a 'work-print' copy for editorial purposes and a master reference copy storing the material.

Although time code is widely used for editing (single-frame accuracy), some organizations count VT control track pulses instead (6–8 frame accuracy) for technical convenience.

In one system, an *editor* reviews the 'work-print' copy, assessing edit-decision points, logging in-out times and audio cues. Freeze frame facilities precise location. An *assembler* using this edit decision list enters the precise frame numbers into the VT editing equipment. Computer-controlled VT equipment locates these points in the master reference tape, dubbing off the selected sequences onto a quad VT machine for broadcast—perhaps with a corresponding back-up helical version. Either reel-to-reel or cartridge (cart) systems are used. The latter providing automatic replay of a series of cart stories from a special recorder/reproducer.

The unaffected original unedited master reference tape remains available for any revised editing of the material (revamping) for later newscast.

During the dubbing process, commentary may be added from an adjacent announce booth; titles can be inserted (multifont electronic character generator); and graphics added (a back-up from an *electronic still store* providing illustrations from digitally-stored graphics in computer memory files).

Fig. 20.3 Field camera mountings
Some typical methods of mounting the TV camera in the field: 1, Roof-top tripod at vantage point. 2, Camera on frame or tower. 3, Lightweight pneumatic pedestal. 4, Lightweight tripod. (3 and 4 may have wheeled bases.) 5, Portable cameras. 6, Van with hatch, roof, rear-platform cameras. 7, Hydraulic platform. 8, Helicopter.

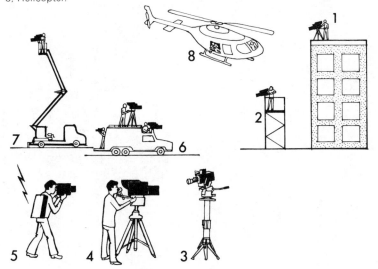

Types of field camera

TV cameras used in the field come into four broad categories:

1. *Larger studio camera*—designed to state-of-the-art parameters, with full broadcast specifications.
2. *Portable broadcast cameras*—in more compact mobile form, often using multi-module configurations that offer various facility options (different lens systems, viewfinders). These are used for both studio and portable field-work.

3. *Lightweight cameras*—designed for ENG/EJ applications, where emphasis is on reduced size and hand-held use.
4. *Small hand-held low-cost cameras*—designed for CCTV industrial applications, but used for certain ENG assignments

Field camera equipment

■ *Mountings* Camera mountings used in field production are varied and adaptable: lightweight pedestals, tripods (fixed or rolling), lightweight dollies, fixed cameras (platforms, towers, roof-sites), hydraulic hoists (cherry pickers), helicopters used as camera platforms (and as microwave relay stations).

Portable cameras may use lightweight tripods, body harness (brace), be shoulder-mounted, or hand-held. The electronics may be contained in a shoulder-slung bag, a back-pack, separately mounted trolley pack, or at the support vehicle.

Fig. 20.4   Portable cameras
Mobile camera designs include: 1, Shoulder-mounted (battery belt supply). 2, Hand-held camera cabled to a back-pack. 3, Body-brace support for camera cabled to control equipment on a trolley-pack (transmitter or VTR). 4, Small hand-held camera attached to shoulder pack VTR.

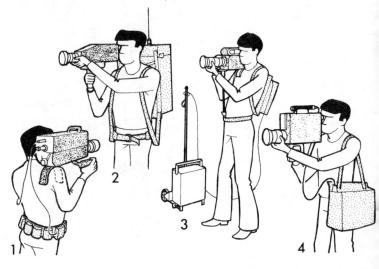

■ *Electronics and power* Lightweight and hand-held camera equipment is usually battery powered (rechargeable battery belt containing silver-zinc/nickel-cadmium cells), with options for the use of an on-board generator, inverter supplied by the vehicle generator, a battery system, or power lines (mains supply).

■ *Camera tubes* Field equipment typically uses plumbicon, saticon, and chalnicon type camera tubes in 1 in or $\frac{2}{3}$ in (25 or

18 mm) formats. In the future, charge-coupled devices (CCD) promise well for further miniaturization.

■ *Lens systems* Although lightweight cameras can support large zoom lenses when pedestal mounted (34:1), lighter lens systems become necessary when cameras are hand-held (6:1 or 10:1). For some purposes, weight-saving fixed-angle (non-zoom) lenses are preferred.

The handling and focusing problems we met earlier when discussing zoom-lens operation (Chapter 3) are inherent. They are particularly troublesome when trying to maintain steady hand-held shots at narrower lens angles.

■ *Automatic video control* Although automatic control circuits help to provide usable pictures under adverse conditions, they have no magic. Any camera system still has inherent limitations. As lighting and scenic tones change, so circuits compensate—but to suit *technical* parameters. A bright light comes into shot, so the camera stops itself down, and lower tonal detail is lost; even if that contains the main subject tones. The artificial effect may be quite fortuitous.

'Hands off' video control usually includes such features as:

1. *Automatic iris* (lens aperture, stop'—adjusting for variations in scenic tones and lighting. When using manual control, a view-finder-screen marker may indicate when exposure is techni-cally cirrect.
2. *Autoblack*—stabilizing the black balance.
3. *Autowhite*—compensating for changes in lighting color temp-erature (readjusting to a chosen 'white' reference in the scene).
4. *Scenic-contrast control*—gamma adjustment for improved shadow detail under high-contrast conditions.

■ *Time-base correction* Complete synchronism is essential if TV pictures are to be transmitted and reproduced without color errors; or even total loss of color, and picture displacement.

A small portable helical VTR offers considerable economy and portability for ENG projects. But although its output produces satisfactory pictures when reproduced on a picture monitor (which adjusts to the varying sync pulses), the video from such a machine is too unstable for use with a precisely synchronized TV system. Particularly when editing or using picture combina-tions (super, insertions) the irregular speed fluctuations, arising from tape travel and mechanical imperfections, would cause color loss and image break-up.

This problem can be overcome by using *digital correction*, which compensates for even substantial errors and produces stable pictures of improved quality from small-format VTRs.

Digital time-base correctors function by storing up the video information from the VTR (notwithstanding the information rate variations) and clocking it out of store again at precisely timed intervals, in perfect synchronism with the station sync-pulse rate. Certain other imperfections may be improved too, such as low color detail and video noise (although not the chroma delay inherent in 'color-under' helical systems). Under poor lighting conditions the digital noise reducer ('snow remover') works wonders to reduce the disturbing background (although low-light lag remains). It may also remove grain from televised film.

21 Video engineering

The disciplines and complexities of television engineering do not normally involve the program maker. He is more concerned with what the equipment can or cannot do, and with using it effectively. However, there are certain engineering areas where a basic understanding will help you to appreciate the TV system's idiosyncracies and to achieve optimum performance.

Video control (shading)

Although scenery and lighting can build up an atmospheric effect, how successfully this is conveyed by the camera will depend on the video equipment and how it is adjusted.

It is really a matter of standards. The very act of producing pictures *at all*, may be sufficient for some purposes—e.g. surveillance. But to produce pictures that are consistently attractive and persuasive requires forethought, skill and patience. To obtain optimum picture quality, you must do more than just point a camera at a scene. Appropriate control is necessary, otherwise shadows can become clogged and detail-less; the lightest tones may merge; successive pictures may jump in brightness; and some shots may look 'washed out', while other are overcontrasty.

In the studio, you can help to ensure good picture quality, by carefully selecting tones, and lighting, wardrobe and so on. But on location, where the situation is less easily adjusted, it may be necessary to avoid shots that produce unacceptable results, such as an overexposed exterior seen outside the window of a room. Alternatively, you may be able to introduce compensatory lighting.

'Picture processing' in the form of video control (shading) is necessary to get the most from the TV system. Ideally all video sources (cameras, film, slides, tape) should be continuously monitored and their pictures suitably adjusted. But this is often neither practical nor economic, so four basic approaches to the control of TV picture quality have evolved. These are preset, automatic, shader and video control operator.

Preset control

Here the equipment is adjusted during a technical *line-up*, and its working limits set (exposure, black level, etc). This involves placing test cards in front of cameras to check equipment performance and adjusting circuits for optimum definition, tonal gradation, geometry, etc. Computerized *auto set-up* is also used.

If at the selected f-stop there is any over or underexposure, then you must adjust studio lighting or change subject tones (e.g. alter clothing). Otherwise you just have to tolerate defects. Lighting with this method generally needs to be even overall and of low contrast. This technique may be quite adequate for simpler presentations and where staffing is kept to a minimum.

Automatic control

In this method, after technical line-up, automatic control circuit are switched in which respond to extreme picture tones. Detectin, a bright area, they will stop the lens down until 'correct exposure is obtained; opening up when the general levels fall. So too, the can adjust the *black level* (set up) of the video signal to ensure tha the darkest areas are reproduced as black. Therefore, any gros errors are avoided.

For many types of presentation, results can prove quite satis factory. But, unfortunately, any automatic compensation can b fooled. The darkest tones in shot may not be black . . . but auto black could set the picture down and shadow detail would be los A person's face should appear of *similar* brightness when shot b two adjacent intercut cameras. But automatic controls react to a picture tones, whatever their significance. With light background to one shot and dark to another, face tones would not match Bring a newspaper up into a shot and the lens stops down, so a fac becomes darker. Have another person move into shot wearing lighter or darker costume and the picture may change.

Automatic systems are useful safety devices under varying o unpredictable light conditions, even if artistic results are quit arbitrary. Set a *preset lens aperture* for a bright interior and the move into a dim interior; we now have a gross underexposure an can see nothing! Set the stop to correctly expose this interior an the bright exterior is now totally overexposed (washed out) The camera tube (or film) cannot accommodate such a brightnes range. We have to select what we want to see best. In these cir cumstances, an *automatic iris* will 'correct' the exposure b opening and closing—however arbitrarily, and produce usabl results. *Skilled* manual adjustment can anticipate, and be artistic ally sensitive; for unlike the film camera, the TV camera enable us to see and immediately correct results.

Shader

This system is widely used in network studios and remote truck (mobile control rooms for OBs). The video engineers, wh service and align the video equipment, sit at the *camera contro units* (CCUs) adjusting the camera lens aperture (remotel controlled), the *black level* (*set-up*, *sit*, *sit-up*), and *video gain* to suit each shot. Camera color balance can also be adjusted.

Video control operator (video control)

Here a specialist video operator remotely controls all the camera equipment at a central video console (vision control desk). He adjusts lens-aperture and black level, with a combination control knob for each channel. Color balance (TARIF) controls are provided for film and caption scanner sources, corresponding controls being available for the camera channels. Video and lighting controls are also closely coordinated. All equipment is lined-up and serviced by video engineers (studio engineers).

Exposure control

The main operational function of video control (shading) is to ensure that the camera tube is suitably exposed—technically and artistically.

The TV camera tube, like film emulsion, can only accept a limited tonal range. If the scenic tonal contrast is restricted and you suitably adjust the lens aperture (*f*-stop, iris), your picture will reproduce these tones effectively.

But many scenes exceed this tonal range, especially if lighting is contrasty, and fall way beyond the camera's limits. Then you must set the *f*-stop to provide optimum reproduction of, what you consider to be, the most important tones. This situation arises for instance, in strongly sunlit exteriors where you cannot expect to reproduce detail and tonal gradation in both highlights and shadows. In the studio, by ensuring that the lightest tones are not overbright nor the shadows too dark, a full tonal range can be reproduced.

Most shots are exposed to provide good reproduction of *faces*; although this may mean that more extreme tones are lost. You may be able to compensate for this loss—for example, by illuminating deep shadows sufficiently to bring them within the system's limits, and so make detail visible.

Black level

The video equipment has two inherent limits within which the video signal must be contained; these are reference white and reference black. Generally, the peak whites in the camera's video signal are adjusted to correspond with the reference white level (by exposure and video gain), while the blackest picture tones are adjusted to meet the reference black level (by a black level control). Beyond these limits, any tones are 'clipped off' electronically and reproduced as solid white or black.

This black level adjustment (*set-up*, *sit*) can be adjusted operationally, to suit the picture. When raised (*setting up/sitting up*), the darkest picture tones 'gray-out'—without revealing any more shadow detail—as all reproduced tones are lifted. On lowering the black level (*setting down/sitting down/batting down on blacks*) the reproduced tones are reduced so that the darkest tones become crushed to a firm, detailless black—without revealing more detail in the lightest tones.

Further controls

■ *Video gain* This adjusts video amplification and can be in creased under adverse light conditions to boost the video—but the expense of greater picture noise.

■ *Gamma* The subtlety or coarseness of tonal gradation is in fluenced by a system's *gamma*. Where it has *unity gamma* gamma of one) the video signal strength is proportional to th subject tones. If a device such as a vidicon camera tube has a lo gamma (below unity e.g. 0·65) the result is thin, reduced ton contrast. A high gamma device (e.g. a color picture tube wit gamma 2·2) provides coarse, exaggerated tonal contrast. Fo tunately these effects can be compensated by the use of gamm correction circuits.

Normally one aims at unity gamma, to avoid color distortio (hue shift) that would otherwise occur. Gamma adjustment m be introduced though, to compensate for otherwise uncorrectab errors when reproducing color film—'TARIF'; or to compensa for extreme tonal contrasts in ENG/EFP (electronic news gathe ing/electronic field production) presentations. In monochrom systems, gamma is sometimes adjusted to deliberately modi tonal quality.

Picture noise increases with gamma correction.

■ *Color balance* Preset adjustment of the video gain and bla level of each camera's red, green, blue channels, enables the vid engineer to ensure color fidelity throughout the tonal range. Th avoids, for example, bluish or greenish blacks, or pinkish white These controls may also be adjusted to improve intercamera col matching, to compensate for light quality changes during a remo telecast (color temperature variations), or for deliberate effect.

Artistic aspects of video control

You can make a picture more attractive or effective by appropria video adjustments:

■ *Surface brightness* This often alters with lighting angle ar camera viewpoint. So in multicamera shooting, changes effective tonal values can arise. (Improved by exposure or bla level adjustments).

■ *Subject tones* Where these exceed the system's limit judicious adjustments can disguise these limitations. For instanc opening the aperture when we need to see shadow detail or r ducing exposure when an overbright area would be intrusive.

■ *Overall lighting* Flat lighting produces pictorially uninterest-ing pictures and skilled lighting directors avoid it. Although un-even lighting can lead to certain problems, such as accidental shadowing, or a performer moving out of his key light, video adjustments can often alleviate matters by compensatory exposure.

■ *Difficult detail* Where we want to see details in very light or very dark-toned subjects, appropriate video adjustments show them more clearly. For example when looking at a book page, reduced exposure and lower black level improve clarity. Otherwise the page might well block out and appear blank!

■ *Subject brightness changes* Our impression of a subject's brightness varies with the amount of light falling upon it, and with background tones—looking lighter against dark backgrounds and darker against light tones. So in intercut shots we could become overaware of subject brightness changes. Again, judicious control can compensate for this subjective effect.

■ *Long and close shots* Distant shots generally appear more dynamic if they have strong tonal contrasts, while closer shots are coarsened by contrast. When a series of viewpoints are intercut in a continuous scene, it can prove more practical to simulate these changes by subtle adjustments to black level (setting down long shots), than to adjust lighting.

Table 21.1 Inherent picture defects

Line beating (strobing)	Localized flickering regularly encountered with close horizontal stripe patterns in clothing, fabrics (checks, stripes), line engravings, close mesh, etc. Can be reduced by altering the shot size or by slight defocusing. Effect is due to subject lines coinciding with picture scanning lines. Interference effects (cross-color) can arise too, producing a color fringing effect on fine stripes (Table 21.3).
Stroboscopic effects	When the movement rate of rotating wheels or pulsating light, coincide with the TV picture scanning rate of 60 per sec USA (50 per sec Europe) they will appear stationary. Slightly faster or slower speeds can cause wheels to apparently rotate forwards or backwards.
Interlace break-up	When viewed closely, a TV picture's interlaced line structure breaks up as we look up or down the frame. Line structure becomes coarser and vertical definition halves. Evident on fast crawl titles and extensive tilting ('panning-up') shots.
Horizontal break-up	During panning or fast horizontal movement, strong vertical lines or detail breaks up into displaced sections. Due to the inherent time lag between scanning the odd and even fields of the picture (1/60 or 1/50 sec).
Highlight flicker	Sometimes visible on very bright highlights, particularly when viewed in dark surroundings; although the picture repetition rate is too fast to be detected under normal conditions.

■ *Some additional adjustments*   Video control can enhance the pictorial impact considerably in various ways:
Controlled overexposure can be used to simulate dazzling sunshine. *Underexposure* in night scenes prevents overbright highlights. *Increased tonal contrast* emphasizes harshness (squalid environments). *Reduced tonal contrast* to aid high-key treatment, misty effects. Deliberate *white-crushing* to obtain even-toned white areas; deliberate *black-crushing* (the most regular technique) to merge darkest tones to a solid black—consolidating background tone for black backgrounds to title cards, or for black drapes.

■ *Color values*   These can be modified to correct *subjective effects*—due to background hues influencing apparent subject colors. *Deliberate color bias* can also be introduced for such effects as moonlight, firelight and gaslight.

The television camera tube

Although the electronic intricacies of your TV camera tubes are really only the concern of the video engineer, their performance can have a direct influence on staging practices and production opportunities. Some of the shortcomings are inherent; others aggravated or emphasized by the way you use them.

As technology has advanced, new types of camera tube have evolved. The excellent Image Orthicon (IO) pick-up tube that was the mainstay of monochrome TV for many years, became obsolete due to the technical difficulties in applying it to color systems. Smaller, simpler camera tubes developed, each with particular merits.

■ *The Vidicon*   (Antimony trisulphide target). This is a cheap, reliable, long-lived, simply adjusted camera tube that is widely used in closed-circuit TV systems (CCTV) as well as in film and caption (slide) scanners. Its low sensitivity (needing high light levels) and *lag* (image smearing on movement) restrict its applications. Its spurious background shading can also prove troublesome.

■ *The lead oxide camera tube*   (Plumbicon—Philips; Leddicon —EEC; Vistacon—RCA). This is currently the major pick-up tube in color TV cameras; offering a higher sensitivity than the Vidicon, a more constant tonal response (unity gamma) and low spurious signals (less shading). Its response at the red end of the spectrum was poor (now improved in 'extended red' versions) and picture noise was 'high in the reds'. It has a tendency to lag, although less than the vidicon and is considerably improved by *light bias*—a technique where the camera tube target is internally illuminated. The tube overloads to create *blooming (puddling)*—

grossly enlarged white blocked-off areas in overbright regions (skies) and disturbing *comet tails* behind moving highlights. Later tube designs, using highlight overload protection (HOP) or anti comet tail (ACT) guns, have largely overcome this problem. *Separate target mesh tubes* have improved detail resolution, high-lights and geometry.

■ *The Saticon* (Hitachi/NHK; RCA). An extremely promising camera tube, with a target layer of selenium arsenic tellurium which is more stable and long-lived than the lead oxide target. It provides higher definition (resolution) and has a more uniform spectral response. Its performance in 18 mm form has proved comparable with larger cameras using 25 mm plumbicons. Image lag is noticeably better than vidicons; with a white comet tail rather than the usual red or green.

■ *Further tube designs* These brief technicalities do remind us that the electronic camera is still evolving. Under optimum conditions of suitable light level and contrast, results can be

Table 21.2   Basic electronic adjustments—line-up

	While the camera shoots special geometric charts and tonal step wedges, picture and waveform monitors are scrutinized to check the following:
Electronic focus	Focuses the camera tube's scanning beam for sharpest picture detail.
Beam current	Controls the scanning beam strength. This is set to prevent lightest picture tones blocking off to detailless white areas (i.e. to discharge target's white elements). Too high a beam loses picture clarity, giving excess video noise. Too low a setting and no video is generated.
Target volts	Set for optimum tonal contrast. If too low, camera needs more light (larger *f*-stops or higher illumination) to compensate for weaker video—but image lag (trailing) is less. In the *vidicon*, target voltage adjustment is used as manual or automatic sensitivity control—instead of continual aperture adjustment.
Black level (sit)	Adjustment relating darkest picture tones to system's video level representing black. May be automatically and/or manually maintained. Virtually moves all reproduced picture tones up or down the tonal scale.
Video gain	Ensures video amplification similar for each camera channel (1 volt overall).
Alignment	Camera tube beam's scan pattern adjustment (geometry).
Scan width and height	Adjusts scan limits on camera tube and obtains 4:3 aspect ratio.
Vertical/horizontal shifts	Enables tube's scanned area to be centralized.
High peaker/aperture correction	Circuitry enhancing apparent picture sharpness by emphasizing higher video frequencies. (But increases video noise.)
Neg/pos switch	Picture tones (hues) as in scene, or 'phase-reversed' for negative effect.
Local/remote switch	Switches lift, black level and target controls from CCU to remote control point.

superb. When conditions are less than ideal, the program materia
may be important enough to disregard even appallingly ba
picture quality (under ENG conditions). But we always need t
be aware of the final image quality.

Further camera tube designs have emerged including th
Chalnicon, Pasecon, Newvicon and Hi Sensicon, silicon diod
array target, diode gun tube; all aimed at the portable and/or non
broadcast future.

Tubeless image conversion devices such as *charge-couple
devices* (CCD) or *silicon diode arrays* are a promise for vide
cameras of the future. The charge-coupled device is a solid-stat
image sensor that uses a small silicon chip containing thousands c
individual elements, each independently charged by a part of th
lens image. This postage-stamp device is rugged, consumes lo
power and is potentially cheap.

Synchronization

When operating most switchers, you have to consider whether o
not an incoming source is *synchronous*. If it is, you can mi
superimpose, or electronically insert pictures. If it is not (*non
sync*), you may only be able to switch (cut) them. In a colo
system, they may be technically unacceptable.

As you saw in Chapter 2, the TV scanning process is kept i
step from camera to picture tube, by specially generated *synchron
izing (timing) pulses*.

Although some monochrome studios employ an inexac
random interlace system (mainly closed-circuit industrial users)
broadcast quality requires precision synchronizing equipmen
(sync. pulse generators, SPGs). The synchronizing pulses are fe
to the entire studio or station video equipment. Therefore, a
these video sources are *synchronous* (scanning in unison), an
your switcher can accept any one (film island, caption machine
camera) and intercut, mix, or superimpose it with others.

External video sources (remotes) on the other hand, operat
from their own synchronizing generators. Although these ar
producing identical types of pulses, they will not be exactly i
step with the stations, pulses (*out of phase*). If you feed such
non-sync source into most production switchers, you will obtai
totally unacceptable pictures (having displacement, tearing, rol
extreme color errors).

Where the production switcher design permits, you can *cut* t
non-sync sources (*hot cuts*), but still cannot mix or combine thei
pictures with other local video sources. Only certain switche
designs (digital systems) allow you to combine 'sync' and 'non
syn' sources.

■ *Synchronizing systems*   To overcome this dilemma severa
electronic processes have been devised, which precisely match an
synchronize the local and remote pulses:

1. *Genlock*—Adjusts timing of local sync pulses into step with those of a remote non-sync source.
2. *Natlock/slavelock*—Compares the timing of local sync pulses with those from a remote source and sends corrective coded signals to adjust remote.
3. *Superlock*—A quick-genlock system, syncing a local source (e.g. local caption machine) to a remote.
4. *Rubidium frequency standards*—Equipment producing highly stable sync pulses. Using this, the remote source can be adjusted to suit studio timing and remain in step for long periods.

Table 21.3   Spurious visual effects

Certain picture effects are created or aggravated by unsuitable staging, lighting, and video operation; while others are inherent.

Picture noise ('snow')	Overall fluctuating grain due to random electronic disturbances. Always present in the TV picture; it becomes particularly noticeable in dark scenes. At worst, develops into a 'snowstorm' overall.
Background	Blemish (spot, patch) on a particular camera's shot. Due to target damage or deterioration. May be a low-grade camera tube.
Burn-in	Spurious after-image of light or bright reflection, temporarily (sometimes permanently) 'stuck on' to all of a camera's subsequent shots. Vidicon and plumbicon tubes are *not immune*, so avoid sustained shots of strong highlights.
Shading	Localized lightening/darkening or color shading of parts of a camera's picture (electronic or optical causes). Most noticeable on plain light-toned surfaces. Reduced by corrective electronic shading signals.
Geometric distortion (non-linearity)	Cramping, stretching or bending of picture due to electronic (scanning/sweep circuits) or optical shortcomings. Includes S-curvature, trapezoidal, barrel and pincushion distortions.
Edge effect	With plumbicon and vidicon tubes, circuits compensating for definition fall-off (*aperture correction*) can produce false black edging round high-contrast borders—even crispening distant detail that would naturally be soft focus. Does not sharpen low-contrast borders.
Exposure time (lag)	Smearing or trailing seen behind movement, particularly on high-contrast subjects in low-key scenes. Evident on vidicon film channels during night scenes. In color systems, colored trailing or hue changes may be seen.
Streaking	Black or white streaks extending across the picture beside high-contrast subjects (stair treads, venetian blinds). Usually due to overload or low frequency losses in video channel.
Cross-luminance	Bands of crawling dots seen at sharp color transitions. This is due to chrominance information intruding into luminance signal.
Cross-color	A 'rainbow' interference effect seen on fine stripe or check patterns caused by luminance information intruding into chrominance signal.
Blooming (puddling)	Camera tube overload causing overbright areas to be reproduced as grossly enlarged areas of blocked-off white. Reduced by tube design: anticomet tail gun (ACT), highlight overload protection (HOP), or automatic beam optimizer circuits dynamic beam control circuits.

5. *Time-base correctors* (TBC)—Equipment using analogue or digital techniques (Chapter 12 and 20) to provide automatic compensation for synchronizing inaccuracies on VTR replay.

6. *Field and framestore synchronizers*—Digital memory circuits allowing video information to be written into the store at any rate, and read out at a stable synchronous speed (Chapter 20).

7. *Standard converters*—Used primarily to convert TV signals to another standard (e.g. NTSC to PAL); may also be used to ensure source synchronism.

■ *Composite signals—and switcher operation* A *composite signal* consists of the tonal video (picture) information together with inserted sync pulses at the end of each scanned line and field.

Table 21.4 Picture monitor checks

	Picture quality can only be judged on correctly adjusted, high-grade monitors, without spill light or reflections diluting the picture. Here are general methods of checking performance. Allow time for warm-up before adjustments.
Gray scale (all picture monitors)	Feed a *gray scale* (from camera, slide, or electronic generator) to all monitors. Adjust BRIGHTNESS (BRILLIANCE) to ensure *black* merges with blank screen-face. Adjust CONTRAST for maximum white brightness without line defocusing and good half-tone reproduction. Ensure all monitors' pictures are similar. Replace gray scale (tonal step wedge) with a monochrome picture containing good tonal variations. Compare all monitors' gradations and highlight brightness.
Color monitors	As above. Examine gray scale reproduction for traces of color cast throughout tonal range. They should be neutral grays (good color tracking). Circuit adjustments may be necessary. Recheck using same color picture on all monitors. The picture should contain facial tones and pastel hues (desaturated and not primary colors, or color bars).
Picture size/shape	Check that the screens show *all* of the TV picture (no edge cut-off) in correct 4 by 3 proportions.
Focus	Scanning lines should be sharp over entire picture, reproducing maximum detail on camera test chart.
Linearity	Picture should not show serious geometrical distortions (cramping, stretching). Check linearity with cross-hatch (grille, square lattice) generator—this produces thin white lattice on black background.
Color defects	Check reproduction of cross-hatch for color fringing (*convergence errors*) along pattern. In practice some spurious color edges may be unavoidable. Check overall for: *color balance* (warm or cool?), *desaturated color* (washed-out, low contrast?), *oversaturated color* (overstrong color, high contrast?), hue shift (predominant overall color), and color shading (*purity* errors).
Black level	Picture *black* should be constant; not varying with picture tones, or reproducing as gray in low-key scenes or when faded to black. (Variations are due to equipment design or fault condition.)
Off-air pictures	Probably appear noisier and less sharp than direct off-line pictures. They may have displaced multiple images (ghosts, reflections) and ringing, due to multipath reception of transmission; and/or interference patterns (herringbone, stripes, edge tearing).

Depending on local facility design, the video system may be supplied with either a full composite signal or a *non-composite* version (video information only, no sync pulses).

Table 21.5 Video engineering terms

Synchronizing pulses	Sync pulses from the master control sync pulse generator (SPG) keep the entire TV system in step—from camera to receiver (preventing picture tearing, roll, etc.).
Line (horizontal) drive pulses	Synchronize line scanning rate.
Field (vertical) drive pulses	Synchronize field scanning rate. (Twice each total picture or frame).
Mixed syncs	The total, combined sync pulse information used to synchronize the system.
Blanking	Special pulses used to suppress video output from the camera, to enable sync pulses to be inserted. The scanning process retraces to the beginning of the line or field during blanking.
Timing	Video signals take measurable time to travel along distribution cables inside the studio complex. If such cable-connected sources (e.g. film, VTR) at different distances from studios/master control are not suitably compensated, their pictures arrive with slight horizontal displacement or cut-off. (Important consideration for multisource picture combinations.)
Color information	
Color subcarrier	A constant carrier frequency that is modulated to encode information.
Reference color burst	A brief burst of the above subcarrier, introduced into the synchronizing information to stabilize color reproduction.
Coder (encoder)	Circuitry (matrix) converting a camera's red, green and blue video signals into a single luminance/chrominance encoded form (NTSC, PAL, or SECAM) for distribution and transmission.
Decoder	Circuitry (matrix) converting the coded video back into separate RGB video to drive the receiver or monitor circuits.
Equipment	
Distribution amplifier (DA)	Video amplifier boosting signal strength prior to distribution, to compensate for subsequent losses.
Processing amplifier/stabilizing amplifier	Corrects errors developing in the video signal. Used to reshape, reinsert, or separate sync pulses from the composite signal (i.e. video plus syncs).
Image enhancer (IA)	Circuitry used to improve apparent picture definition, reduce video noise.
Termination	To match a picture monitor to its video cable, a termination resistance is introduced at its video input. This avoids signal losses, distortions and reflections. Only *one* termination is required for a succession of looped-thro monitors.
Color bar generator	An electronically generated color test pattern, comprising vertical bars displaying white, yellow, cyan, green, magenta, red, blue, and black. Used for video system checks.
Standards converter	Electronic equipment used to enable pictures televised by one TV system (e.g. 525 line NTSC—USA, JAPAN), to be reproduced by another incompatible system (e.g. 625 line PAL or SECAM).

Certain production switchers are self-compensatory, accepting composite and non-composite inputs. Others only permit you to *fade to black* by sliding split A-B faders to out, when controlling *non*-composite video. If this fade-out is attempted with *composite* signals, the switcher loses color. In such designs, you must select a *black-level button*, and mix to this 'non-channel' to simulate a fade-out.

Some switcher designs present severe video overloads when peak-white or light-toned areas are combined through mixing or superimposition. If your switcher does not incorporate protective circuits (processing amplifiers), you must avoid fading up channels fully in such circumstances.

22 Glossary

■ **A & B rolls** Method of assembling recorded material (film or videotape) on two machines running concurrently in sync, to permit fast interselection or combination optical effects. In *A-B film cutting* shots are joined in sequence until a combination shot is required (mixes, supers, inserts), when the additional shots are assembled on the second roll. Opaque blanking is included in each roll, where only the alternative roll's shot is being used. A *checkerboard method* is widely used to provide invisible editing splices between shots (as well as combined transitions) on narrow-gauge films. Here odd number shots are assembled on 'A' roll, and even shots on 'B' roll. The film printer alternately selects between them, the joins being hidden by the opaque leader.

■ **Acoustics (see Fig. 10.5)** Higher Frequency sound waves travel in straight-line paths. So they are easily deflected and reflected by hard surfaces; and are absorbed by porous fibrous materials. Lower frequency sound waves (below 100 Hz) spread widely, so are not impeded by obstacles and are less readily absorbed.

As sound waves meet nearby materials they are selectively absorbed and reflected; the reflected sound's quality being modified according to the surfaces' nature, structures and shapes. Studios are carefully proportioned and their walls covered with suitable absorbent materials (fiber panels, rockwool, mineral wool, seaweed quilting) to prevent unwanted sound coloration and to reduce reverberation (small TV studios 0·3–0·5 sec; larger TV studios 0·7–1 sec). The acoustic characteristics of scenery, furniture, drapes, people, further modify audio quality.

■ **Animation** Creating an illusion of movement—by intercutting stills, using graphics with movable sections, using step-by-step changes in a static subject (objects, models, drawings, photographs, etc.), or control wire activation.

■ **Aperture correction** Electronic correction to compensate for the loss of definition due to inherent spread of the scanning beam in the camera tube.

■ **Audio control (see Figs. 10.21—10.23)** All audio sources (mikes, tape and disc reproducers, etc.) are selectively plugged into a *patch panel*, where each circuit is linked to the *audio control board* or *mixer*. There *input* or *channel switches* select the sources to be controlled. Each audio channel includes a *pre-amplifier* (*pre-amp*) to boost weaker sources and/or *attenuators* (*pads*) to reduce strong audio signals, so matching their relative levels (volumes). Comparisons are made during preliminary *level tests*. The audio signal in each channel is further amplified by a *program* or *channel amplifier*, the audio gain of which is adjusted by a channel fader of 'pot' (potentiometer). Both knob and slider controls are used. (The latter can be simultaneously operated by several fingers, their relative positions being easily discernible.) Channel faders can be used to *fade* individual sources in/out ('pot in/out') as well as *control* their average level. ('Keying-in' by *switching* to a preset pot setting is possible, but program material may be lost by late cuts—'upcutting'.) The relative strengths of several channels (now balanced for each other) can be joined to a *group fader* for combined operation (e.g. grouping various singers' mikes). The outputs of all group faders are controlled by a *master pot* (*line or main-channel control*). *VU* (*volume unit*) *meters* show the signal strength at control points, enabling audio to be maintained within system limits.

■ **Automatic scan tracking (AST)**   System for ensuring that the replay head accurately retraces the recorded track when replaying an helical recording, so avoiding disturbed synchronism. Video playback head position is automatically adjusted as recorded track is sampled for maximum signal. Such *dynamic tracking* also allows full color at varying tape speeds.

■ **Backing track**   A prerecorded track providing musical accompaniment for a performer who, listening on earphones (or a playback loudspeaker), is recorded on his individual mike.

■ **Banding**   Occurs on quad or segmented-head recorders, when a series of wide horizontal bands is seen if head responses are uneven or not automatically compensated.

■ **Bands**   In a videotape recorder, an FM carrier is generated which is then modulated with the video signal. It is this modulated carrier that is recorded on the tape; the video component itself being recovered during replay. However, spurious interference effects can arise in the process (close bands or stripes, noise effects) depending on the FM carrier-frequency band used. A *low band* is suitable only for monochrome. A *high band* is suitable for color, but liable to a fine moiré pattern in colored areas, especially red. *Super high band* considerably reduces the problem.

■ **Beeper**   An identification audio signal sometimes used at the start of a recorded section (videotape, audio tape, or disc band) to aid sound cuing.

■ **Bias current**   An ultrasonic current (e.g. 75–120 kHz) is combined with the audio in the record head; this effectively linearizes the tape's magnetic characteristic. Its strength is adjusted to suit particular tape materials. The bias setting is a compromise between distortion and high note response.

■ **Blimp**   Sound proofing cover for a film camera which prevents camera noise from being picked up by a nearby mike.

■ **Bridge** Words or music introduced to tie together two dissimilar items. *Also* the shorting together of RGB channels on a picture monitor to produce a black-and-white picture.

■ **Bulk eraser** A large electromagnet which demagnetizes an entire tape spool, ensuring that any previously recorded program is completely wiped.

■ **Cans** Earphones; headphones.

■ **CATV** Community antenna TV. The TV signal is distributed over a given area by coaxial cable.

■ **CCIR** Comité Consultatif des Radio Communications (International radio consultative committee).

■ **CCTV** Closed circuit television. A localized non-broadcast TV system for selective viewing.

■ **Cinching (windowing)** Slippage between tape layers on a loosely wound spool, causing physical distortion and even tape damage.

■ **Coding** Electronic process of encoding RGB video signals into an NTSC or other color system. (*Decoder*: circuitry for transforming the encoded signal into its RGB components.)

■ **Color bars** Video test signal presenting a series of vertical bars of fully saturated color: white, yellow, cyan, green, magenta, red, blue, black.

■ **Color characteristics:**
Achromatic values (gray scale)—progressive scale of brightnesses, from black through grays to white.

Brightness—the subjective quantity of light received from a subject. Often used to denote 'luminosity' (US).

Hue—the predominant sensation of color; e.g. red, orange, yellow.

Lightness—perceived brightness of surface colors.

Luminance—the true brightness of a surface. Snow has high luminance, black velvet a low luminance.

Luminosity—perceived brightness of light sources.

Saturation (chroma, intensity, purity)—the extent to which a color has been diluted with white. The pure undiluted hue is fully saturated (100% saturation), e.g. red. Dilution pales the color, e.g. red desaturates to pink.

Shade—a hue mixed with black.

Tint—a hue diluted with white.

Tone—a grayed white.

Value—in the Munsell system, indicates subjective 'brightness'.

■ **Color definition**   The eye cannot readily detect fine color detail. Therefore, to save bandwidth in color TV systems, full detail is transmitted in *luminance* only—4·2 MHz; i.e. 320 lines horizontal resolution—NTSC 525/60 (5·5 MHz. PAL. 625/50). Color information is deliberately restricted: I bandwidth (orange to cyan) 1·6 MHz; Q (green to magenta) bandwidth 0·8 MHz. The effect is of lower definition color 'overprinted' with tonal detail. Colors are transmitted in the proportions 30% red, 59% green, 11% blue, to produce a suitable monochrome tonal balance.

■ **Color temperature**   Defined by reference to a black-body radiator. If such a non-reflective body is heated, the spectral distribution of the resultant light changes with its physical temperature (0 Kelvin = 272°C). This figure gives a good guide to the color quality of light sources. A tungsten halogen lamp (quartz lamp) burns with a color temperature of around 2600–3500 K.

■ **Color under**   The slower tape speed of a small gauge VTR, restricts the maximum information density that can be recorded. So, instead of applying the full encoded signal (e.g. NTSC) direct to the VTR's modulator for recording, its color (chrominance) and brightness/detail (luminance) components have to be separated. This involves *heterodyning* the color signal, converting it down to a lower carrier frequency (hence 'color under'). On replay, filters distinguish between the chrominance and luminance information, but detail is lost in the process (above about 2 MHz). Also uwanted spurious effects arise, such as moiré, edge sparkle, color displacement, detail loss ('cartooning'), and ringing (edge repeats). See *Image enhancement*.

■ **Compatible**   When video or audio signals are generated by one system or piece of equipment and reproduced effectively by another, the second is said to be *compatible*.

■ **Crossed gray scale (chipchart)**   A standard test graphic comprising two opposed horizontal nine-step tonal gray scales (from 3% to 60% reflectance).

■ **Crossing the line**   Discontinuity or position-jumps in the frame, caused by intercutting viewpoints on either side of the 'imaginary line' ('action line').

■ **Cut-back** During editing, returning to the main scene after a series of interpolating close shots.

■ **Date** A system of sound distribution used by PBS (Public Broadcast Service) network, in which audio is digitally encoded for distribution and decoded at the local receiving station. Usable for stereo simulcasts (simultaneous TV and high-fidelity FM stereo radio transmission), the system can relay four audio channels in addition to the mono TV sound.

■ **Decibel (dB)** A logarithmic unit indicating ratios of powers, voltages or currents, and used to express transmission levels, gains and losses. The human ear (like the eye) does not respond proportionally to changes, but follows a 'power law'. A decibel represents the smallest perceptible change in audio level.

■ **Degaussing** Demagnetizing. Tape recording heads must be periodically degaussed to prevent magnetic build-up from increasing tape hiss. Video tape is usually bulk erased before reuse.

■ **Dope sheet** Film cameraman's sheet giving basic story information, location, exposure data, footage, instructions.

■ **Drop-out** Magnetic tape surface defect or damage causing momentary disturbance of reproduced video, resulting in small localized information gaps in a line (black flash, color loss). To disguise this brief loss of FM signal pick-up from the tape, a *drop-out compensator* can insert stored picture information from a corresponding part of the previous line. (This term is also used for the information loss, *noise band* at end of line/frame scanning in helical recording.)

■ **Dubbing sheet** A chart prepared by the film editor for the *dubbing mixer* (sound recordist) showing the exact relationships of the various contributory sound sources in the final mixed track.

■ **Dupe neg** A duplicate negative prepared from a positive print—when the original negative is not available, or to protect the original negative, or for special effects work. By using *color reversal intermediate film*, duplicate negatives of the original color negative can be prepared, so eliminating one printing stage (reduced grain, improved color reproduction and sharpness).

■ **EIAJ** Electronic Industries Association of Japan.

■ **Establish, to** To provide an introductory shot or sound image, to set the scene.

■ **Exposure curve (transfer characteristic)** Only where a

picture system's output is directly proportional to the ligh
input (unity gamma) will tonal reproduction be strictly accurate
Most systems have a linear (straight line) portion to their charac
teristics, but are less responsive to changes at exposure extreme
Lightest tones are therefore progressively compressed at the uppe
end of the exposure curve (knee, shoulder), while shadow deta
merges at the lower end (toe).

■ **FCC**   Federal Communications Commission. USA goverr
mental agency concerned with policy making, and transmitte
technical standards for radio and TV broadcasting.

■ **Flare**   Light shining directly into the lens causing overa
veiling (reducing image contrast), spurious light patches, c
ghost images. Most evident in shadow areas of the picture.

■ **Flash cutting**   A sequence structured from a series of ver
brief shots, each lasting only a few frames.

■ **Fluid head**   Camera panning head using layers of high vis
cosity silicon fluid to dampen head movement. (Adjustable, ver
smooth action.)

■ **Flying head**   In a standard helical scan recorder a fixe
erase head wipes the tape width, so erasing parts of *several* track
simultaneously. It cannot erase one complete track at a time, s
precludes precision editing. To overcome this problem, recorder
designed for *insert editing* have an auxiliary *flying erase hea*
fitted just before the record head (on the head wheel). A *flyin*
reproduce head provides simultaneous confidence playback durin
recording.

■ **Focal length**   The distance from the camera-tube target t
the lens' optical center when the lens is focused at infinity–fa
distance.

■ **Forced processing**   When a film is developed at a highe
temperature or for a longer period than normal, its effective *spee*
(sensitivity) is increased; although contrast, graininess and fo
are emphasized. Reversal films 'push' more successfully tha
negative film material. This technique is resorted to when ligh
levels are too low for normal exposure.

■ **Friction head**   Camera panning head relying on surfac
friction (stiction) of parts to control pan/tilt action. Suffers fron
excess friction at start of movement.

■ **Gamma**   Gamma is a logarithmic measurement of repro
duced tonal contrast, relating light input to the resulting pictur
density. In *photography*, gamma defines the tangent of the maxi

mum slope (straight portion) of the processed emulsion's characteristic curve. In *television*, gamma measurement relates light input to voltage output, or voltage input to light output. The effects of a series of processes combine (cascade). Thus a camera tube having a low gamma (vidicon 0·65) relates to a color picture tube (around 2·2) to produce an overall gamma of 1·43 (0·65 × 2·2), which can be electronically corrected (0·7) for overall unity (1·43 × 0·7) gamma.

■ **Gate**   A metal masking plate with a rectangular hole, against which the film is held stationary (during exposure in the camera to form individual *frames*, or while being projected). Unsteadiness of the image during projection (*gate float*) may be discernible as horizontal weave, or vertical judder (*hop*). This movement, together with extraneous blemishes (sparkle, dirt, 'hair in the gate') often reveals that a televised image originates from film rather than a video camera.

■ **Gearing head**   Camera panning head incorporating gearing to control head movement. Used for large film cameras.

■ **Glitch**   General term for any brief electrical disturbance in a picture.

■ **Highlight overload protection (HOP)/Anti comet-tail gun (ACT)**   Special camera tube design providing enriched scanning beam density (during line blanking periods) to discharge extreme highlights on the target. This prevents blocking off and edge overspill of charges from bright areas, increases contrast handling capabilities; and overcomes spurious 'comet tails' ('blooming') behind moving camera tube's light transfer characteristic.) *Automatic beam optimizer* circuitry has similar functions.

■ **Hyperfocal distance**   When the lens is focused at infinity (∞), focus remains sharp from infinity to a plane near the camera. When focused at this *hyperfocal distance* (H) the scene is sharp from about distance $\frac{1}{2}$H to infinity.

$$\text{H in meters} = \frac{(\text{focal length in cm})^2 \times 100}{\text{lens } f/\text{stop number} \times 0.05}$$

■ **Image enhancement**   Electronic correction and improvement of video signals. This includes processes such as crispening edge sharpness (high frequency synthesis), edge color fringe removal, color displacement correction (chroma/luma delay), reduced noise ('snow remover'), suppressing 'ringing' at edges, and improved small image detail.

■ **Imaginary line (action line or axis, center line, vector optical axis)**   An audience's impressions of space and subject positions within a scene come from various visual clues appearing in the picture. The director must take care not to confuse the concepts by using incompatible viewpoints, avoiding puzzli repositioning or exits.

If you switch across an *imaginary line* established by the actio (direction of look and/or movement of performers), disconcertin position changes (jump cuts) and loss of orientation can occu This is a particular hazard in discontinuous or multicame shooting. No problems arise when dollying across the lin changing the line's direction by regrouping or moving subjects, when altering eye-lines. But you can seldom *cut* across it unol trusively. See also Fig. 15.11.

■ **In vision**   Visible in the picture (in shot); as opposed to 'o' of shot' (out of vision, off camera).

■ **ITFS**   Instructional television fixed service. Transmission o 2500 MHz bad for PTV (Public Television) using short-ran, closed-circuit television transmission.

■ **ITV**   Instructional TV. Program material produced f formal educational systems.

■ **Jump-cut**   Any cut causing a positional change in a subje on the screen.

■ **Lens angle (angle of view)**   The coverage of a lens. Tl horizontal and vertical angles within which the scene is visib The angle from the lens to the left and right borders of the field known as its *horizontal angle*; this is used both as a gener reference or lens identification, and to show shot-width coverag and proportions on scale *plans*. The corresponding angle the le makes to the upper and lower borders of the picture, is its *vertic angle*. This is three-quarters of the horizontal lens angle; due the TV screen's 4 by 3 proportions (format, aspect ratio). Tl vertical angle shows vertical coverage on scale scenic *elevation* anticipating overshoot and showing necessary scenic heights.

■ **Light bias**   'Lag' creates movement-blur resulting in a ectoplasmic smearing (trailing, beam lag) and color fring behind moving high-contrast subjects under low light condition Largely obviated by internally illuminating the camera tul target, e.g. with miniature pea lamps. A similar lag effect ca occur when a camera tube is slow to charge (build-up) durir movement from low tones to a bright surface area.

■ **Lighting measurement:**

Lumen—a unit of light output (luminous flux).

Lux, foot candle—units of incident light intensity falling on surface (illumination).

430

Nit, foot lambert—units of surface brightness; reflected light intensity (luminance).

Nanometer, micron, angstrom—all units specifying wavelengths of light (e.g. identifying spectral colors) and other electromagnetic radiation.

Kelvin—a unit of color temperature.

■ **Lip flap** Lips seen moving before accompanying audio is heard. The result of editing a combined film (picture with displaced sound) in mid-speech.

■ **Liquid gate (wet printing)** Printing system in which the film passes through a liquid at the point of exposure. This system is used mainly for printing from badly scratched originals and gives a final print which is virtually free from blemishes.

■ **Longitudinal video recording (LVR)** System recording 48 video and 96 audio tracks bidirectionally on 8 mm wide videotape moving at 400 cm/s (160 ips) past fixed heads, changing direction and switching tracks every $2\frac{1}{2}$ min.

■ **Matte box** A bellows or box mounted in front of the camera lens. It holds camera mattes (masks), gelatine filters, as well as providing an efficient lens shade (sunshade hood).

■ **MATV (master antenna system)** Distribution system derived from a single antenna, to service any area from a single building to an entire community.

■ **Microprocessors** Miniature computers, using logic circuits with programmed memory instructions. Used as 'intelligent interfaces' to automatically calculate and time VT machine operations—thereby synchronizing switching, assessing machine run-up time, energizing appropriate functions in correct sequence at the right instant.

Advanced videotape editing facilities use microprocessors to provide: random access, random recall, rehearse edit, log edit decisions (store edit-in/edit-out points), make frame accurate edits, trim edit points, shift edit points, autoedit from time code data, autosearch to cuepoint, interswitch between two machines (A–B switching), still frame (freeze frame), animate (record a frame at a time), jog frame (jogging, frame-by-frame search), high-speed shuttle, slow motion, shift audio.

■ **Mired (microreciprocal degrees)** Unit derived to classify a filter, or the color temperature of a light source. It enables easy calculation of the color shift required to modify a light source's color quality. Mired value equals one million divided by the Kelvin value e.g. 2500 K = 400 mired.

■ **Multivision (multiscreen)** A presentation using an assembly of adjoining screens to form a large composite single or multiimage display. Sometimes indicates several images appearing simultaneously in the same frame.

■ **Munsell color scale** System of notation defining color through a set of charts showing *hue* (5 principal and 5 intermediate hues), *value* (brilliance) in a scale of 10 steps from black to white, and *chroma* (saturation).

■ **Narration** Voice-over (explanatory or commentary). Nonsynchronous speech accompanying a picture.

■ **Nemo** Remote pick-up.

■ **NET** National Educational Television. Division of Educational Broadcasting Corporation (EBC).

■ **One and a half heads (1·5 heads) principle** A separate head records the vertical interval information near the lower edge of the tape, thus retaining VIRS, VITS, teletext.

■ **Optical effect** A filmed visual effect, created in an optical printer.

■ **Optical printer** A special film-printing machine in which a filmed image is projected onto the film stock rather than printed by contact with it. The projected image can be independently exposed to provide optical transitions, superimpositions, size changes, freeze frame, step printing, speed and direction changes, skip frame, multiimage, traveling matte, image position changes, image distortions, etc.

■ **Out-take** A shot or scene rejected during editing.

■ **Overcranking** Filming at a higher speed than normal, to produce a slow motion effect when projected at normal speed (24 fps).

■ **Overlapping sound** An introductory editing treatment used when flat cuts join two dissimilar scenes. The sounds of the next scene are heard before the vision switches.

■ **Overrun** To exceed a prearranged duration.

■ **Overscanning** Picture monitor adjusted so that the picture edges are lost beyond the picture tube surround mask.

■ **Padding (fill)** Improvised action/dialogue introduced when a program is underrunning, or *stretching* available material to fill the alloted time.

■ **PBS** Public Broadcasting Service. Network for public television supported by the Corporation for Public Broadcasting (CPB).

■ **Perspective distortion** Perspective effects strongly influence one's interpretation of space. Where the camera's lens angle and your viewing angle to the screen are comparable, you see 'natural perspective', showing proportions and spatial impressions similar to those of everyday life. When a *camera moves* towards a subject to obtain a larger image (i.e. a closer shot), perspective remains consistent, the proportions of subjects are compatible, their relative positions and spacing changing as you approach.

Zooming, on the other hand, simply alters the overall magnification of the lens image, so these natural effects do not result. Although by reducing the lens angle you can fill the screen with a distant subject—so suggesting that it is close—the resultant image is generally dissimilar from the impression you would actually get at a close viewpoint. Instead, subjects show characteristics you normally associate with *distance*: depth in the picture appears foreshortened, and there is often considerable overlapping. Sizes do not diminish naturally with distance. You interpret such incompatibilities as *spatial compression*. A *spatial exaggeration* effect arises when a very wide camera lens angle is used. (The illusion of natural perspective can be restored by changing your viewing distance proportionally to the lens angle used.)

■ **Picture quality** A subjective grading scale may be used to assess video or audio quality (CCIR/CCITT).

Quality	Grade	Impairment
Excellent	5	Imperceptible defects
Good	4	Perceptible, but not annoying
Fair	3	Slightly annoying
Poor	2	Annoying
Bad	1	Very annoying

■ **Pilot tone** VTR development in which a special frequency is recorded (1·5 times the color subcarrier) to provide constant monitoring and correction of the system's chrominance, as well as checking scanning synchronizing errors for improved time base correction.

■ **Pixillation** Jerky fast motion effect created by regularly omitting frames from film shot at normal speed. *Also* an animation effect obtained by conjoining a series of still frames.

433

■ **PLUGE** Electronically generated image used to adjust picture monitors to a uniform standard. Picture comprises a vertical white bar, an area of reference black, and two reference tones slightly higher and lower than the reference by 2·5 IEEE/IRE units (Institute of Electrical and Electronic Engineers). The monitor is adjusted so that the darker patch merges with reference background, the brighter patch still remaining visible.

■ **Post syncing** The process of recording sound (speech, sound effects) to synchronize with an existing picture—as when replacing original dialogue with a different language, closely following the original lip movements; or when fitting convincing new sound to a mute print.

■ **PTV** Public television (ETV, Educational TV). Non-commercial TV, citizens group TV.

■ **Simultaneous contrast (spatial induction)** A visual illusion in which the apparent tone, color and size of an area is modified by its background. A light-toned area appears lighter and larger against a dark background; a dark area even darker against a light one. Similar interaction arises in color, where an area may look cool (bluish) against a dark background, and warm (red/yellow) against light tones. A colored background may modify the appearance of the foreground subject, often biasing its color towards a complementary hue. Juxtapose red and blue, and the red will seem quite orange while the blue has a greenish tinge. A strongly colored subject may similarly influence a neutral background tone.

■ **Slides**

Standard 35 mm slide format—mount 50×50 mm (2×2 in) mount; aperture 36×24 mm; aspect ratio 3:2.
Superslide format—similar mount size; aperture 40×40 mm ($1·6 \times 1·6$ in).

■ **Slide scanner** An optical projector or flying-spot scanner used to televise transparencies. Plastic-mounted slides are fitted into pairs of drums, discs, or trays (typical capacities of 36–40). Selection may be sequential (instantaneous) or random (delayed) from the production switcher position. Scanners do not usually permit size and position adjustment of the image. Ingenious and comprehensive systems of slide storage and retrieval have been developed using computerized digital techniques. Storage capacities vary (several thousand maximum) and facilities may include filed descriptions/titles, the opportunity for 'browse' inspection of a simultaneous 64-slide display, segment-selection from stored slides, compression, frame-position adjustment, animation by inter-slide switching.

■ **Staircase (step wedge)** Video test signal generated to provide a series of distinct brightness steps from black to white.

■ **Standupper** Shot of a reporter addressing the camera at a location.

■ **Steadicam** A special body brace supporting an arm with a free-floating gimbel, to which a camera is fixed.

■ **Stringer** Local freelance cameraman or reporters.

■ **Supplementary lens** Power of supplementary lens in diopters $= \dfrac{1000}{\text{focal length in mm}}$ e.g. lens of 250 mm focal length, has power of 4 diopters.

■ **Synchronizer** A film-editing device enabling the synchronism of the picture and separate soundtrack(s) to be maintained as the sections are run forward and backward. It consists of two or more *gangs* (sprocket wheels) mounted on a common revolving shaft. Sound heads permit the replay of audio tracks, while the picture is visible on a viewer screen.

■ **Teletext (Oracle, Ceefax, Antiope)** Systems for the digital transmission of written information and diagrams during vertical interval period—for selective display of data on visual display units or home TV receiver.

■ **Television systems** All color TV pictures are distributed or transmitted in coded form (not as separate RGB signals).

NTSC (National Television System Committee) The three primary color video signals (RGB) are transformed by a coder into a *luminance* (brightness) signal conveying the tonal values, and two special *chrominance* or color difference signals (I and Q). I and Q signals are used to simultaneously modulate the chrominance subcarrier; its strength (amplitude) varying with a color's saturation at that moment, and its phase changing with the hue (quadrature modulation). The resultant complex transmitted wave therefore contains both chrominance and luminance information. Color errors can arise in the path from transmitter to receiver (phase errors). Corrective receiver circuits increasingly use *VIR* signals to regulate these errors. Overall bandwidth 4·2 MHz. Color subcarrier 3·58 MHz.

PAL (Phase alternating line) Basically a similar system, but one of the color difference signals is reversed in phase on alternate lines, so that slight color errors due to phase shift are averaged out. (Greater errors become visible as *Hanover bars*.) Overall bandwidth 5·5 MHz. Color subcarrier 4·43 MHz.

SECAM (Sequential color and memory) Similarly derived, but here the chrominance signals are transmitted one at a time sequentially, on alternate lines of each field to average out phase errors. (A delay line stores and reinserts color information).

All three systems can exhibit 'herringbone' (moiré) or 'crawling dot' patterns.

■ **Time base correction** Any disturbance of synchronizing pulses in a color system can cause such picture defects as jitter, frame roll (vertical), skew or hooking at the top of frame, scalloping at edges, hue and saturation errors. (Helical scan recorders are very susceptible to such irregularities.) Ingenious circuitry can reduce such scanning errors by using delay lines (as in analogue *time base correctors*, TBC), or by digital storage of a full frame of video information (as in *frame synchronizers*). Sophisticated digital correctors have overcome certain problems in video switching and remote source synchronizing, and in satellite transmission.

■ **Translator** Equipment used to extend the service area of a TV station, by receiving the transmitted signal, boosting it and retransmitting at increased power.

■ **Transmission numbers (T)** The *f*-stop number is based on physical aperture size, and disregards light transmission losses and reflections between lens elements. *T-stops* are based on the actual amount of light passed by the lens system. In practice, these systems are comparable.

■ **Traveling matte** A film printing process in which a matte or mask silhouette is derived from a subject image, to block out a selected area of another picture. This permits the subject (or other material) to be introduced into that area.

■ **Traveling shot** Any shots from a moving camera (e.g. tracking shot).

■ **UHF** Ultra high frequency. TV transmission band *above* Channel 13, 470–890 MHz. (UK bands IV, V.)

■ **Undercranking** Filming at a lower speed than normal to produce fast motion when projected at normal speed (24 fps), or to provide increased exposure on a static scene where light levels are low.

■ **Underscanning** TV picture monitor adjusted so that extreme edges of the picture are visible on the screen.

■ **Up-cut** Late switching causing loss of words or picture at the start of a film or VT insert.

■ **VHF** Very high frequency. TV transmission band covering Channels 2–13, 54–216 MHz. (UK bands I, II).

■ **Video/film animation** Two basic methods are used. 1, A vertical TV camera shoots animation table on which successive drawings (cels) are displayed on a horizontal light box. (With two-frame animation, cels are changed after every other exposure.) These are recorded frame-by-frame on a VTR or video disc. 2, Another system projects the recorded video picture of live action (a frame at a time) onto a glass tracing screen, where it is traced on a cel to derive an animated cartoon.

■ **Videotape standards**

Quadruplex systems

Quad 1 formats Tape width 50·8 mm (2 in); transverse video tracks (0·25 mm wide) with guard bands (0·15 mm) between. Tracks: Full range audio track, video, cue track (low-grade audio, time-code), control. Tape speed 38 cm/s (15 in/s). Writing speed 39·46 m/s at 525/60 Hz (41·22 m/s at 625/50 Hz).

Quad 2 format Derived from Quad 1, with video tracks 0·15 mm wide, guard tracks 0·04 mm between. Half tape speed 19 cm/s (7·5 in/s), same writing speed. FM frequencies increased (Super High Band) providing less moiré interference. Pilot tone recorded to detect amplitude/frequency errors (velocity and saturation error correction).

Continuous/non-segmented helical scan systems—full-field-per-track

Ampex 1 inch format (VPR1) Tape width 25·4 mm. Tape speed 25·4 cm/s (10 in/s). Omega wrap 345°. Writing speed: 21·4 m/s (842 in/s). Separate video head for simultaneous replay (confidence) head. Broadcast quality slow/stop motion (using AST). Zero to fast motion for editing. Video drum heads: video erase, record, replay. Tracks: control, address/cue, video, two audio. No segmentation errors (banding). Vertical interval information (vertical blanking period) is *not* recorded.

Sony 1 inch format (BVH-1000) Similar to above, but extra video head records syncs and other vertical interval information. Writing speed: 10·2 m/s (400 in/s). Video drum heads; video erase, sync erase, video rep, sync rep, video rec, sync rec. Tracks: two audio, video, control, sync, address/cue.

Type 'C' format Development of Ampex and Sony formats, with separate simultaneous replay heads and sync heads. Tape speed 24·4 cm/s (9·61 in/s). Writing speed: 25·6 m/s at 525/60 Hz (21·39 m/s at 625/50 Hz). Video drum heads: video erase, sync erase, video rep, sync rep, video rec, sync rec. Recorded track 400 mm (16 in) long. Bandwidth 5 MHz at 525/60 Hz (6 MHz at 625/50 Hz). Vertical interval period is *not* recorded.

Type 'A' format Old one inch continuous field system (AMPEX); primarily industrial.

Type 'D' format Continuous field, one head NEC system (Nippon Electric Co.).

Segmented helical scan systems

IVC 9000 format Tape width 50·8 mm (2 in). Tape speed 20 cm/s (8 in/s). Writing speed: 38·1 m/s (1500 in/s). Video drum 3·3 in diam. Video drum heads: two video rec/rep heads, full width video erase. Tracks: Audio 2, audio cue, audio 1, video, control, address. Super high band.

Type 'B' format (Bosch-Fernseh—BCN) Tape width 25·4 mm (1 in). Tape speed 24·4 cm/s (9·6 in/s). Recorded track 80 mm (3 in) long. Video drum 2 in diam. Writing speed: 24 m/s (954 in/s). Tracks: Audio, control, audio, video. Entire vertical interval information is recorded. Video drum fitted with two video erase heads and two video rec/rep heads.

HBU format Modified U-matic video cassette operating with segmented scanning at triple normal tape speed. Writing speed: 30·5 m/s (1200 in/s). 7–10 MHz high-band color. Mainly derived for editing applications (video store).

Closed-circuit VCR (video-cassette recorder) systems These are primarily for domestic format. (Recording durations: 1 hour to +3 hours.)

Videotape formats In the traditional open-reel (reel-to-reel) format, the tape is stored on a single reel and fed onto an empty one (take-up reel) on the machine. Later designs use a *cartridge (cart) format* where a single enclosed reel supports an endless tape loop, or a *cassette* enclosing supply and take-up reels in a complete package, for insertion into the videorecorder. Although distinct formats, they are sometimes confused.

■ **Viewing conditions** Preferred viewing conditions recommended by CCIR include—viewing distance 4–6 times picture height; peak screen luminance 60–80 cd/m²; blank screen 0·02 of peak white.

■ **VIRS (Vertical Interval Reference Signals)** These special signals are inserted into the vertical interval period (i.e. while the scanning process returns to top left screen after tracing a complete field). They fall outside the normal picture area and are transmitted just before picture information begins (visible as a dot pattern along the top edge of frame). They can provide reference data to enable receiver screen tint and hue to be adjusted automatically to levels specified at the source. Also, they can provide identification of program source, point of origination, time and date of original broadcast.

438

■ **VITS (Vertical Interval Test Signals)** These specially inserted signals (see VIRS) are used for technical measurement, monitoring, performance correction of transmission systems.

■ **Volume indicators** Indicators used in monitoring audio signals, to ensure that they do not exceed maximum or minimum system limits (avoiding overload distortion and high background noise respectively). The *VU meter* is universally used. Its decibel and modulation scales (volume units) show system limits. Although simple, its continual rapid needle movement can be confusing and tiring to watch. The needle's peak readings are less precise for fluctuating audio than for steady tone. The meter is inaccurate for the loud transient sounds that easily overload and distort the system. Normal range used − 20 to 0 dB. Speech − 14 to 8 dB (true undistorted maximum modulation reading around 50%).

Peak program meters (PPMs) have special circuits responding to volume peaks. Their indication is equally accurate for steady, transient, or continuous complex audio. Normal range used: 2 to 6 (each step on meter is 4 dBs).

Visual displays, using a line of LEDs (light emitting diodes) or patterns inserted into TV monitor picture, are also used to indicate volume range. See also Fig. 10.22.

■ **Waveform monitor (WFM)** A small oscilloscope tube displaying a fluctuating line which traces the variations of the video signal, sync pulses, and insertion pulses. Individual picture lines can be selected and examined (line strobe).

■ **Wild shooting** Shooting mute pictures for a sound film; appropriate sound being added later.

■ **Wild track** Sound recorded independently of vision (non-sync sound).

■ **Work print (cutting copy)** The initial check prints of the camera original (dailies, rushes) are edited together to form the work print, from which initial editing decisions are made. A *rough-cut* then a *fine-cut* are developed. Identifying numbers every 20 frames along the film edge (edge-numbers) aid in conforming camera original to the final edited work print.

■ **Wrap (wrap up)** To finish activities.

Bibliography

TV production

Bretz, R.   *Techniques of television production.* McGraw-Hill, New York.

Chester, G., Garrison, G. R., Willis, E. E.   *Television and radio.* Meredith Corpn., New York.

Lewis, C.   *The TV director/interpreter.* Focal Press, Focal/Hastings House, New York.

Millerson, G.   *Effective television production.* Focal Press, London; Focal/Hastings House, New York.

Stasheff, E., & Bretz, R. Gartley, J. & L.   *The television program.* Hill & Wang, New York.

Zettl, H.   *Television production handbook.* Wadsworth Pub. Co. Inc., Belmont, California.

Staging/scenic design

Millerson, G.   *Basic TV staging.* Focal Press, London; Focal/Hastings House, New York.

Wade, R. J.   *Designing for TV.* Pelligrini & Cudahy, New York.

Wade, R. J.   *Staging TV programs and commercials.* Focal Press, London; Focal/Hastings House, New York.

TV lighting

Millerson, G.   *TV lighting methods.* Focal Press, London; Focal/Hastings House, New York.

Millerson, G.   *The technique of lighting for television and motion pictures.* Focal Press, London; Focal/Hastings House, New York.

Television sound

Alkin, G.   *TV sound operations.* Focal Press, London; Focal/Hastings House, New York.

Nisbett, A.   *The use of microphones.* Focal Press, London; Focal/Hastings House, New York.

Nisbett, A.   *The technique of the sound studio.* Focal Press, London; Focal/Hastings House, New York.

TV camerawork

Jones, P.   *The technique of the television cameraman.* Focal Press, London; Focal/Hastings House, New York.

Millerson, G.   *TV camera operation.* Focal Press, London; Focal/Hastings House, New York.

Make-up

Buckman, H.   *Film and television make-up*. Pitman, London.
Kehoe, V.   *The technique of film and television make-up*. Focal
   Press, London; Focal/Hastings House, New York.

Film

Burder, J. *16 mm film cutting*. Focal Press, London; Focal/
   Hastings House, New York.
Burder, J. *The technique of editing 16 mm film*. Focal Press;
   London; Focal/Hastings House, New York.
Happé, L. B.   *Your film and the lab*. Focal Press, London;
   Focal/Hastings House, New York.
Reisz, K. & Millar, G.   *The technique of film editing*. Focal
   Press, London; Focal/Hastings House, New York.

Television studio

Birmingham, A., and others.   *The small TV studio*. Focal Press,
   London; Focal/Hastings House, New York.

Special effects

Fielding, R.   *The technique of special effects cinematography*.
   Focal Press, London; Focal/Hastings House, New York.
Wilkie, B.   *The technique of special effects in television*. Focal
   Press, London; Focal/Hastings House, New York.
Wilkie, B.   *Creating special effects for TV and films*. Focal Press
   London; Focal/Hastings House, New York.

Color

Evans, R. M.   *An introduction to color*. Wiley, N.Y.

Technical

Ennes, H. E.   *Television broadcasting*. Foulsham-Sams.
Spottiswoode, R. (Ed.)   *Focal encyclopedia of film and
   television*. Focal Press, London; Focal/Hastings House, New
   York.

Index

A and B rolls   423
Acoustics   180, 203, 204, 423
Ambience   279
Ambiguity   133, 303
Angle, lens   *see* Lens angle
Angle of view   31, 34, 35, 430
Animated graphics   267–274
Animation   401, 423, 437
Announce booth   20, 23, 407
Aperture correction   423
Aperture, lens   *see* f-stop
Appearance   389
Architectural units   172
Arcing   74
Area lighting   145
Arranging shots   293
Aspect ratio   93
Assembly, A/B roll   241
Assembly, checkerboard   241
Attention, focusing   297–301, 365, 366
Attitude, subject   92
Audio   202
Audio analysis   362–365
Audio control   17, 18, 20, 21, 22, 219–221, 423
Audio disc formats   229
Audio dub   260
Audio engineer   19–22, 220
Audio quality   202, 203, 223
Audio range   203, 204
Audio sources   17
Audio tape   18, 20–22, 224
Audio tape formats   229
Audio tape recorder   224, 225, 238
Audio tape reproduction   228
Audio-visual relationship   367
Automatic scan tracking   424, 437
Auto set-up   411

Back cuing   346
Backdrop   175
Background, black   121
Background generator   386
Background, influence of   103, 104
Background insertion   392
Background lighting   146, 147
Backgrounds   170, 175
Backing   175
Backing track   424
Back light   140, 142–144, 147
Back projection   *see* Rear projection
Back timing   347
Balance, color   414
Balance, lighting   143
Balance, pictorial   87–89, 108
Balance, sound   221
Banding   424
Bands   424
Barndoor   152, 155
Base light   139
Battens   155
Beeper   424
Bias current   231, 424
Black level   26, 378, 379, 413, 420
Blimp   424

Blocking   333–335
Boom, sound   18, 22, 211–217
Border generator   16, 265, 386, 390
Brace   171
Breakdown sheet   326
Bridge   425
Brightness   425
Budget   316
Bulk erase   224, 425, 427

Calculating shots   328–331
Cameo   148, 175
Camera card   327, 330
Camera control unit   22, 23
Camera crane   52–54
Camera, field   407, 408
Camera height   70–72
Camera mountings   *see* Mountings
Camera movement   *see* Movement
Camera parts   31, 32, 56
Camera tube   24, 31, 34, 37, 48, 408, 416–418
Cans   425
Canted shots   375
Caption scanner   16
Cartridge, audio   17, 22, 229
Cartridge, video   112
Cassette, audio   17, 22, 229
Cassette, video   438
CATV   425
CCIR   425
CCTV   425
Ceilings   179
Character generator   16, 21, 264, 407
Cheated substitute   311–313
Cheating   96
Chiaroscuro   148
Chroma   30
Chroma key   16, 170, 176, 188, 382, 384, 388, 390, 393, 396, 399
Chrominance   25, 26
Cinching   425
Clarity   302
Close shot   74, 75
Coder   421
Coding   425
Color balance   414
Color bars   421, 425
Colored light   162
Color principles   28, 425, 426
Color separation overlay   *see* Chroma key
Color, subjective effects   98, 100, 101
Color synthesizer   16, 237, 390
Color temperature   137, 162, 426
Color TV principles   24–28
Color under   426
Compatible   426
Compatibility, visual   391–396
Complementary picture   379
Composition, dynamic   85, 99, 103, 104, 107
Composition, pictorial   57, 81
Compression, audio   17
Concentration   303–305

Confusing techniques   302
Continuity   234, 286, 309
Continuity, compositional   95–97
Continuity control room   23
Contouring, scenic   176
Contrast range   29, 30, 190
Contrast ratio   30, 160
Cookie   155
Costume   17, 191, 193
Craning   73
Cross cutting   96
Crossed gray scale   426
Crossing the line   *see* Imaginary line
Crowd   312
Cuing   342, 343
Cut   63, 113–115, 126, 131
Cutaways   111
Cut-outs   170, 173
Cutting, continuity   112
Cutting, dynamic   112, 113
Cutting rate   127, 128
Cutting, relational   112, 113
Cutting rhythm   128
Cyclic motion   401
Cycling   387, 390
Cyclorama   151, 162, 167, 170, 174

Date   427
De-beaming   379
Decibel   427
Decoder   421, 425
Dedicated VTR   112
Deep focus   74
Defects, picture   415
Definition   29, 426
Degaussing   427
Deliberate disruption   310
Depth of field   43–47, 60, 62, 63, 65, 74, 75
Detail   28, 29
Diffuser   155
Diffusion disc   80, 373
Digital frame store   250
Digital systems   399
Digital techniques   112
Digital telecine   242
Digital time-base correction   409, 410
Dimmers   157
Director   17, 19–21, 35, 275–277
Disappearance   389
Disc, audio   17, 18
Disc, diffusion   80
Disc reproduction   228
Disc, video   *see* Video disc
Discontinuous recording   112
Discontinuous shooting   96, 284, 285
Display screen   388
Dissolve   *see* Mix
Dissolve, defocus   120
Dissolve, half-lap   121
Dissolve, ripple   120
Distortion, audio   226
Distortion, geometric   41, 62, 74
Distortion, perspective   *see* Perspective
Distortion, visual   372, 375, 400

Distractions   312
Dolby   231
Dollies, camera   54, 55
Dollying   *see* Movement, camera
Dope sheet   427
Double re-entry   386, 390
Downstream key   390
Drapes   172, 175
Drop out   228, 427
Dry run   333
Dubbing sheet   427
Dupe neg   427
Dynamic composition   *see* Composition
Dynamic range   220

Echo   226, 228
Edge light   139
Editing, audio tape   224, 225
Editing controller   257
Editing, film   240
Editing, nature of   110, 113
Editing, off-line   258
Editing, on-line   258
Editing, TV   110, 111
Editing, videotape   110–112, 254–257
Effects, audio   226, 227
Effects projector   *see* Projector
Effects, video   129, 337
EFP   403
EIAJ   427
EJ   403
Electronic effects   129–377
Electronic insertion   170, 176, 181, 265, 390
Electronic still store   407, 434
Elevations   169
Emphasis, sound   359–360
ENG   49, 55, 403–407, 409
Equalization, audio   *see* Filter
Erase, bulk   224, 425
ESG   403
Establish, to   427
Exposure   29, 46, 133, 161, 413, 427
Eyelines   300, 395

Fact sheet   326
Fade   116–120
False perspective   184
Fast motion   400–401, 436
FCC   428
Feedback, video   379
Field   25
Field transmission   *see* Remotes
Filler   140, 143, 144, 147
Fill light   *see* Filler
Film chain   *see* Film scanner
Film cuing   343
Film editing   *see* Editing
Film equipment   238
Film formats   240
Film in TV   236–238, 243
Film lighting   238
Film prints   240, 245, 246
Film recording   244

Film reproduction   241–245
Film scanner   16–18, 22, 23, 242, 243
Film sound   239
Filming process   232, 233, 235
Filter, color   374
Filter, corrective   372
Filter, fog   373
Filter, haze   373
Filter, low contrast   373
Filter, neutral density   48, 371
Filter, night   373
Filter, polarizing   374
Filter, star   80, 372
Filter, U.V.   373
Filters, audio   222, 223, 226
Filters, camera   371
Fire flicker   162
Fishpole   210, 211
Fixed focal length   31
Flag   155
Flare   428
Flashback   119, 120, 309
Flash cutting   428
Flat   170, 171
Floor height   180, 181
Floor manager   18, 338–340
Floor, studio   17
Floor treatment   178, 179
Fluid head   428
Flying head   428
Flying-spot scanners   242–243
Focal length   31, 32, 34, 35, 56, 428
Focus control   33, 43–46
Focusing   33, 43–46, 56, 57, 63, 65, 66, 82
Foldback   18, 226
Forced processing   428
Foreground subjects   81, 82, 91, 182
Forward timing   347
Foundation light   139
Four walls   182, 184, 185
Frame   25, 102
Frame jumps   185
Frame synchronizer   420, 436
Framing   57, 74, 82, 85–87
Freeze frame   103, 400, 401, 406
Friction head   428
Front projection   176, 370
Frustrating techniques   302
f-stop   33, 34, 44, 46–48, 63, 64

Gaffer tape   153, 189
Gamma   30, 414, 428–429
Gate   429
Gauze   *see* Scrim
Gearing head   429
Genlock   419
Geometric distortion   74
Glitch   429
Gobo   376
Golden section   83, 85
Graphics   17, 190, 261–274
Gravity   100, 105
Gray scale   29, 30, 420

Ground row   170, 173
Grouping   97

Hard light   134, 137, 152
Headroom   86, 117
Helical scan VT   249–252, 406
Helicopter   55, 407
Highlight overload protection   429
Howlround   227
Hue   30, 425
Hydraulic platform   55
Hyperfocal distance   429

Illusory attraction   84
Illustrating subjects   307
Image enhancement   399, 421, 429
Image rotation   375, 376
Image transfer   246
Imaginary line   185, 290, 291, 426, 429, 430
Imaginative sound   358, 359
Imitation, direct   313
Imitation, indirect   313
Indicators   389
Infill   386, 398
Insertion, complex   397
Insertion, shadow   396
Inset   387, 388
Intercom   21–22, 56, 57, 335, 346–351
Intercutting   114, 127
Interest   93–95, 107–109
Interscene devices   310
Inter-source switching   380
Inverse square law   134
In vision   430
ITFS   430
ITV   430

Jack   171
Jump cuts   117, 311, 430

Kaleidoscope   375
Kelvins   137, 431
Key, downstream   390
Key, edge   390
Key, external   381, 382, 385, 386, 390
Key fill   390
Key, internal   381, 384, 390
Key light   138, 142–146
Key, linear   391, 396
Key shots   82
Keyed insertion   380, 385, 388
Keying amplifier   383
Keying, self   381, 384–386, 388, 390
Keying, sequential   399
Kinescope   *see* Picture tube

Lamp patching   156
Lamp supports   153, 154
Lamps, types of   149, 150
Lateral reversal   108
Lens angle   31, 32, 34, 35, 37–41, 56, 64, 74, 430
Lens hood   32

Lens, narrow angle 31, 33, 35–40, 60, 62–64, 67, 74, 361
Lens, normal angle 35–40, 74
Lens, prismatic 375
Lens types 31, 34
Lens, wide angle 31, 33, 35–40, 60–62, 64, 65
Library shots 236–237, 312
Light bias 430
Light control 156, 157
Light direction 135–139
Light dispersion 134
Light intensity 136
Light levels (intensities) 48, 133, 146, 163
Light measurement 161, 163, 430, 431
Light quality 134, 135, 137
Light sources 149, 150
Lighting 16, 133
Lighting angles 143
Lighting, animated 165
Lighting an object 138, 139
Lighting areas 145
Lighting, atmospheric 166
Lighting balance 143
Lighting control 21, 22
Lighting director 20, 21
Lighting effects 166
Lighting functions 139
Lighting people 140–145
Lighting plot 169
Limbo 148, 175
Limiter, audio 17
Lines, compositional 94, 95
Line-up 411, 417
Lip flap 245, 431
Lip sync 226, 239
Liquid gate 431
Live on tape 111
Location filming 237
Location production unit 403
Long focus lens 31, 37
Longitudinal video recording 431
Luminance 25, 26, 30, 391, 425

M and E track 226
Magicam 393
Make-up 17, 194–201
Master control room 16, 19, 20, 22, 23
Matched dissolves 119
Matte box 431
Matte, camera 376, 382, 383
Matte, electronic 380
Matte, foreground 376, 378, 394
Matte, insertion 377
Matte, key 391
Matte, surround 388–389
Matte, title 385
Matte, traveling 436
MATV 431
Microphone, baton 208
Microphone, desk 209
Microphone, directional response 212–215, 217
Microphone, lip 208, 209

Microphone, parabolic 217
Microphone, personal 207, 218
Microphone shadow 216
Microphone, slung 209, 210
Microphone, stand 209
Microphone, wireless 207
Microphones characteristics 205–207, 212, 217, 218
Microphones, types of 207, 217–219
Microprocessors 431
Miming 226
Mired 431
Mirrors 55, 180, 368–371, 376, 398
Mismatched cuts 117
Mix (dissolve) 63, 117–121, 130, 131
Mobile control room 403
Mobile scenic units 18–189
Molded pieces 177–178
Monitor, picture 19–21, 25, 420
Monitor, sound 18–21
Monitor, waveform 439
Monochrome 24–26, 30, 48
Montage 124
Mountings, camera 31, 49–57, 238, 407, 408
Mountings, microphone 209–211
Movement, camera 38, 60, 63, 66, 72–78
Movement, impressions of 101–105, 107, 108–109
Movement in pictures 103
Movement, parallactic 79
Movement, subject 289, 295
Moving vehicles 187–189
Multicamera shooting 95, 161, 285–288
Multi-hue Chroma key 399
Multilevel synthesis 398
Multiplane techniques 182, 184
Multiple images 375
Multiple use of units 185
Multiplexing 242
Multisplit screen 389
Multivision 432
Munsell 30, 426, 432

Narration 432
Narrow angle lens see Lens
Natural light 133
Negative film transmission 240
Negative picture 379
NEMO 432
NET 432
Neutral backgrounds 175
Noise reduction 399, 429
Non-additive mix 385, 391
Normal lens see Lens
Notan 147
NTSC 26, 421, 426, 435

Objective techniques 73, 296, 297
One and a half heads 432
Ooze function 400
Open settings 175
Optical effect 432

Optical printer 121, 432
Order of shots 124–125
Out of sequence shooting 234
Outline effects 400
Outside broadcasting see Remotes
Out-take 432
Overcranking 432
Overlapping sound 432
Overrun 432
Overscanning 432
Overshoot 74, 180, 182

Pace 300, 301
Padding 432
PAL 26, 421, 435
Panning 66–69
Panning head 31, 32, 56, 67
Parallactic movement see Movement
Parallels 173, 180
Partial settings 185–186
Partial shot 311
Party line 21
Pattern projector see Projector
PBS 433
Peak program meter 221, 439
Pedestal, camera 50–52, 56
Pepper's ghost 369–370
Performers 17, 341, 342, 344–345
Periscope 55
Perspective distortion 36–40, 60, 62, 64, 74–75, 423
Perspective, false 184
Perspective, natural 433
Perspective, sound 222–223
Photo-enlargements 176, 186
Pictorial lighting 147–148, 162
Picture defects 415
Picture quality 433
Picture tube 24–27
Picture usage 352
Pilot tone 433
Pipe grid 154
Pixillation 401, 433
Plan, camera 168
Plan, floor 167
Plan, staging 167, 169, 170
Plan, studio 167, 168
Planned viewpoints 290–292
Planning, production 316–320
Playback 19, 226
PLUGE 434
Plumbicon see Camera tubes
Portable lighting 153, 238
Postsynchronizing 226, 434
Practicals 188
Pre-focus 78
Pre-scoring 226
Preview 19
Prismatic lens 375
Private wire 21
Producer 17, 19
Production approaches 278, 318
Production control room 18–21
Production emphasis 276, 278
Production procedure 317

Production routines   287, 289
Production switcher   *see* Switcher
Production team   275, 277
Production techniques   279–285
Production rhetoric   352–357
Profile piece   170, 173
Program opening   295–296
Projected backgrounds   176, 373
Projector, effects   152
Projector, pattern   152
Projector, scenic   373
Prominence, subject   91
Prompting   344–345
Proportions   80, 82, 83, 92
Props   188–189, 192
PTV   434

Quad overlay   389
Quad split   391
Quadruplex VTR   247–249, 252, 406,
   437

Reaction shots   111, 311
Real time   400
Realism   186, 187
Rear projection   170, 176, 188, 370,
   372, 373
Rediscovery   118
Reflected shapes   166
Reflection, laws of   368–369
Reflectors   139, 155
Reflex projection   170, 176, 371, 374
Regular formats   314, 315
Rehearsal blocking   332–335
Rehearsal, prestudio   332
Rehearsal problems   336–338
Rehearsal, studio   332–338, 340–342
Rehearse/record   335–336
Relational cutting   112
Remote sources   16, 17, 19
Remotes   403–407
Remotes, program handling   404
Reproduced sound   204
Resolution   28, 29
Reverberation   17, 205, 226, 227
Reversal film   240
Reverse-angle shots   290
Reverse motion   401
Reverse shooting   180
Reversed direction   116
Rhythm, visual   89, 128
Rifle microphone   217
Ripple   *see* Weave
Ripple dissolve   120
Ripple distortion   375
Rolling tripod   50
Rostra   *see* Parallels
Rostrum camera   381, 383, 437
Rotating image   375, 376
Rule of thirds   84
Run-down sheet   326
Running order   326

Safety margins   86, 261
Saturation   30, 425

Scale   90, 91, 183
Scene analysis   305
Scene-sync   393
Scenic design   *see* Staging
Scoop   150
Screens, scenic   174
Scrim   175, 187
Script   17, 111, 234, 320–325
SECAM   26, 435
Selectivity of mechanics   279, 281
Sequential insertion   389
Sequential switching   389
Services, contributory   17
Set design   *see* Staging
Set designer   16, 20
Set dressing   188, 189
Set piece   170, 172
Set proportions   181
Settings, size and shape   180
Set-up   26
Shader   19–21, 411–416
Shadow insertion   396
Shadows   164, 165
Shape of picture   93
Shooting off   *see* Overshoot
Shooting rate   235
Shooting ratio   232
Short-focus lens   31, 39
Shot box   33
Shot card   56, 58
Shot organization   289
Shot planning   328–331
Shot plotter   327, 330
Shot rate   128
Shotgun microphone   217
Shots, defining   59
Shots, duration of   126
Shots, order or   124–126
Silence   359
Silhouette   149
Simultaneous contrast   90, 99, 434
Single-camera shooting   234, 284
Skeletal staging   176, 189
Slide scanner   16, 22, 23, 434
Slides   434
Slow motion   401, 432
Soft light   134, 137, 150, 151
Solid pieces   172
Sound analysis   362–365
Sound balance   221
Sound boom   158, 159
Sound control   20–22
Sound, focusing attention   365, 366
Sound, off-screen   360, 361
Sound on film   239
Sound, substituted   361
Sound treatment   362, 366–367
Sound usage   360
Sound/visual relationship   367
Space economies   181, 183
Space, filmic   402
Special effects   17
Special effects generator   16, 381–383,
   388, 391
Spectacle   312

Split field   376, 377
Split screen   121, 370, 392
Spot, follow   152
Spot, fresnel   151
Spotlight   151, 153
Staging   16, 17
Staging area   180
Staging for color   192
Staircase   434
Standards converter   420, 421
Standupper   435
Steadicam   435
Still pictures   103
Stock shot   *see* Library shot
Storyboard   82, 292
Strength, subject   92
Stringer   435
Studio   15, 22
Studio address   17
Styrofoam   177
Subjective effects   99–101
Subjective techniques   73, 76, 77, 83,
   296–297
Superimposition   121, 123, 124, 131,
   377, 385, 387, 422
Supplementary lens   435
Surface brightness   140, 190
Surface detail   176, 190
Surface finish   177, 190–192
Surface tones   141, 160
Switcher   16–21, 110, 128–132, 223,
   223, 418, 420, 422
Synchronizer   435
Synchronizing video   23, 24, 26, 409,
   418–421
Synthesis, color   130, 398
Synthesis, multilevel   398, 399
Synthesizer, audio   17
Synthesizer, color   129, 398
Synthetic sound   230

Take-bar   132
Talent   *see* Performers
Talkback   *see* Intercom
TARIF   237, 243
Technical Director   17, 19–21, 110
Technical planning   318–320
Telecine   *see* Film scanner
Teletext   21, 435
Temporal effects   400
Tension   300
Three-point lighting   138
Tilting   69
Time address   *see* Time code
Time-base correction   409, 410, 436
Time code   258, 259, 406
Time, extended/contracted   402
Time, passing   308–309, 400, 401
Timing   301, 345–347, 421
Title cards   74, 79, 261–268
Title insertion   387
Titling   261–265, 385
Tonal values   190
Tone, effects of   90, 92, 105, 303
Transformations   402

Transitions 94
Translator 436
Transmission number 436
Traveling matte 436
Traveling shot 436
Treatment 294
Tripod 50
Turret lens 31, 34, 41
TV systems 435

UHF 436
Undercranking 436
Underscanning 436
Unity 89
Unplanned production 314
Unrehearsed formats 314, 315
Up-cut 436
User bits 257, 259

Value 30
Vehicles, remotes 403, 404
Vertical interval 24, 438
Video control 17, 19, 21, 22, 242, 243, 409–416
Video control, artistic aspects 414–416
Video control, automatic 409, 412
Video disc 252–253
Video effects 21
Video engineering 411

Video feedback 379
Video gain 414
Video image processing 400
Video operator see Shader
Video signal 24–26
Video sources 16
Video terms 421
Videotape applications 247
Videotape editing see Editing
Videotape formats 252–254
Videotape leader 258
Videotape recording 16–18, 22, 23, 247–260
Videotape recording, discontinuous 112
Videotape standards 437
Videotape Type A standard 250, 437
Videotape Type B standard 250–252, 438
Videotape Type C standard 249–252, 437
Videotape Type D standard 438
Viewfinder 32, 56
Viewing distance 35, 438
Viewpoint, camera 302
Viewpoint, influence of 92, 93, 98, 117
Viewpoints, planned 290–292
VIRS 432, 438
Vision control see Video control
Vision mixer see Switcher

Visual changes 106
Visual effects 368, 419
Visual padding 306–308
Visual variety 286–289
VITS 432, 439
Volume indicator 439
Volume unit 221, 424
VT build up 390, 397, 398
VT cartridge 112
VT cassette 438
VU meter 221, 424, 439

Water ripple 162
Waveform monitor 439
Weave, electronic 379, 383
Weight 87–89, 99
Wide angle see Lens
Wide screen film 245
Wild shot 439
Wild track 237, 439
Wipe 120–123, 381, 382
Work print 439
Wrap 439

Zoom lens 31–34, 41–43, 45, 46, 56, 78–79
Zooming 19, 42, 43, 45, 46, 64, 72, 73, 76, 80, 123, 433